International Conference on Innovative Technology for Sustainable Development (ICITSD-2021)

ITM Web of Conferences Volume 37 (2021)

Online
27 – 29 January 2021

Editors:

R. Jagaeesh Kannan
Piet Kommers
S. Asha
Abdul Quadir Md

ISBN: 978-1-7138-2725-2

Printed from e-media with permission by:

Curran Associates, Inc.
57 Morehouse Lane
Red Hook, NY 12571

Some format issues inherent in the e-media version may also appear in this print version.

For additional information, please contact EDP Sciences – Web of Conferences at the address below.

EDP Sciences – Web of Conferences
17, Avenue du Hoggar
Parc d'Activité de Courtabœuf
BP 112
F-91944 Les Ulis Cedex A
France

Phone: +33 (0) 1 69 18 75 75
Fax: +33 (0) 1 69 28 84 91

contact-edps@webofconferences.org

Additional copies of this publication are available from:

Curran Associates, Inc.
57 Morehouse Lane
Red Hook, NY 12571 USA
Phone: 845-758-0400
Fax: 845-758-2633
Email: curran@proceedings.com
Web: www.proceedings.com

TABLE OF CONTENTS

Preface

The International Conference on Innovative Technology for Sustainable Development - ICITSD-2021 is a joint venture of Vellore Institute of Technology, Chennai and University of Technology, Sydney, Australia. Any education makes real sense only if it reaches the hands of the common man and it is helping him in some ways. The Sustainable Development Goals (SDGs), also known as the Global Goals, were adopted by all United Nations Member States in 2015 as a universal call to action to end poverty, protect the planet and ensure that all people enjoy peace and prosperity by2030.

The 17 SDGs are integrated—that is, they recognize that action in one area will affect outcomes in others and that development must balance social, economic and environmental sustainability. Everyone is needed to reach these ambitious targets. The creativity, know-how, technology, and financial resources from all of society are necessary to achieve the SDGs in every context. This ICITSD conference is one marked step in helping us to contribute towards the global goals, in terms of engineering solutions.

As per the technical committee norms, to assure the quality of the papers; we involved reviewers who are experts in the field across many countries. 150 reviewers involved in reviewing the conference manuscripts, out of which 50 international reviewers were involved in the review process of this conference. It was a rigorous exercise for the technical committee of this conference to assign the manuscript to the right reviewers and get back the reviewers' valuable comments on time. Considering the assurance of the quality of the manuscripts, we limited to 171 unpublished manuscript submissions, after a preliminary technical content check of the manuscripts. After a stringent review process, 60 quality technical papers qualified for presentation and were accepted out of which 25 high-quality papers are selected for the ITM Web of Conference. Hence, the acceptance ratio of this conference is 35%, which inherently portrays the quality of the conference. To list few topics on which the technical papers that were submitted and presented in our the conference was: Artificial Intelligence and Machine learning, Cloud Computing, Data Analytics, Data Security and Privacy for CPS, Block Chain, Virtual/Augmented reality and Internet of Things, on domains like agriculture, education, health care, Sports, Tourism Recommendations, etc.

This book is the collection of remarkable works related to Innovative Technology for sustainable Development that was submitted and presented in the First International Conference on Innovative Technology for Sustainable Development – ICITSD -2021.This prestigious event was hosted by School of Computer Science and Engineering, Vellore Institute of Technology, Chennai Campus, India and Centre for Cyber Physical Systems (CPS) in collaboration with University of Technology Sydney, Australia. The conference hosted the main theme – "Smart and Sustainable Computing" with topics related to computer science and engineering and also to allied topics. Over the three days of January 27th to 29th 2021, ICITSD 2021, brought together a large number of stakeholders, including engineers, faculty members, researchers, students, industry, governmental organizations, private sectors, etc., to learn, share, and build fruitful and long-term collaborations. 171papers were submitted from different countries. Of these papers, only 60 papers were accepted for presentation and publications – Keeping the acceptance ratio to be 35%.The conference would not have been successful without the support of the Program Committee members, Organizing Committee members, keynote/plenary speakers, authors, tutorial presenters, special events organizers, reviewers and the conference attendees.

We greatly appreciate the meticulous reviews provided by the Program Committee members, and additional reviewers and thank the authors for having revised their papers to address the comments and suggestions by the referees. Our sincere thanks to the Steering committee members for their timely help and supervision. We want to thank all the Keynote speakers and plenary speakers of the conference. The conference program was enriched by the outstanding talks by Dr. Jeevan Prakash Kulkarni, Indian Institute of Tropical Meteorology (IITM), Pune, Dr. Jeganathan L, Professor, SCOPE, Vellore Institute of Technology, Chennai, Dr. Mariofanna (Fanny) Milanova, IEEE Senior Member, Professor, Computer Science Department, NVIDIA Certified Instructor for Deep Learning, University of Arkansas Little Rock , AR, USA, Dr. Nabin Sharma PhD, SMIEEE, Co-Director (Intelligent Drone Lab), Faculty of Engineering and Information Technology, University of Technology Sydney, Australia, Dr. Nebojsa Bacanin, Department of Informatics and Computing, Singidunum University, Belgrade, Serbia, Dr Vimal Kumar, Department of Computer Science, University of Waikato, New Zealand, and Dr. Sivakumar V, Centre for Development of Advanced Computing (CDAC), Pune.

We would like to express our sincere thanks to the Chancellor, Vice Presidents, Asst. Vice President, Executive Director, Vice-Chancellor, Pro Vice Chancellor, Registrar, Additional Registrar and other members of VIT University family for extending all facilities and support for the smooth conduct of the conference. We wish to express our thanks to the ITM Publisher for producing and publishing theICITSD2021 proceedings, and especially to Mr. Isabelle Houlbert and for the excellent support. We would also like to thank all those who contributed to the success of ICITSD2021 but whose names may not be listed. We hope you will find this book informative and resourceful.

March 2021

Editors

R.Jagadeesh Kannan
S. Asha
Abdul Quadir Md
Piet Kommers

1st INTERNATIONAL CONFERENCE ON INNOVATIVE TECHNOLOGY FOR SUSTAINABLE DEVELOPMENT

UTS
UNIVERSITY OF TECHNOLOGY SYDNEY

(ICITSD - 2021)

SMART AND SUSTAINABLE COMPUTING

VIT
Vellore Institute of Technology

School of Computer Science and Engineering, VIT Chennai and Centre for Cyber Physical Systems

ICITSD 2021

in collaboration with

University of Technology Sydney, Australia

January 27 - 29, 2021
VELLORE INSTITUTE OF TECHNOLOGY, CHENNAI, INDIA

ICITSD 2021 Highlights

- ICITSD-2021 will comprise the main conference and several co-located workshops and tutorials in collaboration between VIT Chennai and University of Sydney, Australia through online mode
- Eminent speakers from across the globe will share their views on the theme of the conference
- Technical panel comprises of leading academicians and industry experts

Conference Focus Area

The Conference topics include (but are not limited to)

- Robotics and Automation
- Block Chain
- 5G Technology
- Imaging and Computer Vision
- Deep Learning
- IoT and Artificial Intelligence
- Machine Learning and its applications
- Advanced Numerical Algorithms
- Cloud and Fog Computing
- 3D Printing
- Quantum Computing
- Xtended Reality (AR, VR and MR)
- Data Mining and Information Retrieval
- Fuzzy Logic and its applications
- Personalize and Predicted Machine
- Digital Twins
- Software Engineering

Call For Papers

- Inviting original contributions from researchers, unpublished research work, which is not currently under review in another conference or journal
- ICITSD accepts the following two paper types. a) Full paper: 12-15+ pages and b) Short paper: 1-8+ pages

Submission

- All manuscripts should be submitted online at: **https://easychair.org/conferences/?conf=icitsd2021**

All accepted papers will be indexed in "Scopus"*

Important Dates

Event Name	Date
Last Date for Full Paper Submission	December 26, 2020
Notification of Acceptance of Paper	January 2 , 2021
Registration by the accepted authors	January 9, 2021
Final Paper (Camera Ready) Submission	January 9, 2021
Presentation Slides Submission	January 15, 2021
Conference Dates	January 27-29, 2021

Registration Fees (including 18% GST)

Author(s) / Registration	Indian Authors	Foreign Authors
U.G / PG Students	6000 (INR)	180 (USD)
Ph.D Students	6000 (INR)	180 (USD)
Academicians / Faculties	7000 (INR)	240 (USD)
Industry & Research	8000 (INR)	300 (USD)
Listeners	1500 (INR)	50 (USD)

Conference Committee

Chief Patron

Dr. G. Viswanathan
Founder & Chancellor, VIT

Patrons

Mr. Sankar Viswanathan
Vice-President, VIT

Dr. Sekar Viswanathan
Vice-President, VIT

Mr. G V Selvam
Vice-President, VIT

Dr. Sandhya Pentareddy
Executive Director, VIT

Ms. Kadhambari S. Viswanathan
Assistant Vice-President, VIT

Dr. Rambabu Kodali
Vice Chancellor, VIT

Dr. V. S. Kanchana Bhaaskaran
Pro Vice Chancellor, VIT Chennai

Dr. P.K. Manoharan
Additional Registrar, VIT Chennai

General Chairs

Dr. Jagadeesh Kannan R
Dean and Professor
SCOPE, VIT Chennai

Dr. Geetha S
Associate Dean and Professor
SCOPE, VIT Chennai

Dr. Farookh Hussain
Associate Professor and HoD
University of Technology Sydney

Conference Chairs

Dr. Asha S
Associate Professor Senior Grade
SCOPE, VIT Chennai

Dr. Abdul Quadir Md
Assistant Professor Senior Grade
SCOPE, VIT Chennai

Phone: +91-9884774695, +91-9884004139

Conference Mail ID: icitsd@vit.ac.in Conference Website: http://chennai.vit.ac.in/icitsd

Statement of Peer Review

In submitting conference proceedings to ***Web of Conferences***, the editors of the proceedings certify to the publisher that,

1. They adhere to its **Policy on Publishing Integrity** in order to safeguard good scientific practice.
2. All articles have been subjected to peer review administered by the proceedings editors.
3. Reviews have been conducted by expert referees, who have been requested to provide unbiased and constructive comments aimed, whenever possible, at improving the work.
4. Proceedings editors have taken all reasonable steps to ensure the quality of the materials they publish and their decision to accept or reject a paper for publication has been based only on the merits of the work and the relevance of the journal.

Title, Date and Place of the Conference

International Conference on Innovative Technology for Sustainable Development (ICITSD-2021), 27-01-2021 to 29-01-2021, Online

Proceedings Editors

Dr. R. Jagadeesh Kannan

Dr. Piet Kommers

Dr. S. Asha

Dr. Abdul Quadir MD

Date and Editors Signature

10-03-2021

Simple Single and Multi-Facility Location Models using Great Circle Distance

Baskar A[1*]

[1]Panimalar Institute of Technology, Poonamallee, Chennai – 600 123, India

Abstract. Facility location problems (FLP) are widely studied in operations research and supply chain domains. The most common metric used in such problems is the distance between two points, generally Euclidean distance (ED). When points/ locations on the earth surface are considered, ED may not be the appropriate distance metric to analyse with. Hence, while modelling a facility location on the earth, great circle distance (GCD) is preferable for computing optimal location(s). The different demand points may be assigned with different weights based on the importance and requirements. Weiszfeld"s algorithm is employed to locate such an optimal point(s) iteratively. The point is generally termed as "Geometric Median". This paper presents simple models combining GCD, weights and demand points. The algorithm is demonstrated with a single and multi-facility location problems.

1 Introduction

In any supply chain, the location of a supply point has a great influence on the system parameters like transportation cost, the time required and system efficiency [1]. In addition to the above, other factors like population, economy, and literacy may also play some roles in fixing the final facility location. In the case of collection point(s), the parameters that require attention include proximity to the raw materials, prices of them, infrastructure including manpower. It is generally acknowledged that the cost of logistics is very high in India. Some estimates put the cost at about 13 per cent of GDP, which is higher than the US (9%) and Germany (8%). A study conducted by Assocham-Resurgent India (2016) concluded that India can save around Rs 3.33 lakh crores if logistics costs reduce from 14 per cent to 9 per cent of GDP. Reduced logistics costs would bring down prices of products. In any logistics chain, the transportation services form a third of the cost. Among all modes of transport, roads carry about 60 per cent of the freight cargo in India [2]. Hence, minimising the total distance of any supply chain is important. When we refer to the 'distance' in general, Euclidean Distance (ED) is considered.

According to Zarinbal [3], Euclidean and Rectilinear distances account for more than 63 per cent of distance functions used in location problems. However, since the demand points are distributed over a spherical surface, using Great Circle Distance (GCD) which is based on spherical trigonometry is more relevant [4-5]. The discrepancy between the distances measured based on Euclidean and those based on spherical trigonometry becomes greater, the further apart the locations are from each other [6].

The points on the earth surface are represented by geodetic coordinates (latitudes and longitudes). The GCD between any two points on a sphere can be computed using the Haversine formula [7].

According to ReVelle and Eiselt [8], the location model has four basic features:

- The customers those are located at points or routes.
- The facilities that must be located.
- The geographic space or surface where the customers and facilities are located.
- The metric that indicates the distance, cost, or time between the locations of clients and facilities.

The geographic space and the distance metric play important roles in locating the facilities when the locations of customers remain the same. Most works have considered the space for the FLP to be flat.

The accuracy of the distances for the facility location model depends on the space feature.

As discussed in the study by Shih [9], if the location points are widely separated, the difference in driving distance between the flat surface assumption (i.e. Euclidean distance) and the spherical surface assumption (i.e. great circle distance) may be significant, leading to important variations in the optimal location of the facility.

The use of "Great circle distances" finds its place in marine applications also like exploring natural resources including oil rigs and mobile drilling units. In computing the optimal location, the geometric median; the relative distance, population, GDP or number of trips are considered as the 'weights'. Considering the nature of the problem, this can be well-thought-out to be a Fermat-Weber problem [10].

A single facility Fermat-Weber problem requires finding a point in a real coordinate space of dimension '***n***' (R^n) which minimizes the sum of weighted distances to '***m***' given data points. In a facility location problem, the data points are the customers (demand points) or venders (supply points). The weighted

*Corresponding author: a.baaskar@gmail.com

distance between the new facility and a customer/ vendor is a measure of the cost to transport goods or provide services by the facility to satisfy the known requirements [11].

$$\textit{Minimize } G(y) = \sum_{j=1}^{m} f_j(d(y, p_j)) \qquad (1)$$

Where, $p_j = (p_{j1}, p_{j2}, \ldots, p_{jn})^T$ is the known position of jth data point.

$y = (y_1, y_2, \ldots, y_n)^T$ is the position of the unknown new facility which has to be evaluated.

$d(y, p)$ is the distance function to measure the distance separating the two points y and p; $y, p \in R^n$.

$f_j(d(y, p_j))$ is the cost function of the distance function in the interval $(0, +\infty)$, $j = 1, 2, \ldots, m$.

$$f_j(d(y, p_j)) = w_j \cdot d(y, p_j);\ j = 1, 2, \ldots, m.$$

Where, w_j is the weighted positive constant and $d(y, p_j)$ is the norm.

Analytical solutions are available to evaluate such a 'Fermat-Weber point' for three non-collinear points [12]. Finding a geometric median of a set of data points is an optimization problem, which has no analytical solution for more than four points. In such cases, the Weiszfeld [13] algorithm has been effectively used by researchers for decades. It is an iterative procedure for solving the minisum Fermat-Weber problem.

The optimal facility location problem on a sphere has been a subject of interest for many researchers. This problem was first studied by Drezner and Wesolowsky [14] and by Katz and Cooper [15] where Weiszfeld-like algorithms were proposed. A gradient algorithm was proposed by Xue [16] to solve the spherical facility location problem. The author claimed that the algorithm always converges to a global minimizer of the spherical facility location problem. To compute lower bounds, four ways were proposed and compared by Hansen et. al. [17] within a branch-and-bound scheme for the Weber problem on a sphere.

The Wiggle Factor (WF) is a form of approximate correction factor used to estimate the route distances for road transport. It is defined as the ratio between the real distance travelled by road and the straight line between the two points. It is frequently used to calculate the fuel costs (which represent approximately half of the total truck costs) and hence, a certain degree of accuracy is required. The "curvature of the road" or Wiggle Factor was proposed by Cooper in 1983 [18] and he determined a value of 1.2 for UK roads. An improved methodology applied to Spanish road transport was proposed by Domínguez-Caamaño et al. [19]. They used two different WF: the first WF (1.36) describes mainly road infrastructure in rural areas while the second WF (1.29) characterizes high-capacity roads (typically motor-ways). Similarly, the total distance obtained using any model needs to be multiplied by the WF which may be suitably taken between 1.2 and 1.35.

According to the National Ocean Service, National Oceanic and Atmospheric Administration, U.S. Department of Commerce; the Earth is an irregularly shaped ellipsoid [20]. Cazabal-Valencia et al. [21] claim that minimum total coverage distance is achieved with the facility location problem considering the ellipsoid as the model for the Earth's surface. This distance is smaller than the total coverage distance achieved under the spherical model. The model will be analysed in that direction also using ellipsoidal distance. This also needs further investigation. Several works are available in the literature [22-24] in this area that use both Euclidean distance and Great Circle Distance.

This paper presents two simple models to estimate the optimal facility location point(s) on the earth surface from a data set with known geodetic coordinates (latitudes and longitudes).

2 Great Circle Distance (GCD)

The most popular distance metric is the straight line distance or Euclidean distance.

Euclidean Distance (ED) in 3D space between points (x1, y1, z1) and (x2, y2, z2) can be computed using a simple formula.

$$\text{Euclidean Distance, ED} = \sqrt{(x2 - x1)^2 + (y2 - y1)^2 + (z2 - z1)^2} \qquad (2)$$

But for spherical locations, we prefer great circle distance instead of Euclidean distance. The great-circle distance (GCD) is the shortest distance between two points on the surface of a sphere, in our case earth. Earth is assumed to be a perfect sphere with a radius, R = 6371.009 km for analysis, even though it is not so. Any location, x_j can be represented by its geodetic coordinates (latitude, longitude) on the earth surface. In spherical trigonometry, GCD measurement is based on spherical triangles [Figure 1] whose sides form arcs of great circles [25].

The shortest distance between any two points on the earth's surface lies along the arc of the great circle joining these two points.

GCD = Radius of Earth (R) * Internal Spherical Angle in radians (θ). (3)

Internal Spherical Angle, θ between the approximate supply point (P) and a demand point (Q) is estimated using the relationship,

$$\theta = 2\sin^{-1}.\sqrt{(\sin^2\Delta\alpha + \cos\alpha_1 * \cos\alpha_2 * \sin^2\Delta\beta)} \qquad (4)$$

Where:

$\Delta\alpha$- Half difference in Latitudes = $(\alpha_1 + \alpha_2)/2$

α_1 – Latitude of the first point (P), radians

α_2 – Latitude of the second point (Q), radians

$\Delta\beta$ – Half difference in Longitudes = $(\beta_1 + \beta_2)/2$

β_1 - Longitude of the first point (P), radians

β_2 – Longitude of the second point (Q), radians

These formulae work fine for all points in the earth surface except those points on the earth directly opposite to each other.

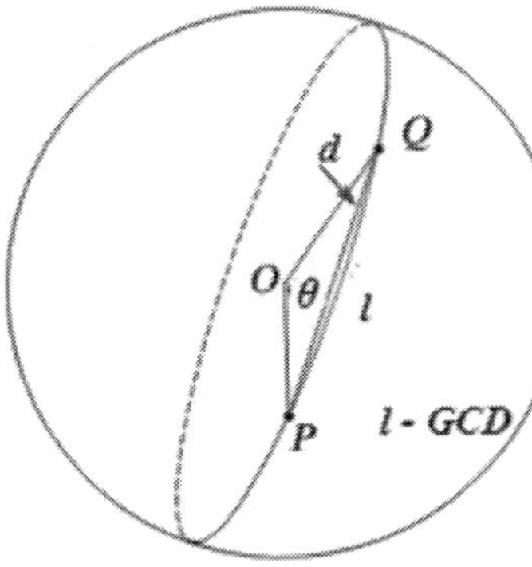

Figure 1 Great Circle Distance

3 Weiszfeld's Algorithm

For a given set of data points, the 'Mass Centre' refers to a point with coordinates that are weighted arithmetic average of the corresponding coordinates of the data set. It is nothing but the weighted mean of the given data points.

$$\text{Mass Centre, } y_0 = \frac{\sum_{j=1}^{m} w_j p_j}{\sum_{j=1}^{m} w_j} \qquad (5)$$

Where w_j is the weighted positive constant associated with the data point p_j and m is the number of data points. Weiszfeld's algorithm is used for improving the 'mass centre' towards the 'Exact Centre' or 'Fermat-Weber Point'. The exact centre refers to a point that has 'total minimum weighted distance to all points. Weiszfeld's algorithm is an iterative procedure used to find the 'Fermat-Weber Point' for '*m*' data points. The point is also known as 'Geometric Median'.

The distance may be of any. The straight line or Euclidean distance can be used for shorter distances. For more number of points on the sphere which are fairly separated, great circle distance gives accurate results. The 'Geometric Median' is estimated iteratively from the 'Mass Centre' by the Weiszfeld's algorithm using the formula,

$$y_{(i+1)} = \frac{\sum_{j=1}^{m} \frac{w_j \cdot p_j}{\|p_j - y_i\|}}{\sum_{j=1}^{m} \frac{w_j}{\|p_j - y_i\|}} \qquad (6)$$

'y_i' refers to the present facility location, '$y_{(i+1)}$" will be the new location and each data point „j" is represented by „p_j". „w_j" is the weight associated with „p_j" and „i" refers the iteration number. The iteration can be stopped if the result reaches the specific "stopping criterion".

4 Estimation of Optimal Facility Location and Results

Given a set of „*m*" data points, the procedure for estimating the optimal location is given in several steps:

Steps:

1. Estimation of geodetic coordinates (latitudes and longitudes) of all points in radians with signs.(South latitudes and west longitudes are expressed in negative values)

2. Assignment of weights, w_j associated with each data point, if required.

3. Computing the "Mass Centre" as the initial approximate centre.

4. Estimation of internal spherical angles, θ between the approximate centre and other points.

5. Computation of Great Circle Distances (GCD) between the approximate centre and other points.

6. Finding the next approximate Centre using Weiszfeld"s algorithm (considering the weights).

7. Finding the final optimal facility location point in subsequent iterations by repeating steps 4, 5 and 6.

The iteration can be stopped if there is no change in two consecutive centres or the stopping criteria is reached.

Here, the two norm represents the GCD, not ED.

5 Validation with Known Results

The Weiszfeld's algorithm is a time tested algorithm to estimate the optimal point iteratively for a given set of data points. Also, the GCD is the minimum distance between two points on a sphere. This paper combines both the concepts in obtaining optimal facility location(s).

It is required to check the coding before the analysis.

Two problems considering weights for each point used by Cooper and Katz [26] were analysed. The results obtained perfectly match with Cooper and Katz both in terms of magnitudes and number of iterations. Euclidean distance is used as the distance metric here.

The other part of the algorithm is estimating the GCD between two points of known latitudes and longitudes. The algorithm is tested using the data points available in the literature [4] and the results are matching.

Also, the GCD between New York (40.7128°, 74.006°) and London (51.5074°, 0.1278°) is reported as 5570 km by a web calculator [27] (https://www.movable-type.co.uk/scripts/latlong.html) whereas, the MATLAB program reports 5570.2 km.

Two faraway locations; Tangier, Morocco (35.7595° N, 5.8340° W (or) 35.7595°, -5.8340°) and Auckland, New Zealand (36.8483° S, 174.7625° E (or) -36.8483°, 174.7625°) were also considered. The GCD reported by the online calculator is 19880 km whereas, the MATLAB program reports 19882.77186 km. The variation is very less.

In the 8 data points problem studied by Mwemezi and Huang [4], the total distance reported is 1983.439 km whereas, the proposed algorithm which takes the geometric median as the facility location reports 1979.12 km. Both the algorithms use GCD but, Mwemezi and Huang considered the mass centre as the facility location. The proposed algorithm iteratively reaches the optimum point using Weiszfeld's algorithm.

6 Example Problem

There are 8 warehouses (demand points) in a country for a particular company to which consumer goods are to be supplied monthly. For the warehouses numbered 1 and 6, two trips are to be made in a month. The company wants to have a centralized supply point from which the supplies are to be made. Locate the centralized supply point, given the locations of the warehouses. The geodetic coordinates of the warehouses are collected and given.

The geodetic coordinates are converted from degree, minutes and seconds format to decimal format (60 seconds = 1 minute, 60 minutes = 1 degree).

Subsequently, the values are converted to radians using the expression,

$$\text{Radians} = (\text{Degrees} * \pi)/180 \tag{7}$$

Weights are assigned to each demand point, in this case, the number of monthly trips to be made to each point. The next step is to compute the "mass centre" as shown in Table 1. The initial approximate facility location, the 'mass centre' is estimated as the arithmetic mean of the coordinates of the demand points.

Table 1 Computation of Mass Centre

Point No.	Latitude, degrees	Longitude, degrees	Latitude, decimal	Longitude, decimal	Latitude, radians	Longitude, radians	Weights
1	N5°30'45"	E15°30'30"	5.5125	15.5083	0.0962	0.2707	2
2	S15°15'15"	E5°30'45"	-15.2542	5.5125	-0.2662	0.0962	1
3	N40°15'45"	W35°15'45"	40.2625	-35.2625	0.7027	-0.6155	1
4	S25°30'45"	E25°15'30"	-25.5125	25.2583	-0.4453	0.4408	1
5	S35°45'45"	E10°45'45"	-35.7625	10.7625	-0.6242	0.1878	1
6	N20°30'30"	E40°30'15"	20.5083	40.5042	0.3579	0.7069	2
7	N10°45'30"	E30°30'30"	10.7583	30.5083	0.1878	0.53254	1
8	S30°30'30"	W20°30'30"	-30.5083	-20.5083	-0.5325	-0.3579	1
				Mass Centre	**-0.06545**	**0.1576925**	

The mass centre is (-0.06545 0.1576925) radians.

First Iteration:

Interior spherical angle, $\boldsymbol{\theta}$ is computed between any demand point and the Mass Centre (a total of 8 angles are computed).

Great Circle Distance = earth radius * interior spherical angle (a total of 8 GCDs are computed).

The weighted distances are analysed using Weiszfeld's algorithm to obtain the next approximate centre (latitude, longitude) in radians.

Subsequent Iterations:

Interior spherical angle, $\boldsymbol{\theta}$ is computed between any demand point and the New Approximate Centre (a total of 8 angles are computed).

Great Circle Distance = earth radius * interior spherical angle (a total of 8 GCDs are computed).

The weighted distances are analysed using Weiszfeld's algorithm to obtain the next approximate centre (latitude, longitude) in radians.

The same procedure is repeated until there is no change in the new centre (or) the terminating condition is reached. The final supply point is the optimal facility location point.

Codes are generated in MATLAB 2012b and run in an i5 PC with 4 GB RAM.

The terminating condition is the difference in distances less than 1 km. That is, if the total distance between any two consecutive iterations is less than 1 km, the algorithm will stop. The analysis results are presented in Table 2.

Table 2 Optimal Supply Point and Great Circle Distances

Point No.	Latitude (radians)	Longitude (radians)	Optimal Supply Point (Latitude, Longitude)	Great Circle Distance from Optimal point, km
1	0.0962	0.2707		7.8343
2	-0.2662	0.0962		2553.6
3	0.7027	-0.6155		6362.2
4	-0.4453	0.4408	(0.095125, 0.271249) radians	3599.3
5	-0.6242	0.1878		4609.7
6	0.3579	0.7069	(5.4503, 15.5415) degrees	3172.5
7	0.1878	0.53254		1749.5
8	-0.5325	-0.3579		5539.6
			Total Distance	27594.32

The optimal supply point is located at ((0.095125, 0.271249) radians or (5.4503, 15.5415) degrees, latitude and longitude respectively. The total distance is 27594.32 km.

As the weights represent the number of trips between the demand points and the optimal supply point, those weights are to be multiplied with the corresponding GCDs and added to get the total minimum distance.

In our case, Total minimum distance = 27594.32 + 7.8343 (first demand point) + 3172.5 (sixth demand point) = 30774.65 km. This is the minimum possible distance for the given problem.

If the number of warehouses (demand points) is high and more than one supply point is to be established, instead of one we can go for two or more facility locations using the same Weiszfeld's algorithm.

Clustering technique should be applied in such case. If two optimal supply points are to be established, two initial centres are assumed randomly instead of one. The demand points will be attached to the nearest centre one by one. Proceeding in the same way, next approximate centres will be evaluated using the Weiszfeld algorithm, and again the demand points will be attached to the nearest centre one by one. This continues till two optimal centres are evaluated.

In the following section, a real-world application is discussed.

7 Estimation of Multi-Facility Locations and Results

In this section, a slightly different type of problem is considered. The population data of [28] 661 districts (2011) and 34 states and union territories (UTs) (2011) in India are considered and the geodetic data of the head-quarters of these districts/ states are collected from a single source [29]. It is assumed that the head-quarters represent the entire district/ state/ UT.

For the analysis using districts' data; equal weights of 1 each (GCD), % area share, %population share (2011) and literacy fraction (2011) are taken as the weights. Similarly, for the states and UTs analysis; %population share (2001 Census), %population share (2011 Census), %real GDP share (2004-05) and %real GDP share (2014-15) are the weights.

Case I, Districts wise Analysis:

The same procedure is followed to estimate the single and two optimal facility location(s) and the final results are shown in Tables 3 and 4.

Table 3 Districts wise Analysis – One Optimal Facility Location

S.No.	Parameter (Weight)	Latitude, deg.	Longitude, deg.	Total GCD, km	Location
1.	GCD (Equal Weights)	23.7881	80.7853	549066	Salaiya Sihora, Madhya Pradesh
2.	Area (% Share)	22.5975	79.1146	563381	Jhirpi, Madhya Pradesh
3.	Population (% Share 2011 Census)	23.3910	80.3029	550762	Dasharman, Madhya Pradesh
4.	Literacy (Fraction, 2011 Census)	23.6397	80.6799	549345	Chandia, Madhya Pradesh

Table 4 Districts wise Analysis – Two Optimal Facility Locations

S.No.	Parameter (Weight)	No. of Districts in each Cluster	Latitude, deg.	Longitude, deg.	Total GCD, km	Location
1.	GCD (Equal Weights)	441	26.1034	82.4589	293605	Biraili, Uttar Pradesh
		220	16.8811	77.2954	122473	Yadgir, Karnataka
2.	Area (% Share)	450	25.7642	80.6633	308465	Jukra, Uttar Pradesh
		211	17.3778	76.9419	117020	Injihal, Karnataka
3.	Population (% Share 2011 Census)	441	25.9183	82.4377	293699	Fattupur, Uttar Pradesh
		220	16.9309	76.6658	123533	Mutkod, Karnataka
4.	Literacy (Fraction, 2011 Census)	442	26.1068	82.4211	294371	Bhoopatipur, Uttar Pradesh
		219	16.5218	77.0806	122125	Tumkur, Karnataka

The GCD will be always greater than ED. The maximum radius from the centre and an extreme location in India can be maximum of 1600 km. If two major cities in India; Chennai and New Delhi are considered; the GCD = 1760.1 km and ED = 1745.4 km. That is, ED is less than GCD by 14.7 km (or) 0.835%. If Chennai and Nagercoil are considered; GCD = 624.47 km and ED = 621.23 km. That is, ED is less than GCD by 3.24 km (or) 0.519%.

The locations, number of districts in each cluster and total distance vary depends on the weight. The total GCD is minimum when equal weights are considered. For a single facility location, all the points are located in Madhya Pradesh.

For two facility locations, one cluster centre is located in Uttar Pradesh covering northern districts and the other one in Karnataka covering southern districts. More number of districts is covered in Cluster No. 1 and hence, it is preferable to go for more number of clusters,

Case II, States and UTs wise Analysis: Movements of Centres

Here, only less number of data points are considered (number of states and UTs). This part shows another important use of this model to analyse the movements of "Centres" over a period of time. Movement of centre implies the shift in activities.

The results are presented in Table 5. When the population is considered for the years 2001 and 2011, the points are not the same. The population optimal centre is shifted (blue dots above the text India) in the north-eastern direction by 20.953 km. Similarly, the %real GDP shares are analysed for the years 2004-05 and 2014-15 and it is observed that the point gets shifted (blue dots below the text India) towards the south-west direction by 44.689 km (Figures 2 and 3), that is about 2.13 times more than the population point.

Population points are above the economic (GDP) points. The accuracy will increase with an increase in the number of data points. Table 6 shows the variation in centres for different number of data points; number of districts and number of states.

Table 5 Movements of Optimal Single-Facility Locations - States and UTs wise Analysis

S.No.	Parameter (Weight)	Latitude, deg.	Longitude, deg.	Location	Distance between Centres, km
1	Population (% Share 2001 Census)	23.5216	80.1738	Chhapra, Madhya Pradesh	20.953
2.	Population (% Share 2011 Census)	23.7061	80.2158	Dhoori, Madhya Pradesh	
3.	Real GDP (% Share 2004-05	21.8824	78.3004	Betul, Madhya Pradesh	44.689
4.	Real GDP (% Share 20014-15	21.6968	77.9165	Betul, Madhya Pradesh	

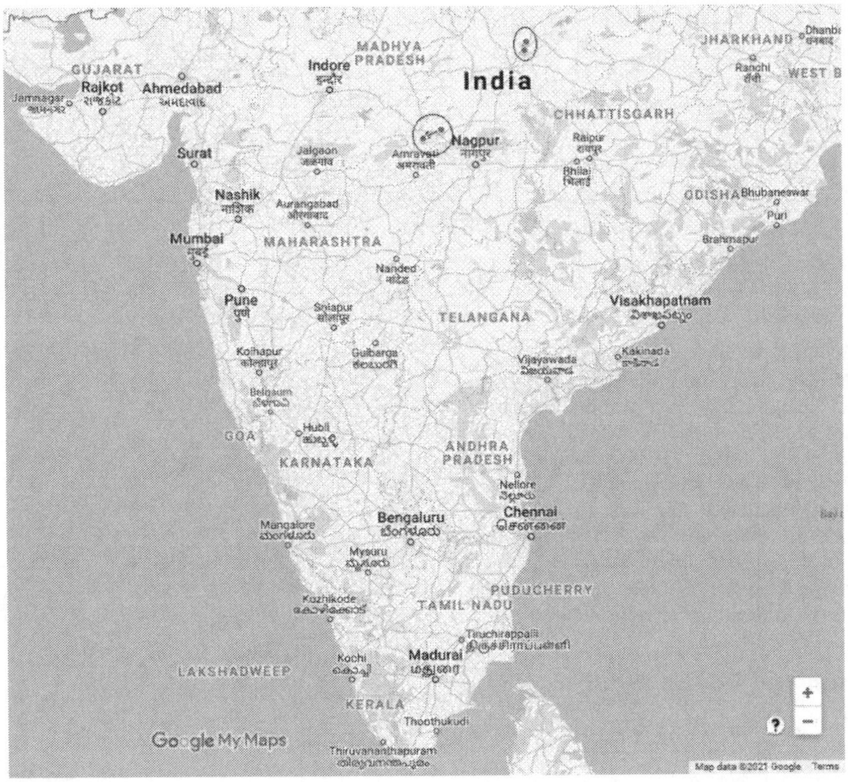

Figure 2 Movements of Population and Economic Centres in India

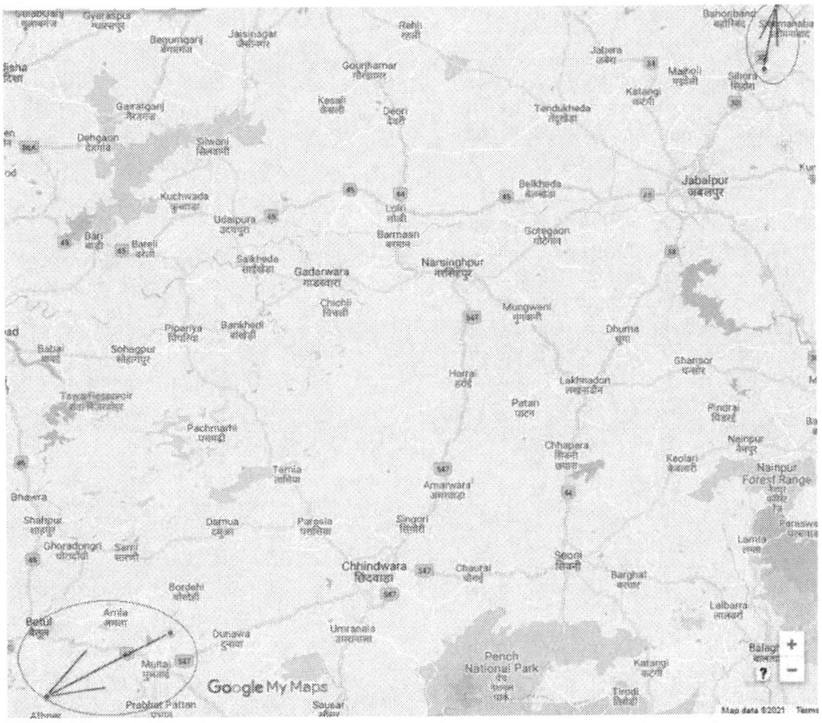

Figure 3 Movements of Population and Economic Centres in India (Enlarged View)

Table 6 Distance between Optimal Centres: States and Districts Analysis

S.No	Parameter (Weight)	Optimal Centre, (Lat, Lon): States	Optimal Centre, (Lat, Lon): Districts	Distance between Centres, km
1.	Distance (Equal Weights)	22.298, 81.943	23.788, 80.785	203.71
2.	Population (% Share 2011)	23.706, 80.216	23.391, 80.303	36.173

The points differ by 203.71 km if no weights are considered and by 36.173 km if %population share is considered as the weight.

The location found using any model lies below the earth centre. If any model uses the earth-centric earth focussed (ECEF) coordinates which is in the form of x, y and z coordinates, the location has to be extrapolated to a point at the earth's surface. The advantage of this model is, we need no such conversion as this uses the angles only during the iteration process. For example, the location (23.788, 80.785) lies 3451.0 m below earth surface in the districts wise Analysis – One Optimal Facility Location (Table 3) and the extrapolation also will result in the same values.

8 Conclusion and Future Work

This paper models both single and multi-facility location problems on earth using great circle distance which is more realistic. Weiszfeld's algorithm is used to move the "Mass Centre" towards the optimal location point(s) in a few iterations. The concept is demonstrated using an eight data points problem for computing a single facility location. The number of monthly trips is taken as the weights.

The information about the population, area, real GDP, literacy rate of India is considered as the weights and analysed for both single and double facility locations. It is shown that the locations and total minimum distances vary with the weights.

Another important use of this model is demonstrated using the population and real GDP data of states and UTs over a period of time. This shows the movement of the population and economic activities in the country, which can be extended for any weight parameter. These models will help for fixing the optimal location centres, analyse the movement of activities over a period of time that helps in management decisions.

References

1. D.R. Sule, *Logistics of facility location and allocation* (CRC Press, 2001)
2. https://www.thehindubusinessline.com/opinion/addressing-indias-high-logistics-costs/article9732208.ece [Accessed on 09-09-2020]
3. M. Zarinbal, *Distance functions in location problems*, In Facility Location (Physica, Heidelberg, 2009)
4. J.J. Mwemezi, Y. Huang, N. Y. Sci. J. **4**(7), 21-8 (2011)
5. C. Carter, *Great circle distances* (2002)
6. D. Ross, *Master Math: Trigonometry* (Nelson Education, 2009)
7. J.E. Bell, S.E. Griffis, W.A. Cunningham III, J.A. Eberlan JA, Omega **39**(2), 151-8 (2011)
8. C.S. ReVelle, H.A. Eiselt, Eur. J. Oper. Res. **165**(1), 1-9 (2005)
9. H. Shih, American Journal of Operations Research 5(05), 450-492 (2015). DOI: 10.4236/ajor.2015.55037
10. A. Weber, Prog. Hum. Geogr. **6**(1), 120-8 (1982)
11. J. Brimberg, R.F. Love, Comput. Math. Appl. **37**(8), 87-97 (1999)
12. A. Baskar, Int. J. Math. Oper. Res. **10**(2), 150-66 (2017)
13. E. Weiszfeld, Tohoku. Math. J., First Series, **43**, 355-86 (1937)
14. Z. Drezner, G.O. Wesolowsky, J. Oper. Res. Soc. **29**(10), 997-1004 (1978)
15. I.N. Katz, L. Cooper, Comput. Math. Appl. **6**(2), 175-96 (1980)
16. G.L. Xue, Comput. Math. Appl. **27**(6), 37-50 (1994)
17. P. Hansen, B. Jaumard, S. Krau, Location Science, **3**(4), 217-37 (1995)
18. J.C. Cooper, J. Oper. Res. Soc. **34**(5), 419-24 (1983)
19. P. Domínguez-Caamaño, J.A. Benavides, J.C. Prado, Braz. J. Oper. Prod. Manag. **13**(1), 52-6 (2016)
20. https://oceanservice.noaa.gov/facts/earth-round.html [Accessed on 09-09-2020]
21. L. Cazabal-Valencia, S.O. Caballero-Morales, J.L. Martínez-Flores, Int. J. Eng. Bus. Manage. **8**:1847979016668979 (2016)
22. A. Baskar, *Locating the Economic and other Performance Centres of Asia using Geodetic Coordinates, Haversine Formula and Weiszfeld's Algorithm*, in IOP Conference Series: Materials Science and Engineering, **912** (6), 062001, (IOP Publishing, 2020)

https://doi.org/10.1051/itmconf/20213701001

23. N. Wichapa, P. Khokhajaikiat, Oper. Res. Perspect. **4**, 39-48 (2017)
24. R. Abyazi-Sani, R. Ghanbari, Comput. Ind. Eng. **93**, 99-109 (2016)
25. P. Das, N.R. Chakraborti, P.K. Chaudhuri, Comput. Optim. Appl. **18**(3), 311-26 (2001)
26. L. Cooper, I.N. Katz, Comput. Math. Appl. **7**(3), 225-34 (1981)
27. https://www.movable-type.co.uk/scripts/latlong.html [Accessed on 30-01-2021]
28. http://www.census2011.co.in/ [Accessed on 10-06-2019]
29. http://www.mapsofindia.com/lat_long/ [Accessed on 21-06-2019]

ITM Web of Conferences **37**, 01002 (2021)
ICITSD-2021
https://doi.org/10.1051/itmconf/20213701002

Remote Monitoring of Heart Patients Using Robotic Process Automation (RPA)

Shwetha R[1] *, and Dr. Kirubanand V B[1]

[1] Department of Computer Science - CHRIST (Deemed to be University), Bengaluru, India

Abstract. After the age of 60 years or older many people are diagnosed with heart related issues to follow up older patients, bedridden, who needs maximum care at their homes implementation of phone monitoring becomes necessary. According to study many heart patients are supposed to have a regular check up to keep a track on their physiological data. Health care is a ground that is rapidly developing in services and technology. A recent development in this region is remote monitoring of patients which has many benefits in a fast-aging world population with increasing health complications. Sensors are used to monitor essentials or vital requirements such as heartbeat, blood pressure, temperature, blood glucose level and many more. Remote monitoring varies for every age group and every aspect. During pandemic times its difficult for many of them to move around, thus remote monitoring of patients helps to have a safer and efficient way to monitor them and also saves time. These new technologies can make a contact-less and monitor illness based on the sensor values. To make remote monitoring more efficient automation places a vital role. In health sector, to make any process automated RPA (Robotic process automation) is used. Using the established application software robots automate the process originally performed by human beings. This paper focuses on easy way to monitor patients who are in a distant place, have a regular check-up without visiting the doctor regularly and in case of emergency contact the doctor within no time.

1 Introduction

The term heart disease refers to a range of conditions which affect our heart and makes a lot of variations in our body. It varies according to the age groups. Common symptoms are chest pain, shortness of breath, numbness, or coldness in legs or arms, discomfort and nausea and many more. 4 out of 5 people who are 70 years and older die due to heart attack. All these comes under CVD (Cardiovascular diseases). Even though advanced treatments are available in the health care, due to certain conditions and lifestyles of many people Cardiovascular diseases are increasing rapidly. Few causes of these diseases are diet, body mass, stress, anxiety medication side effects genetic disorders. Diet: There are good cholesterol and bad cholesterol we need good cholesterol to our body such as HDL with a good count of HDL it lowers the risk of heart related issues, LDL and VLDL are the bad cholesterol which causes major blockages in the valves.

It can be a by-product of what we consume which accumulates in our body and it can be seen in our increasing weights. This can be due to due to bad food habits and drinking habits. Obesity is a condition which accumulates bad cholesterol and makes us vulnerable to heart diseases. It is defined by the World Health Organization as unnecessary fat that accumulate and present a risk to health. Obesity is measured in body mass index (BMI), which is a person's weight (in kilograms) divided by the square of his or her height (in meters). Weight (in kilograms) divided by the square of his or her height (in meters). A person with a BMI of 30 or more is generally considered obese. A person with a BMI equal to or more than 25 is measured over weight [1,2]. Rest of the issues such as anxiety, side effects are inevitable. Most of these causes can be diminished with a simple change in lifestyle.

1.1 Remote Patient Monitoring

Remote monitoring and automation is a process that can help reduce some of the burden off of the network administrators. It keeps a track on all the activities. Automation software works more efficiently than a human being. This improves overall efficiency and maximizes their up time. It is an extremely non-intrusive way to gather data and information about patients. In a pandemic situation like covid-19 it is difficult to go to a check-up regularly and spend more time on it. The patients who are just discharged after a heart surgery will have to get a check-up regularly, this automated remote monitoring will be useful to those patients which helps in monitoring our heart rate with a wearable sensor and analyse the blood pressure and notify the doctors if there are any abnormalities. This reduces time and also it is safe as the patient can get a quick check up in their desired location.

*Corresponding author : shwetha98.r@gmail.com

 https://doi.org/10.1051/itmconf/20213701002

1.2 Why RPA?

Robotic process automation is the only automation technique used in the medical field. This is more reliable and gives more accurate results. Human errors can be reduced and thus has more accuracy. This can save a lot of time and during any major emergencies it can notify the doctors early can possibly save a person's life. It accelerates time to value it where traditional way takes few days create test and deploy. RPA takes just few hours. Increases the throughput. This way we can experience more and better outputs in less time. RPA software can manipulate collected data to generate analytic that offer clinical staff valuable insights to help them make more accurate diagnoses and offer tailored treatments to patients. This method is only for a regular and basic check-up such as monitoring the pulse rate and blood glucose level. In case any variation in these parameters' patient can be sent to hospitals on doctors' advice. This saves a lot of time for both doctors and patients. This is a better approach than the traditional method. In a pandemic situation it is always better not to go out much. This helps the patient to get a quick check up and the doctors will be notified if there is any abnormalities found.

Table 1.RPA Process

Front Office	Middle Office	Back Office	Regulatory Compliance	Data Intelligence

Primary Objective of this paper is to introduce RPA in remote monitoring. It acts as an add-on to an existing model; RPA is used major in Health Sectors. It can automate our anti-virus solution, patch updates, backup process and create tickets through the web portal that will notify you if something is awry. Using RPA, health care companies can automate the check-in process for patients by using self-service terminals integrated into the facility's database. The patient's conditions could also contribute to how much of a priority they are, this would allow the RPA software to automatically organize the list of patients by priority as they entered the system. Automation helps in reducing risks and gives better accurate results and it diminishes a lot human error.

2. Literature Review

There are many different ways in managing a patient and monitor them remotely. All the techniques are mainly focused on how to make use of technology well and help patients in early detection and prevention of worsening of illnesses. This has a lot of advantages for both patients and hospitals. This can reduce the resources usage and also this is efficient way to keep a check on recently discharged patients. These are the main objective of remote monitoring.

Remote monitoring makes more sense for the patients who are recently discharged.

The management of patients with prolonged heart failure (HF) induces a high burden on health care resources due to the regular review, visits combined with repeated hospitalizations due to cardiac decomposition.[2] Early identification of HF deterioration is crucial to prevent HF-related hospitalizations, potentially improve overall survival and quality of life and lower the burden on health care resources. Remote monitoring of chronic HF patients can aid in the detection of HF weakening; therefore, several remote monitoring strategies have been developed.[3]A recently updated Cochran's review included 41 randomized controlled trials (RCTs) investigating the use of structured telephone support or non-invasive tele-monitoring compared with standard HF care [3]. This review helped in getting a into a model where we can make the process of remote monitoring more efficient. As most of them prefer a tele-monitoring first. This increases the growth of the automation of this process too.

With the help of wearable sensors, the patient can test his heart rate and blood pressure by himself. Multiple large multicentre prospective clinical studies and RCTs have examined multiple non-invasive remote monitoring strategies, ranging from indication and body weight monitoring to complex and intensive strategies including body weight, blood pressure, electrocardiography and peripheral capillary oxygen saturation. Standalone systems allow patients, usually in their own homes, to send non-invasively measured data to their health care team, by either telephony-based systems or the internet. Heart rate variability (HRV) discusses the variation (over time) in the period concerning the consecutive heartbeats in a person, which reflects his physiological signs that modulate his heart rhythm. HRV is an indication of the current heart-related abnormalities and can warn about the imminent cardiac diseases.

[4] RPM via regularly scheduled structured telephone contact between patients and Doctors/health care providers or electronic transfer of physiological data using remote access technology via remote external, wearable, or implantable electronic devices is a growing modality to manage patients with chronic HF. [7] Here the electronic transfer of data made more. Various technologies such as actual ECG monitors or textile- based wearable systems are used to get these data. Remote monitoring represents a novel intensified follow-up strategy of heart failure management.[6] The doctors can monitor the deterioration or maintenance of the patient while also advising the paramedics who are physically with the patients as necessary. Doctors can work on two consecutive things and attend more patients. An application for Android OS, which is widely applied in portable devices all over the world, was developed by our team. Its main tasks include data processing reporting of test results to the user test scheduling during the day dialling of automatic emergency calls or sending an SMS in case of critical user conditions [5] this application also provides controls and Bluetooth connections for easy access and report generation for
the patients.

This method helps in monitoring elderly and patients with heart issues. Mobile Application is easier

https://doi.org/10.1051/itmconf/20213701002

to handle. The engineering varies from sensors attached to body to sensors attached to the environment and new innovations allow contact less monitoring which requires only the patient to be present within a few meters from the sensor. System can measure body temperature, ECG, saturation levels, air flow in lungs oxygen, and galvanic skin response. [10] Personalized algorithms are used to make the predictions equal to the expectations. Although majorly to detect abnormalities threshold algorithms are used. AI in health care include implementing ML approaches to track individual patient health, deploying advanced analytic to scale-invariant feature transform through large data sets to realize clinical skillfulness.[9] In all the above-mentioned papers authors have made a survey on couple of patients and given their results. All the RPM (remote patient monitoring) techniques here focuses on early detection. This is easier for recently discharged patients, giving them a quick check up once a day and reporting to the doctor instantly. RPA is more accurate than human workers and runs at a higher throughput. They can greatly benefit from RPA thanks to its advantages which could be even more powerful if machine learning is applied to the RPA processes.

2.1 Existing models

All the existing models have their own add-on to this method, each method helps patient analyse their reports this also keeps a transparency between doctors and patients. The most suitable method to perform our implementation was using wearable sensors and web application to show the data and send it to doctors as well. This method consists of wearable sensor and a device and a web server. Here patient can wear sensor and check his heart rate or pulse rate and that data will be stored in the web application. The data stored will be sent immediately to the doctors and notify the details of the patient. This method transmits real-time data wireless from wearable devices worn by the patient to second tier of the system via Bluetooth low energy. Zephyr BT is a wearable sensor which has been used to extract heart rate information of the patients requiring continuous monitoring.

As this consists of low energy, we need not worry about the other technical issues such as battery and usage. This records the heart rate of the patient at that particular time and stores that information in a web application this can be done in an android application as well. To even check the progress this application can be used. It can keep a track on everyday status and this gives a clear idea to patient to diagnose themselves. Many Machine Learning algorithms are implemented to get better accuracy and get better predictions. Each technique follows a pattern and a different algorithm. Similarly, many methods have been implemented so far and have used threshold algorithms in order to have a track on the patient heart beat or pulse rate. Similarly, many methods have been implemented so far and have used threshold algorithms in order to have a track on the patient heart beat or pulse rate. This way we can monitor heartbeat of recently discharged patients and follow the instructions of the doctor on a phone call that is using tele-monitoring system. Below mentioned are few models which works similarly to our proposed method.

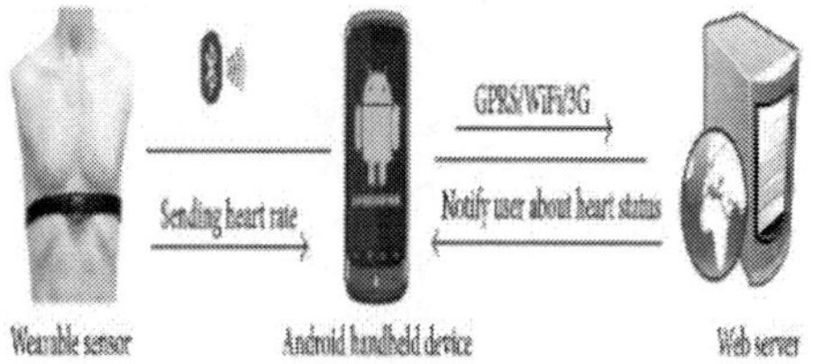

Fig. 2. Working of existing model-1

Above Figure-2 shows the working of how this method works, it is reading the values using sensor and sending heart rate using Bluetooth and that value is sent to web.

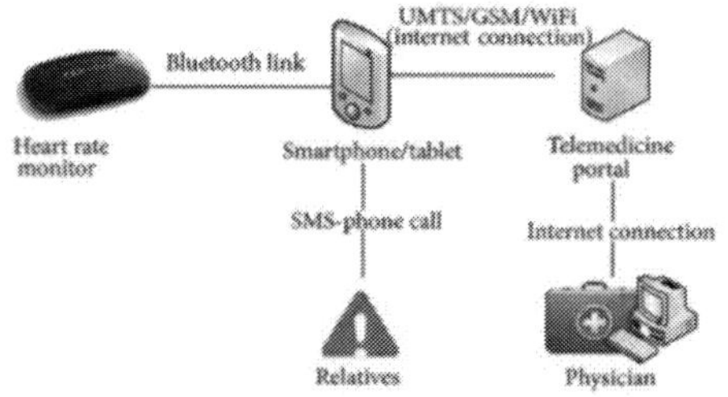

Fig. 3. Working of existing model-2

The working sequence of the existing model 2 is given in Figure 3. This is one more method where RPA can be introduced even in this technique patients heart rate is recorded and that data is stored in smart phones. Using telemedicine portal, a doctor will be notified. Here Heart rate monitor is again a sensor which is linked to a phone using Bluetooth the data will be printed. This is exactly how it works in smart watch where it takes data from human body based on their activity and data will be displayed in a format on a mobile application. Both the systems are based on a simple, inexpensive, and wearable device whose functionalities are comparable to those of more expensive and sophisticated clinical devices.

3. Proposed Work

3.1 Data Collection

This Data set is a real time sensor data which consists of 120 patients diagnosed of heart conditions based on various medical factors such as Blood Pressure, Temperature, Pulse, Glucose level. The attributes in the data set are Patient ID, Patient Name, Gender, Age, Temperature, Blood Pressure, Pulse, Blood Glucose and Heart Rate. Many operations were performed on this data set. Initially a co-relation of each feature in data set was calculated. It was found that heart rate,

temperature and blood glucose level are co-related. The co relation explains these values are directly proportional. When Temperature increases the Heart Rate, temperature and Blood Pressure also increases.
These indications help to analyse the data set better and perform functions on it.

KNN algorithm is used to set target classes which are approximately of same size. KNN is the most efficient algorithm used for pattern recognition. Using this algorithm, we are estimating if the person is suffering from any abnormalities or not. This was done based on the target classes which was male and female.
While processing this data, converting few explicit variables into dummy variables and scale all the values before training the ML models are necessary.

KNN model was trained and different K values and the scores with K neighbors and different K values were recorded. This helps in understanding the prediction of heart issues on that particular patient. Further on this Robotic Process automation implementations and its working can be seen. This prediction can help in a lot of ways we can even increase the value of K and try it on N values which is gives more understanding on it. The graph below is the representation of K Neighbours Classifier scores for different K values.
With the resultant data of the trained KNN model, we will be estimating the extent of the threat when it exceeds the threshold value, the Doctors, Patients or caretakers will be notified of the diagnosed health condition instantly thus reducing the fatality rate because of negligence or delay in treatment protocol.

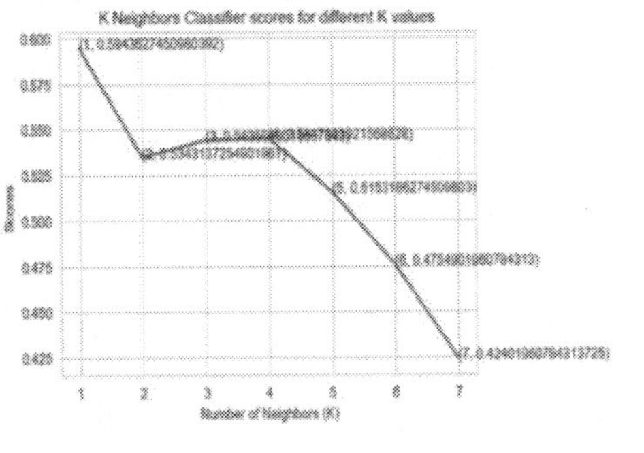

Fig. 1. Representation of K Neighbours Classifier scores for different K values.

4. Implementation

After collecting suitable existing models, Automation can be implemented very easily on web applications or mobile applications. Thus, using wearable sensors data should be collected and the values will be displayed in a web application/excel and those values will run in BOT using UiPath. In this we can collect multiple patient data and check which patient has what abnormalities, we can consider any attribute such as temperature, blood pressure and heart rate. Here we are using BOT to automate the process of getting data and comparing it and sending out the alters. ID of every patient will be unique thus we can search for a particular patient. Most of the work will be done by BOT. It sorts the data and sends the alerts when there were any abnormalities found.
To check for abnormalities and send out triggers we are setting a threshold value and comparing that threshold value with the BOT and send a notification regarding the abnormalities, we can fetch the criteria and perform operation on it. Figure 4 and Figure 5 represents the sample data which consists of criteria such as Heart rate, Blood Pressure, Temperature. Here which ever factor is higher than the threshold value that data will be displayed. In this we have considered heart rate. In the Fig. 4 it shows that heart rate is high we can specify the patient ID and display an alert saying Patient 1234 Heart rate is high. This can be done for multiple values at a time. This method helps the doctor to attend multiple patients at a time.

The same process was performed on the sensor dataset which has reports of 120 patients as mentioned earlier. The same BOT runs and checks for all the 120 patients heart rate and other attributes and whichever value is higher than the threshold value those value will be stored in a new file. This is the file which doctor gets to see first and can immediately contact the patient. Here RPA is automating the process of manual general check-ups for a patient. Certain examples where BOT shows a notification.

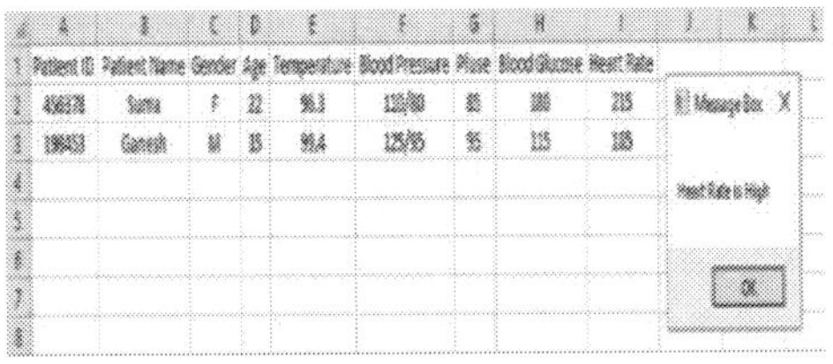

Fig. 4. High heart rate notification.

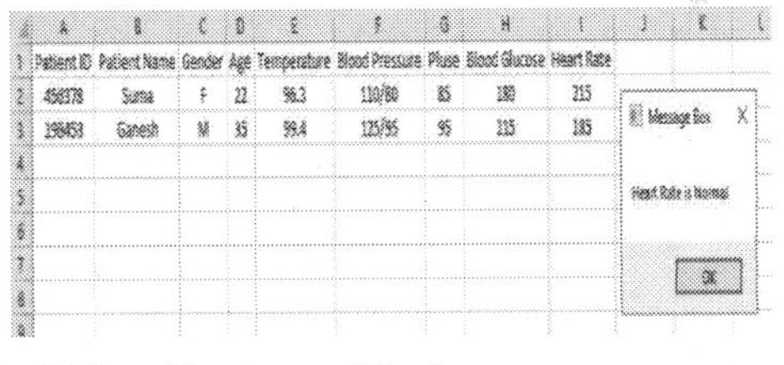

Fig. 5. Normal heart rate notification.

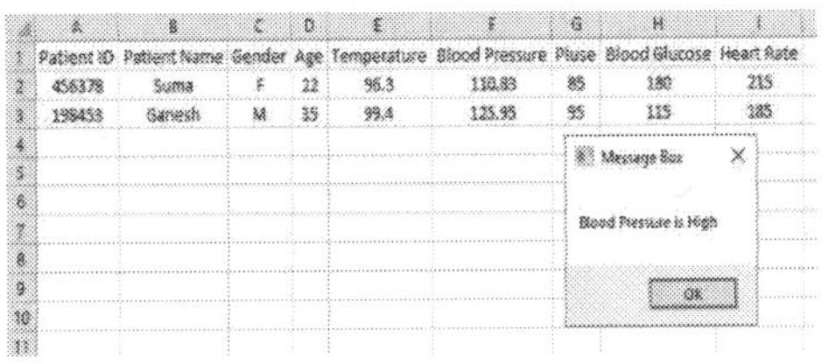

Fig. 6. High Blood Pressure notification.

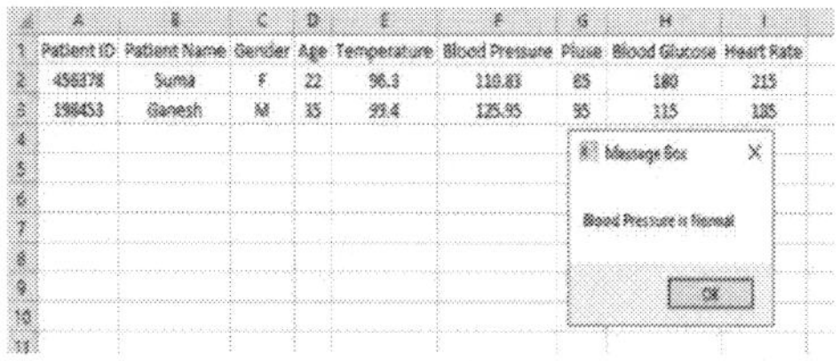

Fig. 7. Normal Blood Pressure Notification.

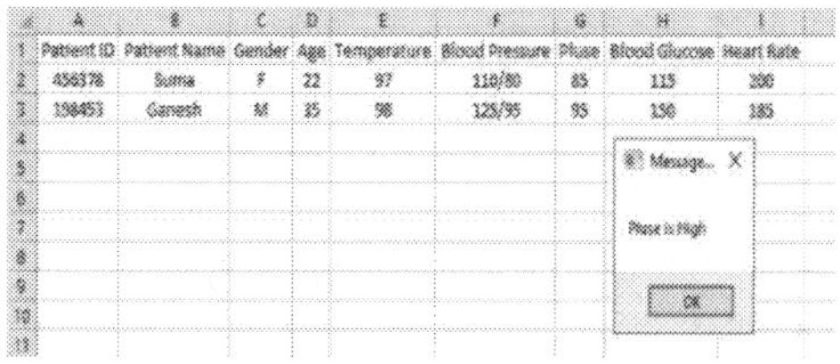

Fig. 8. High Pulse Notification

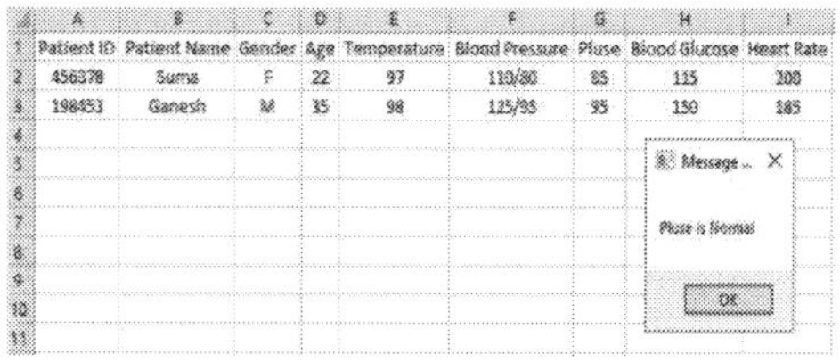

Fig. 9. Normal Pulse Rate Notification

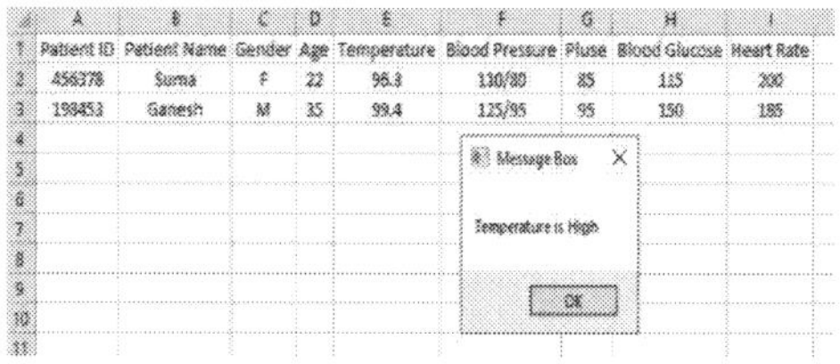

Fig. 10. High Temperature Notification

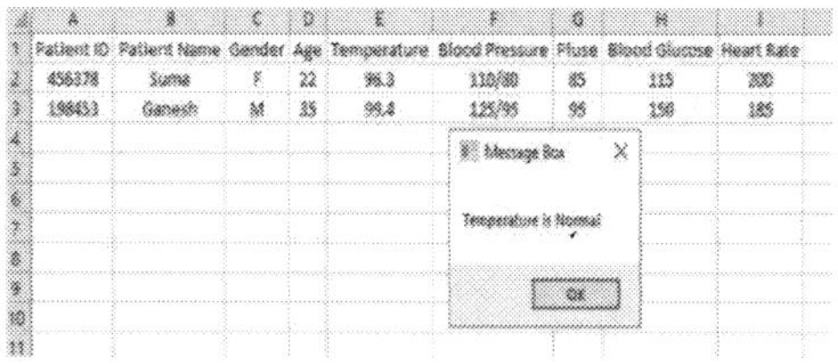

Fig. 11. Normal Temperature Notification

The above given images in Figure 4, 5, 6, 7, 8, 9, 10 and 11 were only for lesser number of patients. Same implementation was applied to a larger number of patients.

5.Results

If any of the values are greater than the threshold value, the comment," Please check on the patient records." will be added by the records. This helps the doctor to contact the patient immediately. The data set is processed and the output will be stored in a different file where only the values greater the threshold value will appear. Age and Heart Rate goes hand in hand. Result Graph represents the age and Heart rate. The person is at more risk with the increasing in age. Age is directly proportional to heart rate.

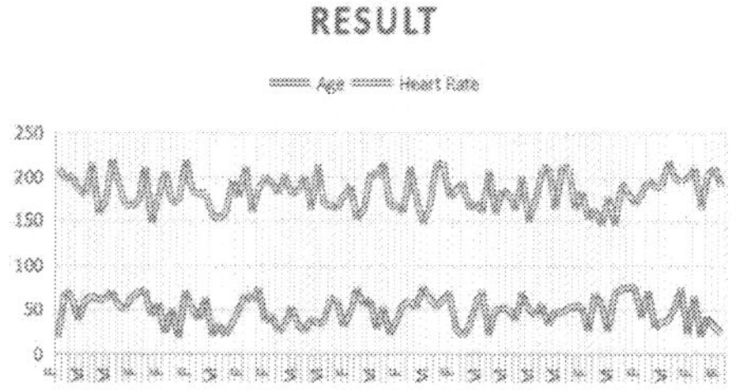

Fig.12. Result Graph

RPA is implemented on most of the business and it is more actively used tool. Many articles have been published about the usage of RPA in health care sectors. RPA in health care results in tracking and tracing many standardized, patient specific routes. It enhances the data confidentiality as well.

5.1 Advantages and Disadvantages

There are many advantages in this, Robotic Process Automation helps in faster and better execution and also gives more accurate results. It can be considered as an evolved version of present system. Implementing RPA on present or existing methods save a lot of time and money. It provides a better and efficient way to keep a track on the patient details. Here there is no need of any manual entry of the details so it also reduces human errors. This is very helpful for busy doctors who can just take a quick glance of multiple patients at a time and have a telecommunication mode soon. As far as patient is concerned, he can go to doctor only when there is abnormalities and only when doctor contacts them. This way we can save time and help elderly or mentally challenged people to have a treatment at home also get a good monitoring system. This reduces a lot of risks.

Disadvantages can be regarding signals and bandwidth patients should always have a stable internet connection and should also take care of their sensors and their batteries, any fault in any of these may result in a wrong output. This can again put the patient at risk in case of emergency.

5.2 Comparison with traditional and automation methods

With traditional method only one task can be performed at a time. For instance, only one patient data can be seen and the doctor should manually check for abnormal entries. But, using RPA we can perform multiple tasks at a time. All the process can go hand in hand without any disturbances. Doctors can prioritize the patient list according to the alerts available. Doctors can check many patient data at once and the check-up will be done much faster than the usual check-ups. patients need not wait for a long period of time. In traditional methods each patient to should visit the doctor and get a clarification, this is very risky during pandemic like covid-19. And also, patient has to go through many other safety measures before getting into hospital. Instead for all the general check-ups patient can do it at home with these methods and doctors can notify if there is emergency.

6. Conclusion

This research work concludes that Application of Robotic Process Automation helps the patients and doctors to get a report sooner and also reduces time to travel to hospital each time for general check-up. Instead, they can go to doctor only when it's necessary. This method is cost efficient and accurate compared to other traditional methods.

RPA check that all process steps are traceable, documented and tracked, as well as consistently organized in well-structured log offers a quick solution. Software robots can provide very much needed helping hand to the industry facing new requirement as times are constantly changing.

Acknowledgement

The author honestly thank the research guide for their continued support and motivation.

References

1. Priyanka Kakria, "A Real-Time Health Monitoring System for Remote Cardiac Patients Using Smartphone and Wearable Sensors, "Article Volume 2015.
2. Maurizio Volterrani, Barbara Sposato, Remote monitoring and telemedicine, European Heart Journal Supplements, Volume 21, Issue Supplement, 1 December 2019
3. Brahmbhatt, Darshan Cowie, Martin. (2019). Remote Management of Heart Failure: An Overview of Telemonitoring Technologies. Cardiac Failure Review. 5. 86-92. 10.15420/cfr.2019.5.3.
4. A. F. Hussein, N. A. kumar, M. Burbano-Fernandez, G. Ram ́ırez- Gonza ́lez, E. Abdulhay and V. H. C. De Albuquerque," An Auto- mated Remote Cloud-Based Heart Rate Variability Monitoring System," in IEEE Access, vol. 6, pp. 77055-77064, 2018, doi: 10.1109/AC-CESS.2018.2831209.
5. An Android-Based Heart Monitoring System for the Elderly and for Patients with Heart Disease Paola Pierleoni ,1 Luca Pernini ,1 Alberto Belli,1 and Lorenzo Palma 1 Dipartimento di Ingegneria dell'Informazione, Universita ́ Politecnica delle Marche, Via Brecce Bianche, 60131 Ancona, Italy Aug 2014
6. Mihran Martirosyan, Kadir Caliskan, Dominic A.M.J. Theuns Tamas Szili-Torok (2017) Remote monitoring of heart failure: benefits for therapeutic decision making, Expert Review of Cardiovascular Therapy, 15:7, 503- 515, DOI: 10.1080/14779072.2017.1348229
7. A Meta-Analysis of Remote Monitoring of Heart Failure Patients, Journal of the American College of Cardiology, Volume 54, Issue 18, 2009
8. Andrew Steger "What does robotic process automation look like in Health care" HealthTech magazine January 2020
9. Developing an Artificial Intelligence–Enabled Health Care Practice: Rewiring Health Care Professions for Better Care Author links open overlay panel David Wiljer MA, PhD ZakiHakim MSc, MHIa November 2019.
10. Remote patient monitoring: a comprehensive study Lakmini P. Malas inghe1 · Naeem Ramzan1 · Keshav Dahal1

Hardiness sensing for susceptibility using American Fuzzy Lop

Krishna Prakash R[1]*, *Vasudevan* I[2], *Indhuja* I[3], *Thangarasan* T[4] and *Krishnan* C[5]

[1] Kongunadu College of Engineering and Technology, Trichy, India
[2] KSR College of Engineering, Tiruchengode, India
[3] Mahendra College of Engineering, Salem, India
[4] KSR College of Engineering, Tiruchengode, India
[5] Mahendra Institute of Technology, Namakkal, India

Abstract. American Fuzzy Lop is an automated software testing method to give unexceeded values, else random data's as input to computer programs for testing the hardiness process. Existing fuzzing methods will come under carried based type. Pointing to test a certain input that covers a certain region of units to fit under retain part. Still, a susceptibility over a program space may not arrive paraded in all execution that occurs to see the certain program parts; just some program executions will go the location could disclose the susceptibility. In this paper, we introduced a unified fitness metric known as direct board, which can be used for American Fuzzy lop, and that is explicitly pointed toward exploring for test input that can disclose susceptibilities. To improvise the AFL, we have enhanced its methodology to produce a more effective quality fuzzing tool. This causes our method to notice buffer invade as well as the whole number invade susceptibilities. Enhanced AFL is tested with similar benchmark programs to compare it and adding an extension to it called AFLBLEND. By our method, many uncover susceptibility could be found with a given amount of time and faster than the AFL approach by 8%.

Keywords: AFL, American Fuzzy Lop, Fuzzing.

1 Introduction

The unusual behavior of the software with a bug in the code is named software susceptibility [20]. It is done by hostile attackers and they can potentially manip-ulate to attain control over the software during its execution. To uncover such susceptibilities, it automatically pointing out test inputs in the prevailing soft-ware that helps the developers to bring in the software more protected. In the target program, the runtime errors are revealed by using the search-based soft-ware testing (SBST) [1, 2, 3] during the time of execution. Search Based Software Testing (SBST) is a class of methods with a huge number of inputs produced au-tomatically. To uncovering an error in these methods, rely on a fitness metric for test inputs to lead the search towards the test inputs that fulfill a testing objective. Compare to other test inputs with a greater fitness cadent and is closer to ful-filling the testing objective and consequently produces the newer inputs. In this way, the test inputs possible to attain the test objective to get elicited eventually.

Some various practical tools and techniques are imposed with a grey box fuzz-ing [4]. It is a certain kind of SBST approach. On each test inputs, we intend to use the lightweight instrumentation to obtain a profile from the run of the pro-gram and to utilize this profile to evaluate the fitness cadent of the interrelated test inputs. Coverage based fuzzers is a method having the object of uncovering test inputs as promptly as possible and that cause execution to attain as several parts of the program as probable. The other class of grey box fuzzing tool is known as directed fuzzers [11], use a fitness cadent that brings the program exe-cution closer to a defined susceptibility location and fitness cadent mainly gives priority to the inputs. The inconsistent existing is incremented at the location which has high enough value when its location is attained, an integer overflow [19] susceptibility in a program location will be demonstrated only in the test runs. To generate an input we need to be fuzzed further to reach the susceptibility locale that is induced by the test inputs. And it not only attains the susceptibility area but also uncovers the susceptibility. However, the intention of both cover-age-based fuzzers and directed fuzzers is being met when a test run attains the susceptibility area and it would not fuzz the interrelated test input further. This process stimulates our key hypothesis, which leads the search towards the test inputs, which one wants vulnerability specific fitness metric that not only attains the susceptibility areas, but it obtains them along with

* Corresponding author: krishcommon@gmail.com

the execution paths that the program states result cause the susceptibility to be uncovered. We nurture this concept in the rest of the section.

1.1 Causative Example

The illustration flowchart showed in figure 1 performs as a running and motivating representation. And this flowchart initially reads the input from a buffer input (step 2) and character 'b' copies all occurrences in the input from another buffer outC (step 7). The program units are tagged with letters A to F at (step 7) the buffer overflow [12] error remark that this program is susceptible. The buffer outC size is 250 if the input comprises more than 250 characters 'b' of happenings then the outC overflow at this statement. The well-known coverage based grey box fuzzing tool is AFL [6] here we discuss pertain why coverage-based fuzzing [5] may not be enough to find susceptibilities. The branch of pair is outlined as it's a pair of sequentially executable branches of the program e.g.: in (figure 1) AB, AC, BD, BE, CE, and EF, are instances of branch pairs. To each branch pair during an enactment, the branch pair contour counts the number of visits of each branch pair. And it further buzzing off for at least one branch pair bp, AFL evaluates the newly produced test input t, to be capable and maintains. the visit count of the same branch pair by each execution on each other maintained test input is significantly distinct from the visit count of bp by the enactment on t so far. AFL uses the higher bar significant difference and many test inputs dramatically slow down the persuasiveness and efficiency of the search as maintaining too. Approximately speaking if their logarithms are the base 2 are distinct that two counts are significantly varied.

The running example is shown in figure 2(partially) the prospect tree of devel-oped test inputs that may be produced by the AFL [11]. It is used because due to randomization in the transformation operations. A similar tree may not be produced in every run of AFL. the relationships between parent-child are characterized by the edges in the test inputs and the base node is a given seed input "a ". From the middle, tree runs vertically down we may concentrate on a specific path. At the right side of the input, the blue color indicates the branch pair profile and its test inputs. The number of 'b's in the input is illustrated by the number suggested within the round parenthesis for each test input on this path. There is no symbol other than 'b' in the input. For an instance, we can ignore all the redcolored numbers within the boxes, as they are relayed to AFL [11]. It contains six inputs, the first four inputs will be maintained in this path, but the fifth input will not be retained because the contour of the parent and the branch pair contour is not relatively varied from its parent profile (i.e., 131 'b's). Therefore, the fifth input would not be produced the sixth input (I.e.,261'b's). Remark that at statement 7 the sixth input would have been the first one to inflict the buffer to over-flow in this path.

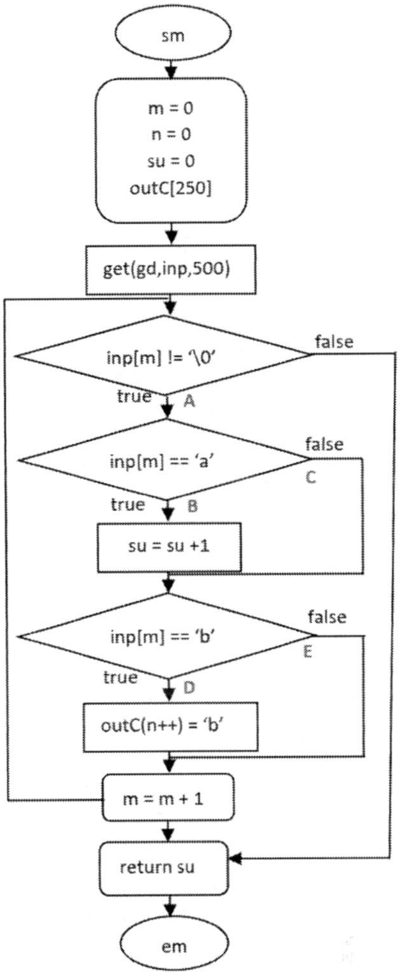

Fig. 1: Example program for buffer overrun susceptibility

In supplementary phrases, by mutating the fourth test input the AFL would have to produce an input comprising more than 250'b's immediately(i.e.,131'b's) and this process can transpire only via a mutation operation that puts in 129 or more 'b's in a single operation. The process is set up in a way that the randomization in AFL is the ratio of inserting a bigger number of identities in a single mutation operation are instantly proportional to the number of mutant test inputs produced by AFL so far. Thus, such "large" mutations strive for a lot of time could elapse earlier.

In mutation operation, the run of AFL ahead in a simple implication here would be to improve the proportion of putting a larger number of bytes. Regard-less, this straightforward idea would originate extensively input files before on, thus each test input duration of running is improving, and holding back the number of mutants is produced per unit time. And these several large mutations may not beneficial also, e.g., ones that insert an extra number of 'a's than 'b's. there-fore, the uncovering susceptibilities of the possibility may not be obtained before.

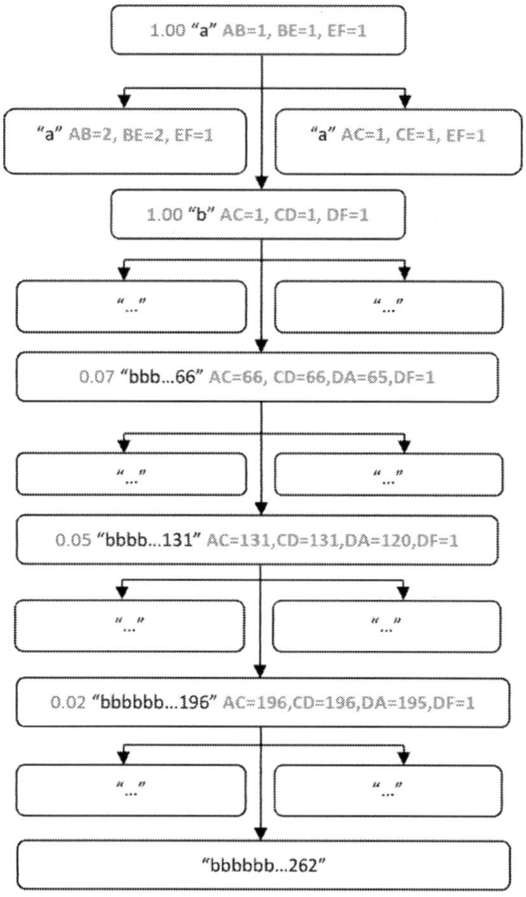

Fig. 2: Production of test input for fuzzing

This issue is not inevitably significant in addressed by even directed fuzzing. In our instance, the first-time (step 7) is breached a directed fuzzer would con-template its matter-of-fact as have been met the first moment. That this would transpire with modest input 'b' itself, and which does not uncover the susceptibility.

2 Susceptibility Spotting Method

In this section, by extending our method to detect susceptibility over buffer run and the methodology used for it.

2.1 Methodology

Susceptibility is a glitch at a susceptibility location s_l in a program P such way on certain carrying out paths will reach s1 as an error occurs, in that way the malicious attackers will lead to having unexpected action and overtake the control to serve their purposes. Susceptibility is of various types, e.g., buffer overflow, integer overruns, reuse after completion, etc. Let S_L be the set of altogether susceptibility locations in the given program P. (This methodology for various susceptibility locations to display several types of susceptibilities.) A chiefboard c ∈ [1,0] is an evaluation of how near a susceptibility is found in susceptibility location s1. The least value represents how close is susceptibility found, as c=0 gives a higher probability of susceptibility.

To explain the functioning of ISNEARER that takes a susceptibility profile N_h from a given input, compares it with N_l and gives an answer containing a Boolean value, and updates the N_l value.

```
bool ISNEARER(Nh)
{
        scp = ∃Sl ∈SL. Nh(Sl) ≺ NL(Sl)
        Nl = min(nL, Nh)
        return scp
}
```

The function ≺ is a comparison of log2 plate; i.e., its second value is above 0.50 then it returns true if the first value is below 0.50; again, if the second value is betwixt 0.25 and 0.50, it delivers true if the first argument is lower than 0.25 and below.

2.1.1 Our Methodology

Our methodology is a combination of the AFL model to produce AFLBLEND[7], where more than two values are tested and involved at the same time using their functions. To accomplish two sets of given test inputs. Set 1: Until getting of dissimilar values compared to previous test values. This belief is acquired from a report conducted from fuzzing technique [13], and is meant to assure that it gives various points near to susceptibility locations are beget. Set 2: Given test input doesn't reach new values, but gets lower chiefboard at susceptibility locations greater than previously acquired test inputs.

Argument 16-23 of algorithm 1 add up the freshly generate test values v_g if both v_g achieves a various branch couple profile than some inputs in V_G, as tried by AFLBLEND's reinforced function CHECKIF, else it achieves less chiefboard than older generated test input for some susceptibility location, as tried through the function ISFINAL, that has already shown in part 2.1.

Later algorithm 1 is concluded, the susceptibility disclose test inputs are obtained by selecting the test inputs in V_G where susceptibility visibility has a zero-entry for any susceptibility location.

3 Initialize chiefboard to Particular Susceptibility Type

In this part, by establishing the idea of the impression of chiefboard to initialize it to a real-time susceptibility.

3.1 Susceptibilities in Overflow Buffer

A buffer overflow comes when a buffer or array is scripted to above its limits. It is dominant susceptibility in the real-world program; for example, it ranks third in the top 10 harmful program errors in the CVE database [8].

Seeing a buffer approaching location s1 of the form "*loc = ….", where loc is a pointer, and consider a particular visit to this location during a run. Let L_c be the value of loc during this call, and let L_h and i be the initial address and size of the allotted buffer within that loc is alleged to choose to during this call. Then define the chiefboard c1 as follow below:

$$C1 = \begin{cases} 0, \text{ if } L_c \geq L_h + i \\ \frac{Lh+i-Lc}{i}, \text{ if } L_h \leq L_c < L_h + i \\ 1, \text{ otherwise} \end{cases}$$

At last, the chiefboard for all the run at s1, i.e $N_h(s_1)$, is outlined as the low value of i_l as explained above among every call to s_l while running.

Algorithm 1: Detecting Susceptibility by Testing
Essential: Main Program MP, and a set of input v
Assure: A tree of various input to test V_G for MP.
Create an abandon tree V_G of various inputs.
$[f_v, i_v]$ = EXECUTE(MP, v.data)
n_r.data = v n_r is a newly tree set
n_r.fit = EVALUATEFIT (f_v)
n_r.reasonToAcuqire = "original"
V_G.setRoot[n_r]
$N_L = \lambda s_l \in S_L.\ 1$
repeat
get n = CHOOSENEXTI(V_G)
get T = LETFUZZPOSSIBLEI(n)
get V_n = ACQUIREOFFSPRING(n,T)
while all v_g in V_n do
    Get $[f_g, i_g]$ = EXECUTE(MP, V_g.value)
    Get N_g = EVALUATEFIT(f_g)
    Get N_h = SetVPROFILE(i_g)
    if CHECKIS(N_g) V (c = ISFINAL(n_h)) then
        v_g.check = N_g
        v_g.causeToAccquire = "original"
        if c then
            v_g.nearest = N_h
            v_g.causeToAccquire = "chieboard"
        endif
        appendnew(V_G, V, v_g)
    endif
end while
until manual termination
return V_G

4 Executions

In this part, we discuss the execution of our method, on finding over the buffer overflow and value overrun susceptibility as mentioned in part 3. In this method the extension is AFL. Where AFL has quite famous for researchers to execute and measure extension to fuzzing [11, 9, 10]. The AFL is a method in C program fuzzers. Additional extension AFL changes to have two sets of test inputs, from set 2 to fuzzing, and evaluate the fuzzing potential for set 2 input if got importantly than susceptibility data than another set 2 test inputs in the tree V_G. The changes were discussed in part 2.

```
void *glb_obuf;

int gld_siz_obuf;

void main() {

        char in[100], out[50], c;

        int i = 0;

        glb_obuf=(void *) obuf;

        glb_size_obuf = 50;

        fgets(in, 100, stdin);

        while((c = in[i++]) != null) {

                _hdr_calc_bov(1,glb_obuf,
(void *) out, glb_size_obuf);

                *out++ = c;        }

}
```

Fig. 3: Program for buffer overflow

5 Valuation Using Benchmark Programs

To evaluate our Buffer overflow susceptibilities (BOS) and Value overrun susceptibilities (VOS) a set of eight benchmark programs recognized by "MIT Benchmark" compounded by "SV-COMP benchmark" [15]. The bug size taken for valuation is mentioned in the x-axis up to 25 in figure 4 likewise 160 in figure 5. This method is compared with AFL and AFLBLEND is our tool extension added to it.

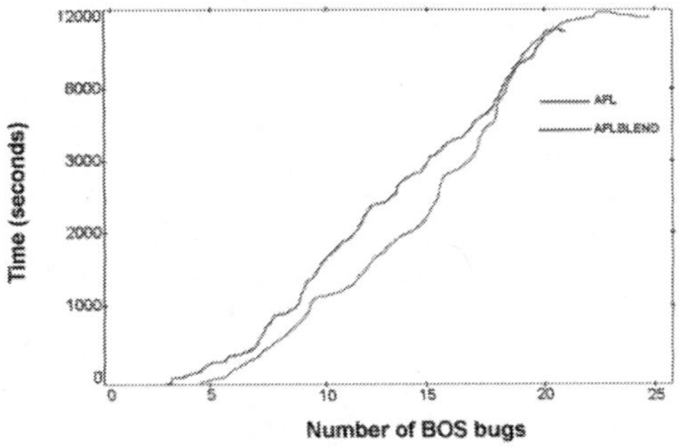

Fig. 4: Results for buffer overflow susceptibilities

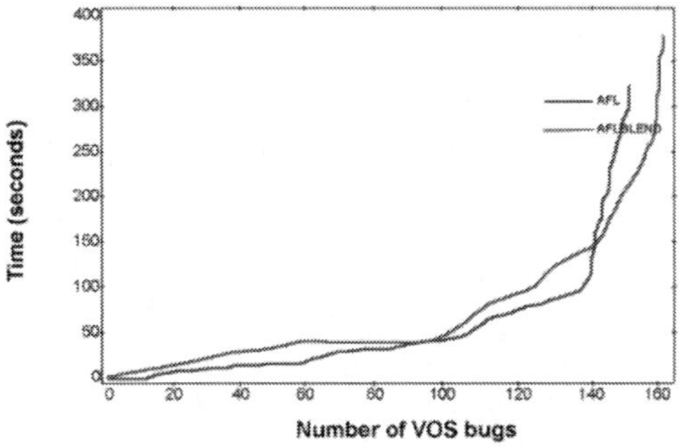

Fig. 5: Results for value overrun susceptibilities

Table 1. Mean value of buffer overrun violations perceived by every tool and results of statistical important tests.

Benchmark		v1	v3	v4	v5	c1	c3	c4	g1
Number of susceptible locations		28	3	4	3	1	1	2	4
Suscep. Location detected	AFL (A)	8	2	3	0	1	0	1	1
	statistically important?	Y	Y	Y	Y	N	Y	N	N
	AFLBLEND (B)	17	3	4	1	1	1	1	1
	statistically important?	Y	Y	Y	Y	N	N	N	N

5.1 Susceptibilities in Overflow Buffer

MIT benchmarks we used are initialized as v1, v3, v4, v5, c1, c3, c4, and g1. These eight MIT benchmarks are injected with 46 susceptible buffer overflow locations. It took nearly 3 hours for each benchmark test and each benchmark has tested 5 times depending on the time and randomness susceptible given to each program. Since it is a strict evaluating method competed with SV_COMP and Test_COMP [16, 17], evaluation of each tool depends on each benchmark. Figure 5 and Table 1 shows the overall result of benchmark results.

Figure 5 and table 1 shows the overall result of benchmark results.

5.1.1 Value Overrun Susceptibilities

These primary benchmarks test for SV_COMP will use over 177 values to check the susceptibilities. Since the susceptibilities are smaller in size each tool will test over 900 seconds for each benchmark. The values given are likely one negative number, one zero number, and one large positive number. Each set will run over 5 times on every benchmark and figure 6 shows the overall result in the decided format and found that AFLBLEND is on top compared to the traditional AFL method.

6 Conclusions

By addressing the problem to expose the susceptibilities that can be found while testing software programs and to detect susceptibilities location in the coding part we implement the AFLBLEND method by deriving core part from AFL. We proposed a method that fits right from the first-time susceptibility specific fitness metric to call and continue test input that comes near to exposing susceptibilities. A valuation of our AFLBLEND tool on normally used benchmarks program generates that our method finds multiple times many susceptibilities or detects them in a short period compared to AFL also with recently added directed fuzzer.

The main restriction over this concept that while running on benchmark program we have run with limited values of inputs, not in large scale values [18] to the program. To address this issue, we might need to migrate our method to have Whitebox Fuzzing [21] on Low Level Virtual Machine (LLVM) so that it will enhance memory availability for tracking chiefboard during program execution. To keep track of chiefboard values a memory address checker method such as Address Sanitizer (ASAN) [14] can be used to accurately keep track of it. These works could consistently improve to move for a large set of database tests and challenging type of susceptibilities.

7 References

1. Gordon Fraser & Andrea Arcuri (2012) in Whole test suite generation IEEE Transactions on Software Engineering 39, 2 (2012), 276–291.
2. Phil McMinn. (2004) in Search-based software test data generation: a survey. Software Testing, Verification and reliability 14, 2 (2004), 105–156.
3. Joachim Wegener, André Baresel, & Harmen Sthamer. (2001) in Evolutionary test environment for automatic structural testing. Information and Software Technology 43, 14 (2001), 841–854.
4. Jared DeMott, Richard Enbody, & William F Punch. (2007) in Revolutionizing thefield of grey-box attack surface testing with evolutionary fuzzing. BlackHat and Defcon (2007).
5. Sanjay Rawat, Vivek Jain, Ashish Kumar, Lucian Cojocar, Cristiano Giuffrida, & Herbert Bos. (2017) in VUzzer: Application-aware Evolutionary Fuzzing. In USENIX security. 1–14.
6. Michael Zalewski. [n.d.]. American Fuzzy Lop. http://lcamtuf.coredump.cx/afl/
7. Mark Harman, S Afshin Mansouri, & Yuanyuan Zhang. (2012) in Search-based software engineering: Trends, techniques and applications. ACM Computing Surveys (CSUR) 45, 1 (2012), 11.
8. CVE database (2017) in CVE Details - Vulnerabilities by type. Technical Report.

https://www.cvedetails.com/vulnerabilities-by-types.php

9. Caroline Lemieux, Rohan Padhye, Koushik Sen, & Dawn Song. (2018) in Perffuzz: Automatically generating pathological inputs. In Proceedings of the 27th ACM SIGSOFT International Symposium on Software Testing and Analysis. ACM, 254– 265.
10. Nick Stephens, John Grosen, Christopher Salls, Andrew Dutcher, Ruoyu Wang, Jacopo Corbetta, Yan Shoshitaishvili, Christopher Kruegel, & Giovanni Vigna. (2016) in Driller: Augmenting fuzzing through selective symbolic execution. In Proceedings Network & Distributed System Security Symposium(NDSS). 1–16.
11. Raveendra Kumar Medicherla, Raghavan Komondoor, & Abhik Roychoudhury (2020) in International Conference on Software Engineering (ICSEW'20), 513-520.
12. Misha Zitser, Richard Lippmann, & Tim Leek. (2004) in Testing Static Analysis Tools Using Exploitable Buffer Overflows from Open Source Code. In Proceedings of the 12th ACM SIGSOFT Twelfth International Symposium on Foundations of Software Engineering (FSE) (Newport Beach, CA, USA). New York, NY, USA, 97–106.
13. George Klees, Andrew Ruef, Benji Cooper, Shiyi Wei, & Michael Hicks. (2018) in Evaluating fuzz testing. In Proceedings of the 2018 ACM SIGSAC Conference on Computer and Communications Security. ACM, 2123–2138.
14. Konstantin Serebryany, Derek Bruening, Alexander Potapenko, & Dmitriy Vyukov. (2012) in Address Sanitizer: A Fast Address Sanity Checker. In USENIX Annual Technical Conference. 309–318.
15. Dirk Beyer. (2019) in Automatic verification of C and Java programs: SV-COMP 2019. In International Conference on Tools and Algorithms for the Construction and Analysis of Systems. Springer, 133–155.
16. Dirk Beyer. (2019) in International competition on software testing (Test-Comp). In International Conference on Tools and Algorithms for the Construction and Analysis of Systems. Springer, 167–175.
17. Dirk Beyer & M Erkan Keremoglu. (2011) in CPA checker: A tool for configurable software verification. In International Conference on Computer Aided Verification. Springer, 184–190.
18. Cristian Cadar, Daniel Dunbar, Dawson R Engler & et al. (2008) in KLEE: Unassisted and Automatic Generation of High-Coverage Tests for Complex Systems Programs. In OSDI, Vol. 8. 209–224.
19. Will Dietz, Peng Li, John Regehr, & Vikram Adve. (2015) in Understanding integer overflow in C/C++. ACM Transactions on Software Engineering and Methodology (TOSEM) 25, 1 (2015), 2.
20. Gordon Fraser & Andrea Arcuri. (2015) in 1600 faults in 100 projects: automatically finding faults while achieving high coverage with EvoSuite. Empirical software engineering 20, 3 (2015), 611–639.
21. Patrice Godefroid, Michael Y Levin, David A Molnar & et al. (2008) in Automated Whitebox Fuzz Testing. In NDSS, Vol. 8. 151–166.
22. Animesh Basak Chowdhury, Raveendra Kumar Medicherla, and R Venkatesh. 2019. VeriFuzz: Program aware fuzzing. In International Conference on Tools and Algorithms for the Construction and Analysis of Systems. Springer, 244--249.
23. Yaohui Chen, Peng Li, Jun Xu, Shengjian Guo, Rundong Zhou, Yulong Zhang, Long Lu, et al. SAVIOR: Towards Bug-Driven Hybrid Testing. In IEEE Symposium on Security and Privacy (SP), 2020.

Block Mining reward prediction with Polynomial Regression, Long short-term memory, and Prophet API for Ethereum blockchain miners.

Jeyasheela Rakkini Simon[1*], *Geetha* K[2]

[1] School of Computing, SASTRA Deemed University, Tamil Nadu.
[2] School of Computing, SASTRA Deemed University, Tamil Nadu.

Abstract. The Ethereum blockchain is an open-source, decentralized blockchain with functions triggered by smart contract and has voluminous real-time data for analysis using machine learning and deep learning algorithms. Ether is the cryptocurrency of the Ethereum blockchain. Ethereum virtual machine is used to run Turing complete scripts. The data set concerning a block in the Ethereum blockchain with a block number, timestamp, crypto address of the miner, and the block rewards for the miner are explored for K means clustering for clustering miners with a unique crypto address and their rewards. Linear regression and polynomial regression are used for the prediction of the next block reward to the miner. The Long Short-Term Memory (LSTM) algorithm is used to exploit the Ether market data set for predicting the next ether price in the market. Every kind of price and volume for every four hours is taken for prediction. The root mean square error of 34.9% is obtained for linear regression, the silhouette score is 71% for K-means clustering of miners with same rewards, with the optimal number of clusters obtained by Gap statistic method.

1 Introduction

Blockchain is a decentralized ledger of transactions with the traits of decentralization, transparency, and anonymity. Ethereum 2.0 has the proof of stake consensus algorithm, which has energy efficiency, high economic incentives, and opens up high revenue-generating capability for a large number of users. The validator nodes add blocks to Proof of Stake (POS) Ethereum blockchain [1] by staking their cryptocurrency, ether on the network and anyone validator can be randomly selected to propose a block. The other validators agree with this validator who is proposing the block. The reward is for both attesting the other validators when they are proposing the block and for the self proposal of the block also. Their own money or ether is at stake when they behave maliciously and this is quite in contrast to the hectic energy-consuming proof of work consensus protocol in the bitcoin blockchain. If the validator does not propose a block when his turn comes or fails to do attestations when other validators propose a block then he is penalized for that also. The validator's stake amount will be fully debited and the validator will be sent out from the proof of stake network if he performs any re-org attack of transactions. The proof of stake consensus algorithm does not require the miners to invest more in hardware, the initial hardware setup costs, the electricity bills but the monetary stake amount,32 ether (ETH) has to be staked by the Ethereum miner for active participation as a validator in Ethereum mining process. For those miners who find 32 ETH, a significant amount to be staked upon, then they can join staking pools which aggregate the funds of miners and the resultant rewards are distributed proportionally. There is a decentralization trend in PoS mining since anyone can participate with minimized electricity costs and this opens up revenue opportunities for many people. PoS achieves both scalability and security with sharding of the blockchain.

2 Related Work

The quintessential details about the proof of stake consensus algorithm for the Ethereum blockchain is given in [1]. Qin, R. et al. [2] discusses the speculation of individual minersin joining mining pools. Due to increased hardware cost, electricity, and maintenance costs, the chances or the probability of the individual miners who can successfully mine a block is very less, since they

* Corresponding author: jeyasheelarakkini@cse.sastra.edu

have to produce a hash value early and once their hash value is less than or equal to target hash produced by the system, they gain rewards. So, the miners join mining pools and contribute their hash to the overall mining pool. The choice of the mining pools or choosing the right mining pool to join and go ahead with the mining is not a trivial one and this is discussed by the authors. This is modelled as a risk decision process. The pool manager has to distribute the reward got from successfully mining to all miners who have joined the mining pool. The mechanisms for distributing the reward include the pay per share mechanism, pay per last N share mechanism, and the flat proportional mechanism. This paper discusses the computational result for a decision of selecting a mining pool.

Qin, R., Yuan et al. [3] discusses the three levels of participants in the bitcoin mining, namely mining pools, individual miners, and blockchain systems. The author lists the issues based on artificial societies, computational experiments, and parallel systems (ACP) and proposes a research framework. Easley et al. [4] discuss the strategy adopted by miners, the significant role of mining rewards, transaction fees, price, waiting time, and external constraints that influence the participation of users in the blockchain. Kiayias, A. et al. [5] discusses the two simplified forms of the game wherein the first game the miners mine a block and release it immediately, in the second form of the game the miners mine a block, and they announce it but defer to release it. In both forms of the game, the best response and the Nash equilibrium is demonstrated. The best response whether it matches the expected behaviour of the bitcoin designer is introspected. Wang, W.et al.[6] discusses the decentralized nature of the blockchain from the perspective of the individual nodes who form the backbone network of the blockchain, the decentralized consensus that runs on these nodes and they have surveyed the incentive mechanisms and byzantine fault tolerance mechanisms from a game theory perspective.

Li, J.et al.[7] discusses the users bidding transaction fees for faster confirmation of transactions. Generalized Second Price auction method is used for the bidding process. The metrics that are considered are transaction size, the bidding scores submitted by the user, the scores of quality, and the virtual fees of the transactions. These metrics play a vital role in reducing and saving the user's fees. Saad M.et al. [8] explores the features in the bitcoin blockchain network and Ethereum blockchain and how they change over time and how they relate to the demand and supply economics of both the cryptocurrencies. Jang, H. et al. [9] discusses the highly correlated network indicators that are useful in predicting the bitcoin price with high accuracy. A Bayesian neural network is used to predict the price of cryptocurrency that takes in to account a sudden decrease in the hash rate that leads to delay in block publishing time and reduction in network throughput. Poongodi M.et al.[10] exploits the ARIMA model for bitcoin data set which is publicly available on the URL http://coinmarketcap.com for 4 consecutive years to analyze the volatility of bitcoin and how it affects the miners joining mining pools and dropping their solo mining. Catalini, C. et al. [11] discusses the two metrics such as the cost of verification of transactionsand thecost ofnetworking, with the cost of verification of transactions related to the cost of verifying the state of transactions along with their attributes, their previous history, and their current ownership of a digital asset. The cost of networking relates to the cost of operating a decentralized digital market place. Salimitari, M.et al. [12] discusses the group of miners joining a mining pool and giving their share of electricity for mining, and three concepts of dynamic game theory such as Social optimum, Nash equilibrium, and myopic Nash equilibrium is proposed. Vimal, S.et al.[13] discusses the efficiency of transferring p2p files in a decentralized interplanetary file system and awarding the miners who successfully shared their resources.

Gupta, S.S et al. [14] discusses clustering methods are used for fraud detection in cryptocurrency financial transactions. Chawathe, S. S. et al. [15] explores the clustering algorithm for the blockchain data and its impact on miners. The chapter from the book, mastering-bitcoin [16], refers to the process of transactions being validated by miners, a new block formed ever since last block creation, 10 minutes ago in blockchain, and the miners mint new coins for each new block creation and the transaction fees from all transactions that are included in the block. The miner who mines for a block by fetching transactions from a mining pool and by solving the target hash given by the bitcoin system or any blockchain system, that is the proof of work system has the crypto address. The miners have to find a hash that is less than or equal to the target hash and the first miner to find it gets the reward in bitcoins. Here we will consider an Ethereum blockchain. But to our surprise for the data posted in http://xblock.pro/xblock- eth.html [17], a simple k means of clustering the mining rewards based on the crypto address of the users for each progressing block number, a linear regression model for prediction of the next block reward, the polynomial regression model for the prediction of the next block reward for theminer has not yet been analyzed in the previous works. Here we have also taken a dataset with stock market prediction of ether from the same url and we have done prediction with LSTM algorithm and Prophet API for ether with ether market data.

3 Problem Definition

Clustering of miner's crypto address, their rewards per unique miner address and the prediction of miner's rewards in Ethereum blockchain has not been addressed in previous works and we have implemented with the existing K-means clustering of miner's crypto address and rewards. The linear regression and Polynomial regression are used for the prediction of miner's rewards. The LSTM algorithm, Prophet API for prediction of Ether crypto currency in Ether stock market. Here the Ethereum blockchain which has Proof of Stake consensus mining, with Ether currency is first analyzed for mining rewards with the dataset in the website [17] that has attributes, miner crypto address, number of transactions, and the rewards. K means clustering of the miner's crypto address with rewards is implemented and the rewards for 'n'(n=3) optimal number of clusters is implemented below in Section 4. The centroids of the cluster are found, the elbow method of K means clustering, and the scatter plot of the clusters is shown with the linear regression line fitted for the scattered values is shown. The polynomial regression is implemented for the same with the prediction for the next block reward for the miner crypto address and values. The LSTM method and the Facebook Prophet API for the prediction of the stock price of Ether is used here. These methods are plotted and analyzed with statistical measures.

4 Results and discussion

We have implemented a K mean clustering in python, given three attributes, timestamp, miner crypto address, and block rewards in the dataset [17]. Then we did a polynomial regression and compared it with the normal linear regression. The Ordinary Least Square (OLS) regression model fit is shown below in Fig.3. The data is scattered and the accuracy of the model with the statistics is shown below. The timestamp for the miner, the crypto address of the miner with the rewards for successfully mining is shown in the head of the data frame with five values in Fig 1.

	timestamp	miner	VALUES	reward
0	1438269988	0x05a56e2d52c817161883f50c441c3228cfe54d9f	58	5.000000e+18
1	1438270017	0xdd2f1e6e498202e86d8f5442af596580a4f03c2c	59	5.000000e+18
2	1438270048	0x5088d623ba0fcf0131e0897a91734a4d83596aa0	60	5.160000e+18
3	1438270048	0xc8ebccc5f5689fa8659d83713341e5ad19349448	61	3.750000e+18
4	1438270077	0x5088d623ba0fcf0131e0897a91734a4d83596aa0	60	3.130000e+18

(9999, 4)

timestamp	int64
miner	object
VALUES	int64
reward	float64

Fig 1. The head of the Ethereum block dataset.

The crypto address being an object data type, the other attributes such as timestamp, VALUES being integer values and reward being float value respectively. The data set is acquired from the block dataset present in http://xblock.pro/xblock-eth.html[17].

The summary of the data frame is given below in Fig 2. with the distribution of the data frame columns, count (number of non-null observations), mean, standard deviation min, max, 25% (lower), 50% (median), and 75% (upper) of values. The min+(max-min) * percentile is used for calculating percentiles per column values.

	timestamp	VALUES	reward
count	9.999000e+03	9999.000000	9.999000e+03
mean	1.438286e+09	124.072107	4.807288e+18
std	1.330892e+04	70.835563	5.698741e+17
min	1.438270e+09	1.000000	1.250000e+18
25%	1.438275e+09	68.000000	5.000000e+18
50%	1.438282e+09	97.000000	5.000000e+18
75%	1.438296e+09	148.000000	5.000000e+18
max	1.438317e+09	356.000000	5.310000e+18

Fig 2. Summary of the block dataset.

Here we have replaced the miner's crypto address with a unique integer value for each unique crypto address, so the clustering with the unique integer per miner's crypto address is implemented seamlessly. The linear regression results are shown below in Fig 3, which has the dependent variable "reward", the R-squared value, the measure of the symmetry of the data about the mean, F- statistic, T- statistic, P-values, Durbin-Watson measure for autocorrelation and Jarque-Bera measure of skewness.

OLS Regression Results

Dep. Variable:	reward	R-squared:	0.013
Model:	OLS	Adj. R-squared:	0.013
Method:	Least Squares	F-statistic:	47.33
Date:	Sat, 03 Oct 2020	Prob (F-statistic):	4.40e-21
Time:	16:23:32	Log-Likelihood:	-2.9509e+05
No. Observations:	6999	AIC:	5.920e+05
Df Residuals:	6996	BIC:	5.920e+05
Df Model:	2		
Covariance Type:	nonrobust		

	coef	std err	t	P>|t|	[0.025	0.975]
const	-7.699e+21	7.93e+20	-9.706	0.000	-9.25e+21	-6.14e+21
timestamp	6.354e+12	5.51e+11	5.711	0.000	4.27e+12	6.44e+12
VALUES	-4.151e+14	1.03e+14	-4.028	0.000	-6.15e+14	-2.12e+14

Omnibus:	3741.634	Durbin-Watson:	1.966
Prob(Omnibus):	0.000	Jarque-Bera (JB):	23811.179
Skew:	-2.503	Prob(JB):	0.00
Kurtosis:	10.213	Cond. No.	1.70e+14

Fig 3. OLS Regression summary.

The linear regression line fit is shown below in Fig 4. with x-axis as values (integer assigned to unique crypto addresses) and y-axis as rewards of the miners.

https://doi.org/10.1051/itmconf/20213701004

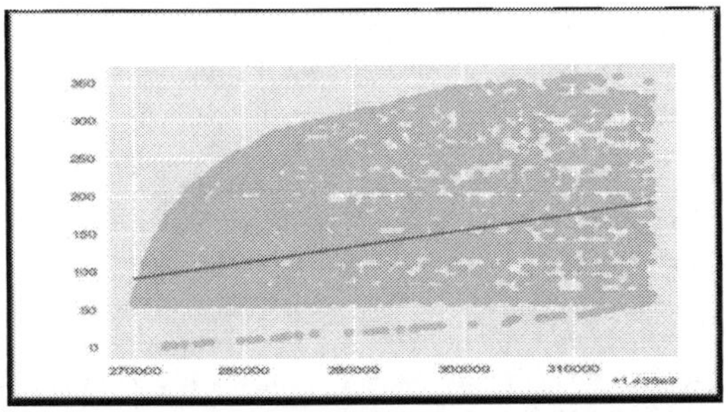

Fig 4. Linear regression fit in a scatter plot.

```
[3.83485401]
[[0.]]
Mean Absolute Error: 0.3014282015259374
Mean Squared Error: 0.12183620349642657
Root Mean Squared Error: 0.34905043116493434
```

Fig 5. Linear regression accuracy metrics.

Then we have done a cluster with K-means for three clusters, to find miners with similar rewards,this is used for analyzing the miners with a similar hash rate of computation. The three clusters are formed with red, green, and cyan colours and are represented in Fig 6.

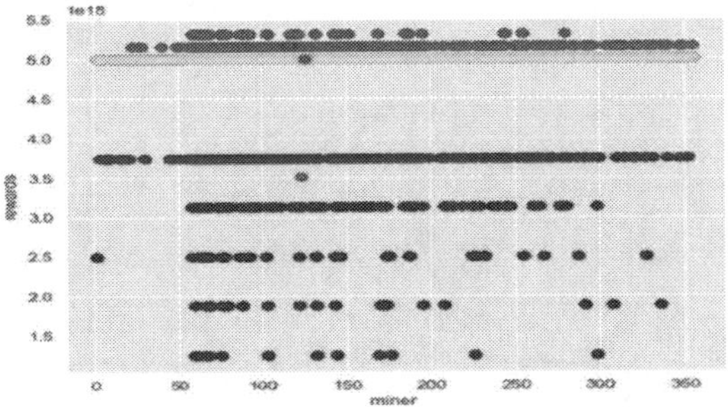

Fig 6. The three clusters.

The cluster of three is formed with values shown below in Fig 7. The number of clusters is user-defined. The centroid of each cluster is determined and the distance of the observations (which are values, rewards) from the cluster is calculated and they are grouped in appropriate clusters.

Cluster	reward	VALUES
1	3.510048e+18	123.567680
2	5.173063e+18	115.926426
3	5.000000e+18	125.676271

Fig.7. Three clusters with reward and values.

The elbow method of k means is shown below in Fig 8. As the number of clusters increases, sum of squared distances decreases in the elbow method.

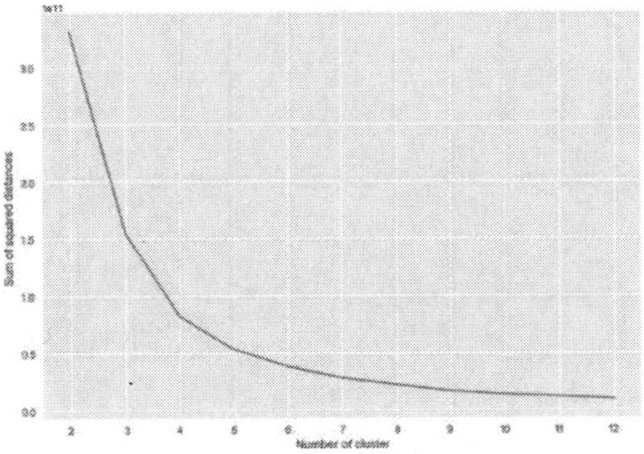

Fig 8. Elbow method of K-means clusters for Ether mining rewards.

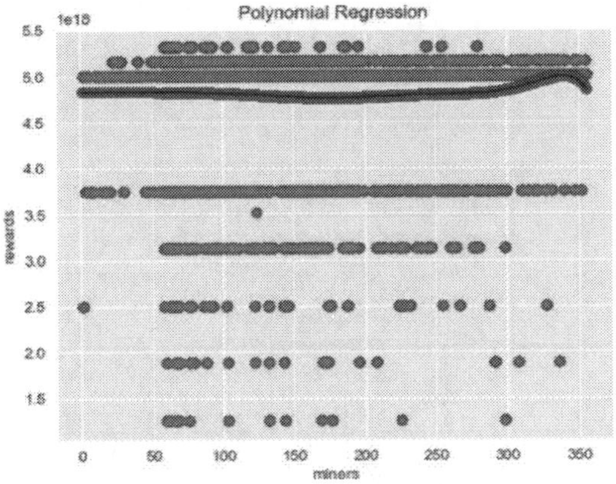

Fig 9. Polynomial Regression with n='3'.

The result of the polynomial regression is shown above in Fig 9. We noticed there was skewness in data and so we did a skew transformation as shown in Fig 10, Fig 11. and Fig 12. The miner's address was more skewed towards the left and the rewards were skewed towards the right. After the removal of the skewness of data with normalization, we have implemented polynomial regression results with a degree of 10.

```
'timestamp' had 'positive' skewness of 0.72
Transformation yielded skewness of 0.00
'VALUES' had 'positive' skewness of 1.26
Transformation yielded skewness of -0.03
'reward' had 'negative' skewness of -2.65
Transformation yielded skewness of nan
In [119]: transformedDF.head(5)
Out[119]:
   timestamp    VALUES         reward
0   0.000539  3.420461  3.708765e+152
1   0.000539  3.432457  3.708765e+152
2   0.000539  3.444234  4.792648e+152
3   0.000539  3.455799  3.489676e+151
4   0.000539  3.444234  2.916962e+150
```

Fig 10. Skewness transformed data set.

 https://doi.org/10.1051/itmconf/20213701004

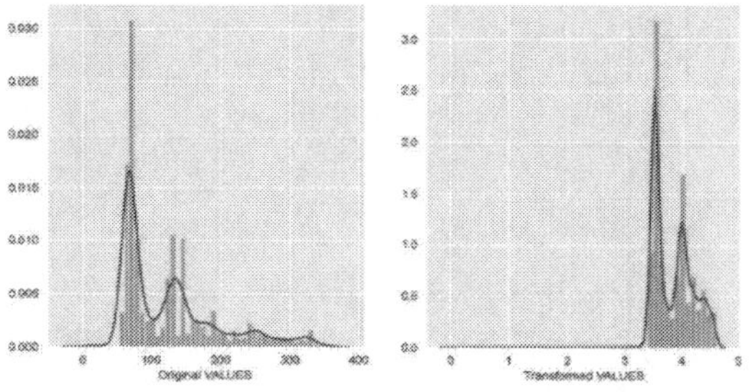

Fig 11. Skewness transformation of the VALUES in Ethereum dataset.

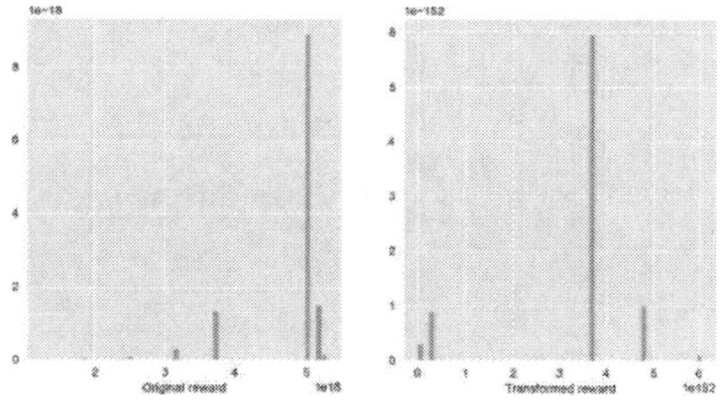

Fig 12. Skewness transformation of the reward attribute in Ethereum dataset.

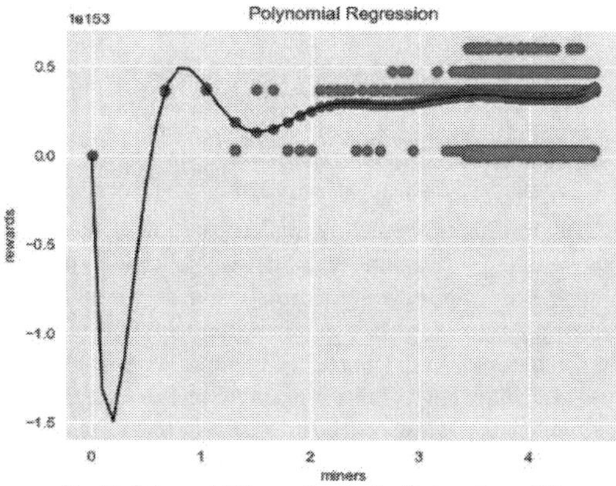

Fig 13. Polynomial Regression for Prediction of next Ethereum block reward.

The root mean square error for polynomial regression is 1.3368300755199987e+152. R2 score, which is a measure of how close are the data to the fitted regression line is 0.005977411651757691. The Polynomial regression with the degree of 10 is shown in Fig 13. The curvilinear fit of the polynomial regression fits the line of best fit so that the best approximation between the independent variable and the dependent variable is implemented.

4.1 Ether price prediction with LSTM

The data set we acquired from http://xblock.pro/mt/ had the following attributes, close: the close price in the period, date: the timestamp at the beginning of this period, high: the highest price in the period, low: the lowest price in the period, open: the open price in the period, quote Volume: the quote volume in the period, base volume: the base volume in the period and weighted Volume: the average price for those base volume and quote volume. The plot of the original values is shown in Fig 14 with x-axis as time in days and y-axis as the price of the Ether.,and the prediction with LSTM for the future values with the given observed values is shown in Fig 15. The original dataset, training data set and testing data set is shown in the graph in Fig 15.

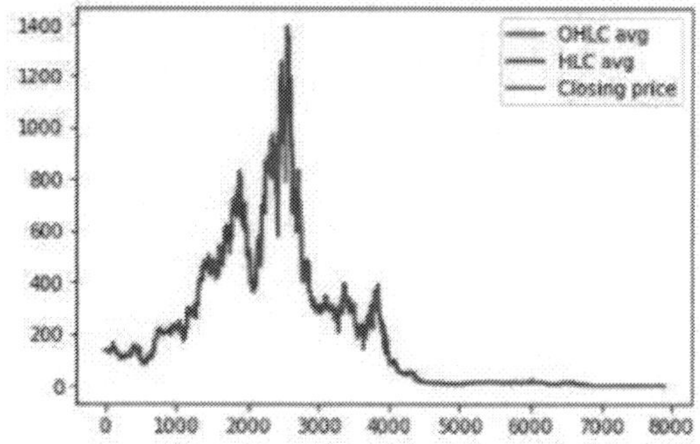

Fig 14. Plot of original values.

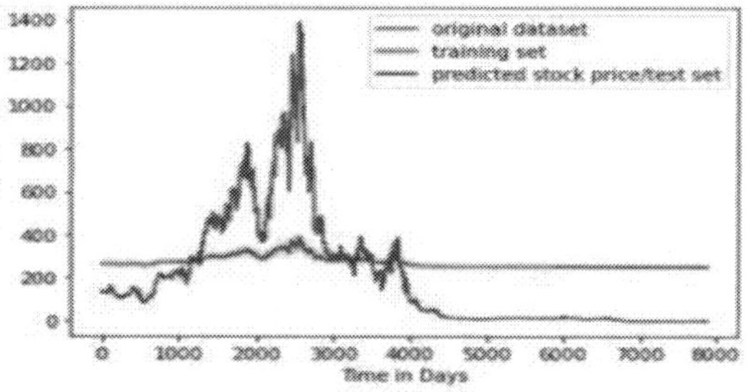

Fig 15. LSTM Prediction with training set and testing set

The LSTM model has predicted lower values than the original values of ether, cryptocurrency with themean square error of 36.45 %. The Ether market prices in the future are predicted by Facebook Prophet API in Fig 16. The column 'ds' has dates for which prediction has to be done using this API. The predicted value is shown in column 'yhat'.'yhat_lower' and 'yhat_upper' represent uncertainty intervals.

	ds	yhat	yhat_lower	yhat_upper
1799	2021-04-12	4521.662794	-30957.902682	40696.046442
1800	2021-04-13	4741.287733	-28437.709043	39877.217167
1801	2021-04-14	4709.327598	-29577.932056	39545.338922
1802	2021-04-15	5000.149948	-29537.530429	42966.354571
1803	2021-04-16	4619.096229	-28157.431980	39048.772218

Fig 16. Prophet API prediction of Ether prices in the future.

The x-axis has the 'ds' column with the year as values and y-axis have ether, cryptocurrency values. The prediction of the ether prices, the trend of the ether price, weekly and yearly is shown in Fig. 17 and Fig. 18 respectively. The trend of the Ether prices, yearly seasonality, weekly seasonality is shown with Prophet API for the time series. The future data frame for a time series of 365 days is taken. Here prediction is done on a data frame with the date and closing values.

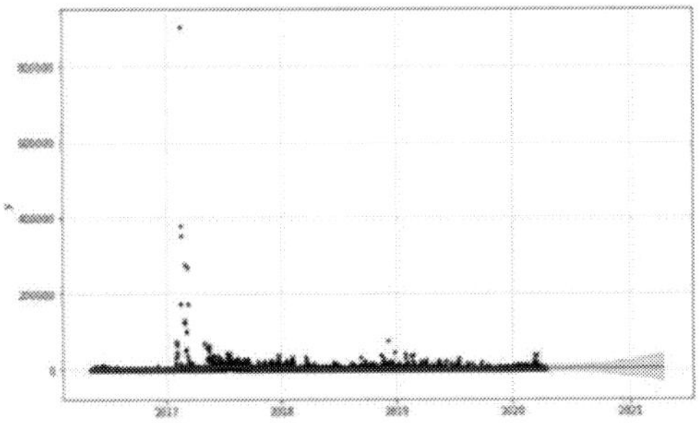

Fig 17. Prophet API prediction of Ether prices.

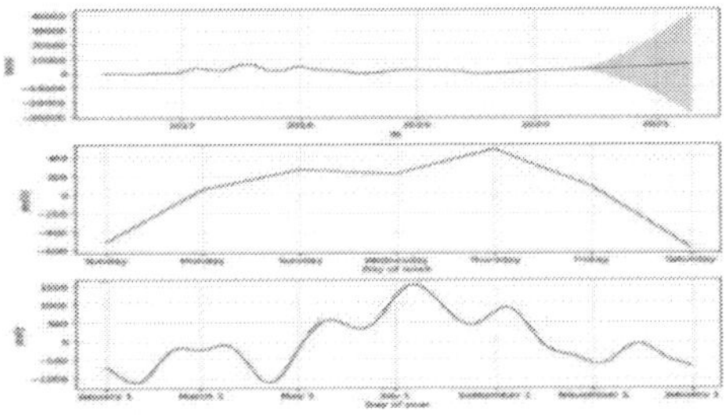

Fig 18. Forecast components yearly, monthly and trend-wise for Ether price.

Thus, Prophet API for forecasting components and future ether values prediction is implemented in python language. The mean absolute error of 1326.71 is obtained with Prophet API. The prediction of Ether price, which is a Time Series problem, is also done with LSTM.

5 Conclusion and Future work

Thus, we have devised a k means clustering algorithm for clustering the unique addresses of the miners based on their rewards. The linear regression and polynomial regression to implement prediction of the mining reward for the next block with the next sequential block number and time period is implemented. The LSTM model to predict the stock price of ether is also implemented. The future work would be the implementation of pattern mining algorithms, centralization versus decentralization traits in Ethereum mining pools with deep learning models.

References

1. https://www.coindesk.com/proof-of-stake.
2. Qin, R., Yuan, Y., & Wang, F. Y. (2018). Research on the selection strategies of blockchain mining pools. *IEEE Transactions on Computational Social Systems*, *5*(3), 748-757.
3. Qin, R., Yuan, Y., Wang, S., & Wang, F. Y. (2018, June). Economic issues in bitcoin mining and blockchain research. In *2018 IEEE Intelligent Vehicles Symposium (IV)* (pp. 268-273). IEEE
4. Easley, D., O'Hara, M., & Basu, S. (2019). From mining to markets: The evolution of bitcoin transaction fees. *Journal of Financial Economics*, *134*(1), 91-109.
5. Kiayias, A., Koutsoupias, E., Kyropoulou, M., & Tselekounis, Y. (2016, July). Blockchain mining games. In *Proceedings of the 2016 ACM Conference on Economics and Computation* (pp. 365-382).
6. Wang, W., Hoang, D. T., Hu, P., Xiong, Z., Niyato, D., Wang, P & Kim, D. I. (2019). A survey on consensus mechanisms and mining strategy management in blockchain networks. *IEEE Access*, *7*, 22328-22370.
7. Li, J., Yuan, Y., & Wang, F. Y. (2019). A novel GSP auction mechanism for ranking Bitcoin transactions in blockchain mining. *Decision Support Systems*, *124*, 113094.
8. Saad, M., Choi, J., Nyang, D., Kim, J., & Mohaisen, A. (2019). Towardcharacterizing blockchain-based cryptocurrencies for highly accurate predictions. *IEEE Systems Journal*, *14*(1), 321-332.
9. Jang, H., & Lee, J. (2017). An empirical study on modeling and prediction of bitcoin prices with bayesian neural networks based on blockchain information. *Ieee Access*, *6*,5427-5437.
10. Poongodi, M., Vijayakumar, V., & Chilamkurti, N. (2020). Bitcoin price prediction using ARIMA model. *International Journal of Internet Technology and Secured Transactions*, *10*(4), 396-406.
11. Catalini, C., & Gans, J. S. (2016). *Some simple economics of the blockchain* (No. w22952). National Bureau of Economic Research.
12. Salimitari, M., Chatterjee, M., Yuksel, M., & Pasiliao, E. (2017, October). Profit maximization for bitcoin pool mining: A prospect theoretic approach. In *2017 IEEE 3rd international conference on collaboration and internet computing (CIC)* (pp. 267- 274). IEEE.
13. Vimal, S., & Srivatsa, S. K. (2019). A new cluster p2p file sharing system based on ipfs and blockchain technology. *Journal of Ambient Intelligence and Humanized Computing*, 1-7.
14. Gupta, S. S. (2017). *Blockchain*. John Wiley & Sons, Inc.

15. Chawathe, S. S. (2019). Clustering blockchain data. In *Clustering Methods for Big Data Analytics* (pp. 43-72). Springer, Cham.
16. https://www.oreilly.com/library/view/mastering-bitcoin/9781491902639/ch08.html.
17. http://xblock.pro/xblock-eth.html.

https://doi.org/10.1051/itmconf/20213701005

Review of Medical Image Synthesis using GAN Techniques

M.Krithika alias AnbuDevi[1,*],Dr.K.Suganthi[1]

[1]*School of Electronics Engineering,Vellore Institute of Technology, Chennai,*
**Corresponding author: krithika.anbudevim2019@vitstudent.ac.in*

Abstract: Generative Adversarial Networks (GANs) is one of the vital efficient methods for generating a massive, high-quality artificial picture. For diagnosing particular diseases in a medical image, a general problem is that it is expensive, usage of high radiation dosage, and time-consuming to collect data. Hence GAN is a deep learning method that has been developed for the image to image translation, i.e. from low-resolution to high-resolution image, for example generating Magnetic resonance image (MRI) from computed tomography image (CT) and 7T from 3T MRI which can be used to obtain multimodal datasets from single modality. In this review paper, different GAN architectures were discussed for medical image analysis.
Keywords: Generative adversarial network, Computerized Tomography, Magnetic Resonance Imaging

1. Introduction

Today, the majority of medical professionals diagnosing diseases with the help of computer aided images. In general, some of the diseases could not be identified with low- resolution images. For example, it is difficult to segment the area of the hippocampus for diagnosing Alzheimer's disease in MRI. So, the deep convolutional network will help to generate the synthetic images along with their segmented images. These synthesized images are in high resolution.

GAN is the deep convolution neural network that was defined by a team of research workers under the guidance of Ian Good fellow. GAN has two competing neural network model. One uses the noise as input and generates samples (and so named generator). The second model which is known as discriminator receives samples from the generator and the training data. According to game theory, the generator is trained to produce an image that looks like a real image, whereas the discriminator is learning to discriminate perfectly from generated data to actual data.

The GAN is trained similarly to the Minimax algorithm from Game Theory and the two networks try to achieve Nash Equilibrium with respect to each other. In medical image synthesis, MRI imaging is obtained from CT imaging. The generator produces the CT image and the discriminator is trained to differentiate the generated CT image with reference to the desired input (CT image) given to it. GAN is the perfect model for generating images and images with better resolution compared to the images generated by MLE (maximum likelihood Estimation) algorithm.

In this review, overview of the GAN principle and some of its variants are explained in section 1 and section 2 respectively. It is followed by a review of different GAN architecture in Section 3.Applications of GAN in medical image synthesis in Section 4. Section 5, 6, and 7 summarizes the review, Future challenges application , and conclusion

2. Structural Variants

GAN is the deep neural framework that contains two models namely the generative (G) and discriminative (D) model. G produces faux samples looks like training sample from the latent variable z, whereas faux samples from G is given to the D. Discriminator find the difference between real data and data generated from G.G tries to satisfy the discriminator with the faux sample, whenever the faux data is found by the discriminator the error is back propagated to the generator. This adversarial learning is formulated in equation (1)

$$\min_G max_D V(G,D)=\min_G max_D\{\mathrm{E}_{a\sim}p_{data}[\log D(a)]+\mathrm{E}_{z\sim}p_{z:}[\log(1-D(G(z))]\}. \quad (1)$$

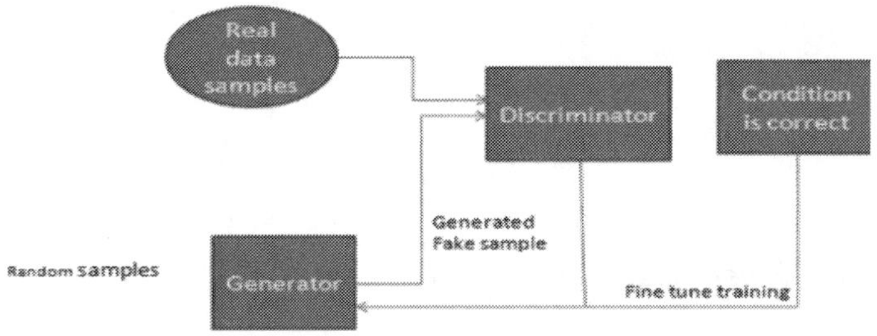

Fig.1.GAN Architecture

whereas

p_{data} (a)-distribution of real data in the data spaceA

$p_z(z)$-distributor of generator on the latent space Z

The generator and discriminator neural model contains the differential function where the weights and bias of two models can be modified to adjust the probability density function through the backpropagation algorithm. Generator satisfies the discriminator with faux data when the PDF of input(p_{data} (a)) is equal to the PDF of the data created by the generator($p_g(a)$)[1].

In this instance it is difficult to differentiate faux and real data hence the discriminator produces the output as 0.5. It shows that D gets confused. The objective of the discriminator is to maximize the two terms log $(1\text{-}D\ (G\ (z)))$ and log(G(a)) i.e. D(z)=0 and D(a)=1 to correctly classify the faux and real data.

The best or optimal generator is given as:

$$D_G^*(a) = \frac{p_{data}(a)}{p_g(a)+p_{data}(a)} \quad (2)$$

Here b=0 is the generated data and b=1 is the real data(b is assumed to be the output) and it is assumed that the probability of both generated and real data are equal. This condition represents that the relative behavior of the two distributions[3] is used by the GAN to pass the collection of data to the generator for generating actual samples. GAN also measures the discriminator discrepancy.

3. Different GAN Architectures

In this section, we introduce the different architectures of GAN like the DCGAN, MGAN, CGAN, Cycle GAN, ACGAN, WGAN which are applied in medical applications.

3.1 Cycle-GAN

Cycle GAN [5] is a technique using the cycle consistent property for training unsupervised image translation models with unpaired images. It translates the original input image into another fake image and again the fake image is converted back into the original input image. The forward and backward generators are named G and F respectively in figure 2. The cycle consistency forces $F\ (G\ (A)) \approx A$ and $G\ (F\ (B)) \approx$B. Both the generators work similarly to an auto-encoder [6]. The proper translation of one image from another image is obtained by combining the cycle consistent loss with the adversarial loss.

In improved Cycle-GAN [7] architecture, gradient consistency loss is adding to improve the boundaries of lung module images in between the MRI images and CT images. This gives accurate results in the process of synthesis and segmentation in generated CT/MRI images.

Cycle-GAN with end to end synthesis and segmentation network can label the image from one modality (MRI) to another targeting modality (CT) [10].Their experimental results suggest that particular segmented CT images are generated from MRI images with multiple organ labels. Tanner et al. [11] used the MR-GAN framework for training paired and unpaired data together with combining various loss functions include adversarial loss, dual cycle-consistent loss, and voxel-wise loss. Even Cycle-GAN is used for DIR(deformable image registration) of lung inhale and exhale volume measurement [8] which results in the lowest mean absolute error between the paired and unpaired images.

A TD-GAN [12] combined with cycle GAN is used for pixel-to-pixel translation in chest X-ray image segmentation. A U-Net-based network [13] was trained using the supervised algorithm on simulated synthetic x-ray which is known as Digitally Reconstructed Radiographs (DRRs) for bone extraction. DRRs are reconstructed in the Cycle GAN with the addition of segmentation loss.

https://doi.org/10.1051/itmconf/20213701005

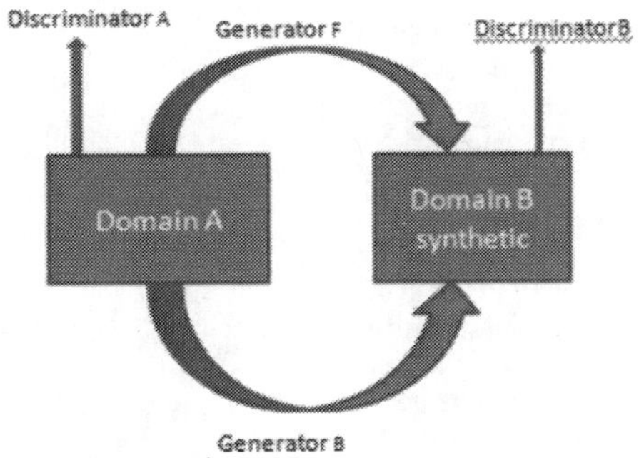

Fig.2.CycleGAN

3.2 MGAN

The MGAN makes use of a pre-trained fixed weights VGG19 network. It is added beginning to both the generative and discriminative network of GAN for extracting feature maps and make the network to preserve the original image content. The generator transfers the style of these feature maps with target texture to an image and the discriminator transforms real or texturized input image with a Fully Convolutional Network (FCN) into VGG19 feature maps.FCN classifies the real and fake image which belongs to patches classified in the VGG19 attributes map space. The generator makes the discriminator satisfied with the faux images, So it is trained to generate images that are similar to the original characteristics of VGG19. A loss function is calculated using VGG which predicts that the real input image content is preserved while transferring the style of the high-level features.

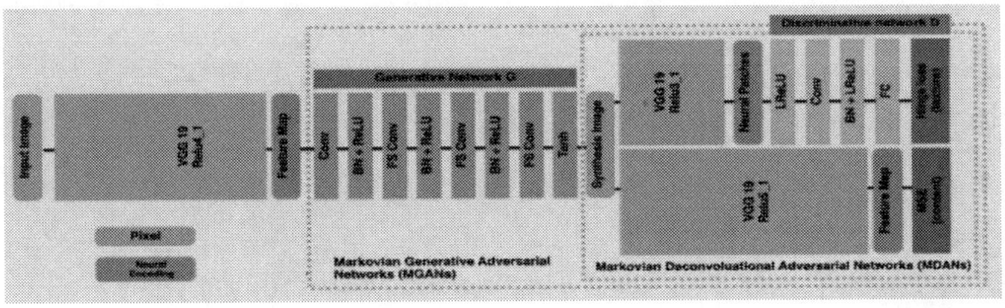

Fig.3.The architecture of MGAN

3.3ACGAN

Auxiliary classifier GAN gives the perfect results with the limited dataset. In this network, side information is reconstructed by the discriminator. Here, the Discriminator model can adapt to multi-class identification and gives the probability distribution of sources and class labels. Samples in the generator have the corresponding class labels (fig.4).

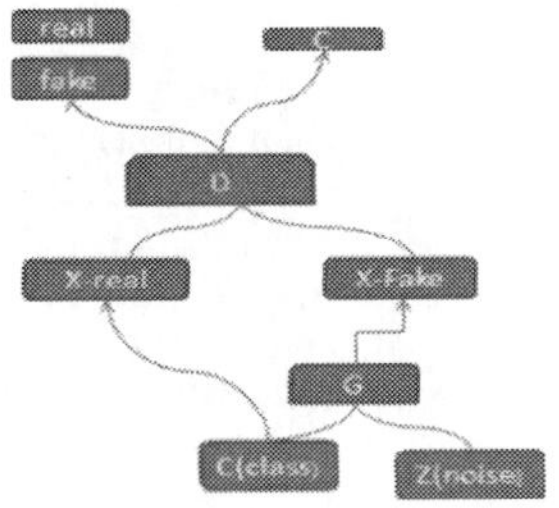

Fig.4.ACGAN

3.4 DCGAN

Generative Adversarial Networks with deep convolutional neural network . [16] is the extension of GAN introduced first. DCGAN can maintain stability in the training process and create high resolution images. DCGAN uses convolutional neural networks with learning phase and generation phase for generating synthetic images. During training process, the random noise is given as the input to the generator and the generator with multi deconvolutional neural network produces the sample images looks like real images. Discriminator tries to differentiate generated images and training set of images [16]. DCGAN uses Batch Norm [17] for normalization of extracted data, and Leaky ReLU [18] for preventing dead gradients.

3.5 Conditional GAN (CGAN) [19]

A latent variable is passed to the generator and the discriminator as shown in figure 6. The generator learns side information conditional distributions, as it can disentangle this from the overall latent space. Class label is given to discriminator hence discriminator predicts only the faux and real data (fig.5)

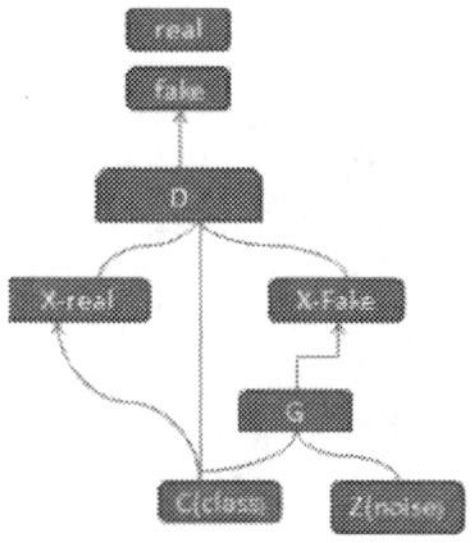

Fig.5.CGAN

3.6 Info GAN

Info GAN [20] is designed in such way to maximize the mutual information and it is trained without label to classify the data with less than 6 % error rate. Learns disentangled features. In info GAN, class label is not given to the discriminator hence discriminator has to predict both the class label and also discriminates the real and faux data (fig.6)

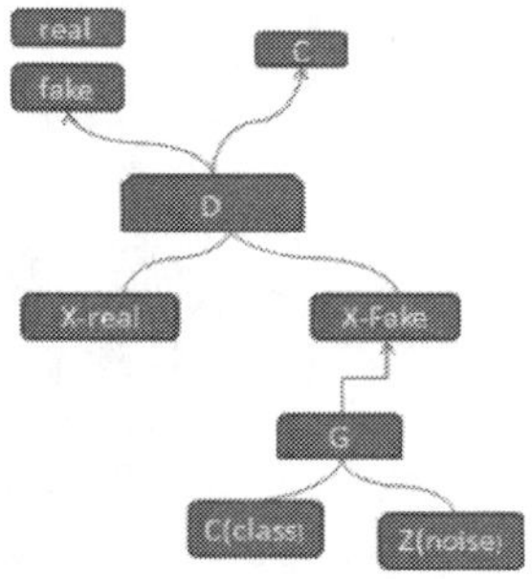

Fig.6. InfoGAN

3.7 WGAN

In the GAN framework, real and generated images data distribution is exactly compared using the divergence called JSD(Jensen-Shannon divergence). This will make a diminished gradient and optimization problem, which results in mode collapse and instability. Hence, Earth Mover (ME) or Wasserstein-1 distance estimation is used in new architecture namely the Wasserstein-GAN (WGAN). Learning process of WGAN estimates the relationship between real and generated distribution in a deep manner. Practically WGAN leads to a slow optimization process. The value function of WGAN is designed using the Kantorovich-Rubinstein duality [23] to obtain

$$\min_{G} max_{D \in D} \mathbb{Z}_{a \sim p_r}[D(a)] - E_a \sim p_g[\mathrm{D}(\hat{a})] \qquad (3)$$

4. Applications of GAN in Medical Image Synthesis

4.1 Synthesis

Medical images can be synthesized with conditional and unconditional image generation techniques to reduce the loss function. The generative property of both supervised and unsupervised GAN has been utilized to synthesis different medical images. In the following, works of all the frameworks will be given. Particularly conditional framework GAN is classified based on the type of imaging modality.

4.1.1 Unconditional Image Synthesis

Many works have proposed recently within the field of unsupervised medical image generation using GANs, which raise data simulation [22] and reduces class imbalance along with data scarcity problem [21]. At first, results have shown that the DCGAN synthesized the images of small patches of prostate benign, prostatic hyperplasia and Prostatic carcinoma [24], digital retinal images [23], or lung cancer nodules [21]. A separate generative model is used to train each metastases class. For liver lesion classification , synthetic samples in addition to heavily augmented data are used to train the GAN which considerably improves a CNN classifier. The synthetic samples are also used to classify skin lesions [25]. Recently, progressive GAN is proposed [26] to synthesize highly realistic-looking images of skin lesions that even skillful dermatologists could not differentiate from a real sample.

4.1.2 Conditional Image Synthesis

To diagnosis certain diseases from the medical image, CT scan images are required. However, CT imaging causes cell damage to the brain and cancer due to radiation exposure. Hence, the3D Fully Convolutional Networks are cascaded for extracting the CT image from MRI images. To improve the resolution of the synthetic images, the adversarial model is trained with some of the loss functions such as image gradient and reconstruction with pixel-wise loss. The cascaded generators are combined to form Auto-Context Model. The main idea of ACM is to train each classifier with the feature data of source image and probability map from the previous classifier gives the context information and allows the GAN for removing redundancy. Use of cycle GANs [27] to change 2D MR to CT pictures without the requirement of co-registered and explicit data pairs. Deep supervision discrimination is proposed to get feasible results in image-to-image translation, which takes advantage of the attributes related to the model of VGG16 which provides exact gradient updates to the generator and also provides real and newly synthetic CT images apart.

4.1.3 Segmentation

Segmentation of various biological structures is an essential process to extract a particular region. The manual segmentation method cannot give accurate results which made it utilize the Deep-Learning networks for analyzing diseases from the medical image. In Deep learning it is hard to optimize the model coefficients, hence there is a lack of annotations training. GAN modifies the CNN's top layers such as CRFs and SSMs [28] in different learning flows to optimize the coefficients.

4.1.4 Reconstruction

Reconstruction is the process of obtaining tomographic images from a projection.L2 weight regularization loss function is used to reduce the overfitting problem [32] for reconstructing the thin slice medical images usual thick slice images.3D-Y-Net-GAN and 3D-DenseU-Net reconstruction framework based GAN is proposed to reconstruct thin slice infant MR image [33].

4.1.5 Registration

Parameter dependency and heavy optimization load is the major disadvantage of the registration process. An improved version of the adversarial image registration method called WGAN with its image transformation capabilities has been used as an image alignment algorithm in MR-TRUS image registration. An unsupervised adversarial registration network does not get information from the ground truth deformation by maximizing the sum of squared difference and cross-correlation to define the similarity between the two pairs of images.

4.1.6 Classification

To diagnosis cardiovascular diseases, two-stage SCGANs(Semi-coupled-GANs) [34] is proposed to completely classify the Left ventricle that covers from basal to apical slices in CMR images. A Combination of DCGAN[30] generated images and real images of Chest X-ray given to DCNN is used to classify abnormalities with the utilization of ReLU activation functions, L2 regularization, and Cross-validation which improves the prediction accuracy of GAN. GAN based synthetic data augmentation [35] plays a major role in liver lesion classification to identify the middle stage of cancer along with the bone and lungs.

Table1.Summary of GAN in Medical Image Synthesis

Type of GAN	Uses of GAN	Ref
DCGAN	Synthesized the images of small patches of prostate benign, prostatic hyperplasia and Prostatic carcinoma	[24]
Progressive GAN	Synthesize highly realistic-looking images of skin lesions	[26]
Cycle GANs	Change 2D MR to CT	[27]
Y-Net-GAN and 3D-DenseU-Net	Reconstruct thin slice infant MR image	[33]
CollaGAN	Synthesis image across different MR sequences	[31]
SCGANs	Completely classify the Left ventricle that covers from basal to apical slices in CMR images.	[34]

5. Future challenges

It is difficult to stabilize the metrics such as Structural similarity index and MAE for quantitative evaluation. Even though it is no need for the visual quality of the image during the reconstruction process. Mode collapse is the most important issue to analyze the similar images. Diminished gradient and non-convergence are also the major problems in GAN. It requires a huge number of training steps to reduce the loss function which makes it difficult to balance both generator and discriminator in a single training. This increases the cost function.

6. Future applications

GAN can be used for Improving radiology work flow and patient care. The advantage of GANs is to perform semi-supervised learning and unsupervised learning. Progressive GAN is applied to medical image analysis for improving resolution. GAN can predict the future video frame and 3D modeling.

7. Conclusion

GAN plays a major role in Cross-modality image synthesis. GAN uses the MR high quality imaging modality that can minimize the data acquisition time by originally forming a sequence of data from a previously taken one. Recently proposed CollaGAN[31] network synthesis image across different MR sequences and find the missing input data with the single generator and discriminator. The availability of huge MR data sets is the reason for the utilization of MR in the GANs framework.

Adversarial training network regularize the texture and shape in the generator output which is important in the reconstruction and segmentation process of medical image processing. This technique significantly reduces the loss function in the segmentation of the liver from 3D CT Volumes. Further, this paper is focused on classification using the data augmentation with domain shift that generates the tiny objects such as nodules, lesions, and cells. Many studies explained that GAN is used to develop the synthesized images to diagnosis neurodegenerative diseases.

References

[1] Ian Goodfellow, Jean Pouget-Abadie, Mehdi Mirza, Bing Xu, David Warde-Farley, Sherjil Ozair, Aaron Courville, and Yoshua Bengio. Generative adversarial nets. In Advances in Neural Information Processing Systems. Curran Associates, 2672–2680. (2014)

[2]X. Mao, Q. Li, H. Xie, R. Y. K. Lau, Z. Wang and S. P. Smolley, "Least Squares Generative Adversarial Networks," IEEE International Conference on Computer Vision (ICCV), Venice, 2017, pp. 2813-2821,doi: 10.1109/ICCV.2017.304, (2017)

[3] Mihaela Rosca, Balaji Lakshminarayanan, David Warde-Farley, and Shakir Mohamed .Variational approaches for auto-encoding generative adversarial networks. arXiv preprint arXiv:1706.04987 (2017).

[4] Lan Lan1, Lei You2, Zeyang Zhang3, Zhiwei Fan4, Weiling Zhao 2 , Nianyin Zeng5 , Yidong Chen 6 and .XiaoboZhou, Generative adversarial networks and its application in biomedical informatics,volume 8,article 154,1-14,Frontiers in public health, Advanced Deep Learning Methods for Biomedical Information Analysis(2020)

[5] Zhu JY, Park T, Isola P, Efros AA. Unpaired image-to-image translation using cycle-consistent adversarial networks. In: IEEE International Conference on Computer Vision. Venice (2017)

[6] Lu X, Tsao Y, Matsuda S, Hori C. Speech enhancement based on deep denoising Auto-Encoder. In: Proceedings of Interspeech. Lyon, pp. 436–40, (2013)

[7] Hiasa Y, Otake Y, Takao M, Matsuoka T, Sato Y. Cross-modality image synthesis from unpaired data using CycleGAN: effects of gradient consistency loss and training data size. In: Goksel O, Oguz I, Gooya A, Burgos N, editors. Simulation and Synthesis in Medical Imaging - Third International Workshop, SASHIMI 2018, Held in Conjunction with MICCAI 2018, Proceedings. Granada: Springer verlag (2018)

[8] Tanner C, Ozdemir F, Profanter R, Vishnevsky V, Konukoglu E, Goksel O. Generative Adversarial Networks for MR-CT deformable image registration. arXiv:1807.07349 (2018).

[9] Huo Y, Xu Z, Bao S, Assad A, Abramson RG, Landman BA. Adversarial synthesis learning enables segmentation without target modality ground truth. arXiv:1712.07695,doi: 10.1109/ISBI.2018.8363790(2017)

[10] Chartsias A, Joyce T, Dharmakumar R, Tsaftaris SA. Adversarial image synthesis for unpaired multi-modal cardiac data. In: Tsaftaris S, Gooya A, Frangi A, Prince J, editors. Simulation and Synthesis in Medical Imaging. SASHIMI 2017. Lecture Notes in Computer Science. Cham: Springer (2017)

[11] . Jin CB, Jung W, Joo S, Park E, Cui X. Deep CT to MR synthesis using paired and unpaired data. arXiv:1805.10790, doi: 10.3390/s19102361(2018)

[12] Yue Z, Miao S, Mansi T, Rui L. Task driven generative modeling for unsupervised domain adaptation: application to x-ray image segmentation,arXiv:1806.07201 (2018).

[13]OphirGozes, Hayit Greenspan, Bone Structures Extraction And Enhancement In Chest Radiographs Via CNN Trained On Synthetic Data, arxiv:2003.10839v1 [Cs.Cv] (2020)

[14] Zhang Z, Lin Y, Zheng Y. Translating and segmenting multimodal medical volumes with cycle- and shape-consistency Generative Adversarial Network.arXiv:1802.09655.doi:10.1109/CVPR.2018.00963 (2018).

[15]M. Arjovsky, S. Chintala, and L. Bottou. Wasserstein gan. arXiv preprint arXiv:1701.07875, (2017)

[16] M. Mehralian and B. Karasfi, "RDCGAN: Unsupervised Representation Learning With Regularized Deep Convolutional Generative Adversarial Networks", 9th Conference on Artificial Intelligence and Robotics and 2nd Asia-Pacific International Symposium, Kish Island, Iran, pp. 31-38, doi: 10.1109/AIAR.2018.8769811(2018)

[17]Ioffe S, Szegedy C , Batch normalization: accelerating deep network training by reducing internal covariate shift. arXiv preprint arXiv:1502.03167(2015)

[18] Maas AL, Hannun AY, Ng AY,Rectifier nonlinearities improve neural network acoustic models. In: Proc. icml. vol 30, p 3(2013)

[19] M. B. Lee, Y. H. Kim and K. R. Park, "Conditional Generative Adversarial Network- Based Data Augmentation for Enhancement of Iris Recognition Accuracy," *in* IEEE Access,vol.7,pp.122134122152,doi:10.1109/ACCESS.2019.2937809(2019)

[20] Xi Chen, Yan Duan, Rein Houthooft, John Schulman, Ilya Sutskever, Pieter Abbeel. InfoGAN: Interpretable Representation Learning by Information Maximizing Generative Adversarial Nets (2016)

[21]Maayan Frid-Adar, Eya lKlang, Michal Amitai, Jacob Goldberger, and Hayit Greenspan, Synthetic data augmentation using GAN for improved liver lesion classification. 2018 IEEE 15th International Symposium on Biomedical Imaging ,pages 289–293, (2018)

[22]Maria J. M. Chuquicusma, Sarfaraz Hussein, Jeremy R Burt, and UlasBagci. How to fool radiologists with generative adversarial networks a visual Turing test for lung cancer diagnosis. IEEE 15th International Symposium on Biomedical Imaging, pages 240–244, (2018)

[23]Pedro Costa, Adrian Galdran, Marialne sMeyer ,Meindert Niemeijer Michael Abr`amo, Ana Maria Mendonc¸a, and Aur´elio Campilho. End-to-end adversarial retinal image synthesis. IEEE transactions on medical imaging, (2017)

[24]Andy Kitchen and Jarrel Seah. Deep generative adversarial neural networks for realistic prostate lesion MRI synthesis. CoRR, abs/1708.00129, (2017)

[25]M. TarekShaban, ChristophBaur, Nassir Navab, and ShadiAlbarqouni. Staingan: Stain style transfer for digital histological images. CoRR, abs/1804.01601, (2018)

[26]TeroKarras,TimoAila,SamuliLaine,andJaakko Lehtinen. Progressive growing of gans for improved quality, stability, and variation. arXiv preprint arXiv:1710.10196, (2017)

[27]Agisilaos Chartsias, Thomas Joyce, Rohan Dharmakumar, and Sotirios A Tsaftaris. Adversarial image synthesis for unpaired multi-modal cardiac data. In International Workshop on Simulation and Synthesis in Medical Imaging, pages 3–13. Springer, (2017)

[28]Liang-Chieh Chen, George Papandreou, Iasonas Kokkinos, Kevin Murphy, and Alan L. Yuille. Deeplab: Semantic image segmentation with deep convolutional nets, atrous convolution, and fully connected crfs. IEEE Transactions on Pattern Analysis and Machine Intelligence, 40:834–848, (2018)

[29]Yan P., Xu S., Rastinehad A.R., Wood B.J. (2018) Adversarial Image Registration with Application for MR and TRUS Image Fusion. In: Shi Y., Suk HI., Liu M. (eds) Machine Learning in Medical Imaging .Lecture Notes in Computer Science, vol 11046. Springer, Cham(2018)

[30]H. Salehinejad, S. Valaee, T. Dowdell, E. Colak and J. Barfett, "Generalization of Deep Neural Networks for Chest Pathology Classification in X-Rays Using Generative Adversarial Networks", IEEE International Conference on Acoustics, Speech and Signal Processing (ICASSP), Calgary,AB,doi:10.1109/ICASSP.2018.8461430(2018)

[31]Lee, D., Moon, W.J., Ye, J.C..Which contrast does matter? towards a deep understanding of MR contrast using collaborative gan. arXiv preprint arXiv:1905.04105(2019)

[32]Zeju Li, Yuanyuan Wang, and Jinhua Yu. Reconstruction of thin-slice medical images using generative adversarial network. In International Workshop on Machine Learning in Medical Imaging, pages 325–333. Springer, (2017)

[33]J. Gu, Z. Li, Y. Wang, H. Yang, Z. Qiao and J. Yu, "Deep Generative Adversarial Networks for Thin-Section Infant MR Image Reconstruction," *in* IEEE Access, vol. 7, pp. 68290-68304,doi: 10.1109/ACCESS.2019.2918926 (2019)

[34]Zhang L., Gooya A., Frangi A.F. Semi-supervised Assessment of Incomplete LV Coverage in Cardiac MRI Using Generative Adversarial Nets. In: Tsaftaris S., Gooya A., Frangi A., Prince J. (eds) Simulation and Synthesis in Medical Imaging. SASHIMI 2017. Lecture Notes in Computer Science, vol 10557. Springer, Cham. (2017)

[35] M. Frid-Adar, E. Klang, M. Amitai, J. Goldberger and H. Greenspan, "Synthetic data augmentation using GAN for improved liver lesion classification," 2018 IEEE 15th International Symposium on Biomedical Imaging Washington,DC,doi: 10.1109/ISBI.2018.8363576, (2018)

ITM Web of Conferences **37**, 01006 (2021) https://doi.org/10.1051/itmconf/20213701006
ICITSD-2021

An intelligent detection and therapeutic device to support sleep apnea in infants

Sindu **Divakaran**[1*] ,*T.***Sudhakar**[1] ,*Sindhiya* **R**[1], *Rimisha* **Gupta**[1], **and***J* **Premkumar**[1]
[1]Department of Biomedical Engineering, Sathyabama Institute of Science and Technology,Chennai, India

Abstract. Among the numerous sleep-disorders breathing patterns encountered by babies, such as intermittent respiration, premature apnea, obstructive sleep apnea wa sconsidered a major cause of concern. Upper airway structure, pulmonary system mechanics, etc. are only a few reasons why the babies are vulnerable to obstructive sleep disorder. An imbalance in the viscoelastic properties of the pharynx, dilators and pressure can lead to airway collapse. Low level of oxygen in blood or hypoxemia is considered a characteristic in infants with severe Obstructive Sleep Apnea (OSA). Invasive treatments like nasopharyngeal tubes, continuous positive airway pressure (CPAP), or tracheostomy are found to be helpful in most cases where infants experience sleep apnea. This paper suggests an appropriate method for long-term monitoring of obstructive sleep apnea in infants and, if any abnormalities are observed, the tool provides continuous airway pressure treatment until the abnormality is stabilized. Resilient propagation algorithm is utilised to train the datasets and produce a relevant output.

1.Introduction:

Pediatric sleep apnea is a sleeping condition where the infants breathing gets blocked partially or completely due to some blockage or constriction of the respiratory airway while sleeping [1]. Obstructive apnea and Central apnea which block the back of nose or throat and improper function of brain controlling breathing respectively are the common types of sleep apnea significantly seen in infants [2, 3]. Mixed apneas are commonly seen in premature infants, where central apnea is followed by obstructive apnea. These blocks usually occur due to over relaxation of the throat muscles[4, 5]. It is hugely caused due to medical conditions during birth, genetic history or any problems associated with mouth and nose[5, 6].

There is always some instability in an infant's breathing which is a regular part of an infant's development [7]. There are instances of healthy infants also experiencing brief central apnea. But these are isolated events occurring during infancy [8, 9]. Movements during sleep or sighs can also cause pause in breath with short duration [10, 11].

Physiologically, babies have more chances of getting sleep apnea than older children. The reported prevalence of infant OSA is difficult to estimate because of differing definitions; but published literature indicates that obstructive incidents in infancy have been characterized by a higher prevalence of PSG than older paediatric age groups. Three diagnosis procedure of questionnaire, Physical Examination and polysomnography were executed for enrolled children between 2 to 15 years. The proportion of low educational level was higher in parents who breast fed their babies than those who didn't. Multivariate analysis showed that snoring for ≥ 3 months, male gender, obesity, breastfeeding, tonsillarhypertrophy, and adenoid hypertrophy were associated with OSAS .

Corresponding Author- *sindudiva@gmail.com

Relation between maternal OSA and fetal growth was analyzed with help of 155 pregnant women of non obese. An accelerated fetal growth is noticed when the fetus experiences an OSA during its third term of growth from this study.

Common symptoms seen during sleep are breathing through mouth, very noisy breathing, cessation in breathing, coughing or choking in sleep, uneasy and disturbed sleep, and sweating [12, 13]. OSA has been linked to behavioral changes and sometimes unexpected infant death [14].

The usually preferred techniques to overcome this condition are anti–gastroesophageal reflux disease treatment,supraglottoplasty, adenoidectomy ,supplemental oxygen ,CPAP/BiPAP and other surgical and non- surgical methods [15, 16].

Our ultimate focus is to reduce the mortality rate of infants dying due to sleep apnea. This is done by carefully monitoring the infant throughout detect sleep apnea. Along with the breathing rate, other important bio signals are also simultaneously measured. This helps to keep a check on the infant's entire health condition [17 , 18]. When sleep apnea is detected in an individual, continuous airway pressure therapy is automatically applied to the infant. This saves a lot of time and prevents the infant from falling to the risk of dying [18, 19].

Lemyre et.al suggests that in preterm babies who are not attached to an endotracheal tube, the Nasal intermittent positive pressure ventilation increases the effectiveness of NCPAP. Preterm babies with breathing problems are fixed to a ventilator that aids in breathing through a tube in the windpipe. Minimally invasive methods like NCPAP and NIPPV assists babies' breathing by using tubes that extend only till the nose, thus minimizing any harm to the lungs. After ventilator extubation, NCPAP and NIPPV can be utilized, thus, reducing the re insertion of endotracheal tube. The baby can breathe more comfortably when a steady pressure is given by the NCPAP to the back of the nose.

Martin et al., used the airway occlusion technique to measure the strength of the Hering-Breuer inflation

 https://doi.org/10.1051/itmconf/20213701006

reflex by using low pressure CPAP before and during the treatment. They observed that CPAP did not alter the inflation reflex in five of twelve preterm infants during the first 14 days. It is mentioned that short inspiratory efforts were noticed on the other 7 infants two times when they were on CPAP and 32 instances when they were taken off it which can be compared to the shortening that occurred two times during 58 occlusions when the seven infants were checked only two hours after birth. The peak inspiratory pressure produced following an occlusion sprang up to 3.4 to 7.1 cm H_2O in preterm infants who showed short responses. The respiratory rate dropped from 61 to 49 breaths/minute. The values obtained put forward the point that the infants were able to adjust to heavier respiratory loads when connected to a CPAP thereby removing the Hering-Breuer deflation reflex.

Subramanian et.al., predicted that time and again disease and death in infants born before 8 months of pregnancy were caused by Respiratory distress syndrome (RDS. Intermittent positive pressure ventilation (IPPV) has been utilized for these babies to breathe. Otherwise, it was suggested to dispense continuous distending pressure of the lung, either as continuous positive pressure to the airway or as continuous negative pressure. The distending pressure was achieved by using any one of the methods given. A negative pressure chamber, face mask continuous positive airway pressure , nasal CPAP and negative pressure for less serious illness and endotracheal CPAP for more severe illness. The consequence was a better improved technique for babies. It was also acknowledged that the rate of pneumothorax can be increased by CDP.

Nasal continuous positive airway pressure (NCPAP) gives a positive pressure to the airways via the nose when an endotracheal tube is not available. A continuous distending airway pressure (CDP) is given continuously during the breath cycle in normally breathing infants. NCPAP is the preferred mode of non-invasive respiratory support in very low birth weight infants, soon after birth and neonatal resuscitation or after mechanical ventilation. The pressure to be used will be provided by the equipment providers. High pressure levels may increase lung volumes and improve respiratory mechanics and gas exchange in under-developed lungs, excessive pressure may directly dam the gas exchange and respiratory mechanics. A single constant pressure level for all very low birth weight infants or a personalized approach that chooses an initial pressure with gradual pressure level increase has to be adopted. Personalized increases may be steered by measuring of gas exchange, work of breathing, radiographic lung expansion.

2. Materials and Methods:

2.1 Hardware Components:

SPO2 sensor probe

The sensor detects the oxygen saturation or the amount of oxygen carrying molecules in the blood. It is a non invasive procedure.

Respiration Sensor

These pressure sensors detect the breath or respiration rate of the infants.

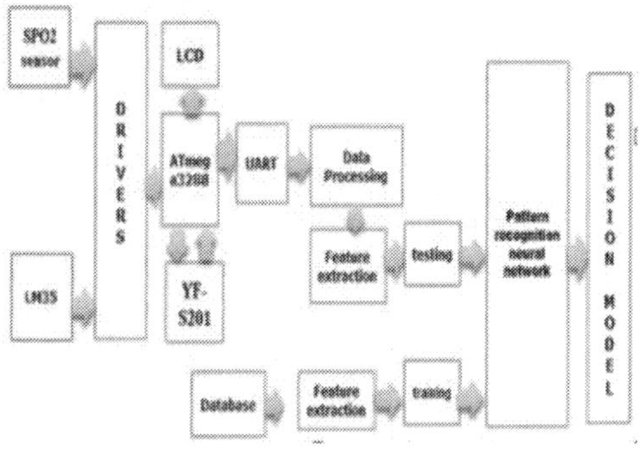

Fig 1. Block diagram of the device

.LM-35/Temperature Sensor

This analog temperature sensor produces an electrical output proportional to Degree Centigrade. They provide good accuracies and need not be calibrated [23].

Flow Sensor/YF-S201

The flow sensor determines the mass flow rate of air entering through it. This is one of the most important parts of the therapy unit [24].

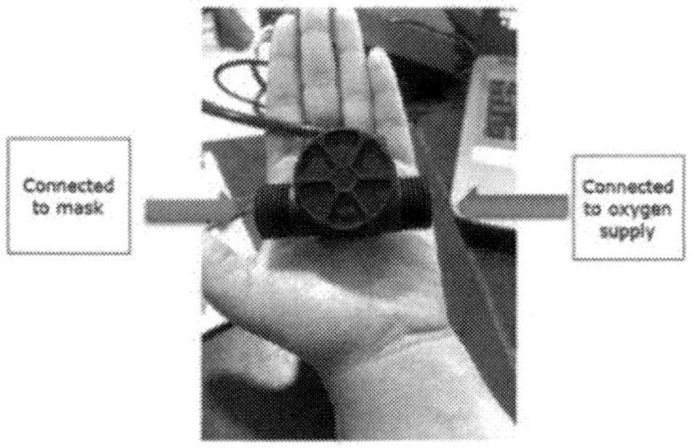

Fig 2: Flow sensor

ATMEGA32

The microcontroller employed is a 8 bit microcontroller developed on CMOS with an automatic voltage regulator based RISC architecture. 1 million instructions per second can be calculated by this microcontroller if cycle frequency is 1MHz

UART

A receiver-transmitter is introduced in the work for asynchronous serial communication which has a configurable data format and speed of transmission . An external driver circuit takes care of the electric signaling levels [25].

LCD

A seven segment LCD display is utilized in this work to display the readings.

2.2 Software Used:

MATLAB
A software language for technical computing. This language does computation, programming in a platform where problems and solutions are expressed in easy to understand mathematical notations [26].

Resilient propagation algorithm
RProp is a gradient descent algorithm that makes use of thee signs of gradients to determine updates. It adapts the step size dynamically for each weight independently. This utilizes different step size for each weight.

The gradient descent variants mostly make use of the sign and the magnitude of the gradient. The gradient points are always directed to its steepest ascent. As we want to observe a minimum, we note the gradient in the opposite direction. This direction is committed to the sign of the gradient.

2.3 Methodology:

The various sensors are placed on the infant to measure parameters such as heart rate, SPO_2, breath rate and temperature. These vital parameters are read continuously and the live data is sent in through a microcontroller through a serial port. This data is then processed and extracted. The values are then compared with the database that contains standard values. The decision model makes a decision and comes to a conclusion whether the condition of the infant is abnormal (sleep apnea) or not. If abnormal, the decision-making model sends a trigger to the flow sensor whose one end is connected to oxygen supply and the other end is connected to the mask on the patient's mouth. Once the trigger is given, the oxygen supply reaches the patient's mask. The therapy unit is stopped only when the condition turns normal.

3. Results and Discussion:

The system is a closed loop system. The live data from the patient is obtained continuously. The data is transferred to the system through a serial port. Tons of data are received every minute and hence only the data that is needed is segregated from the whole lot. The data is then tested and sent to the pattern recognition neural network. The data from the database is trained simultaneously and sent to the pattern recognition neural network. The live data and the pre-existing data are compared and the decision model takes a decision.

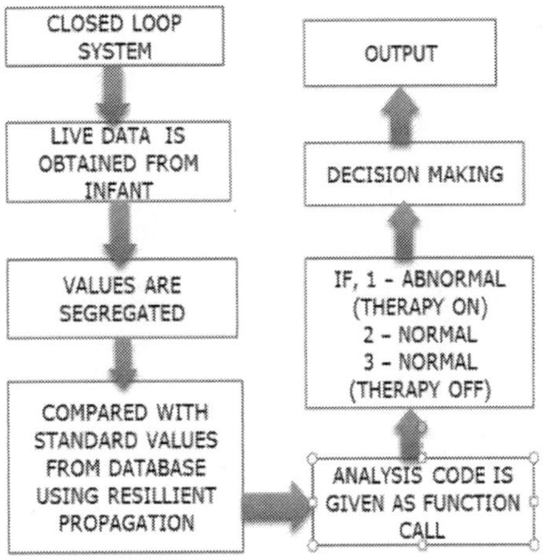

Fig 3: Flow Diagram of process

When the program is run, a pop up box appears with either numbers '1', '2' or '3'. '1' denotes abnormality and the therapy is ON, '2' denotes that the condition is normal and '3' denotes that the condition has come back to normal and the therapy is OFF.

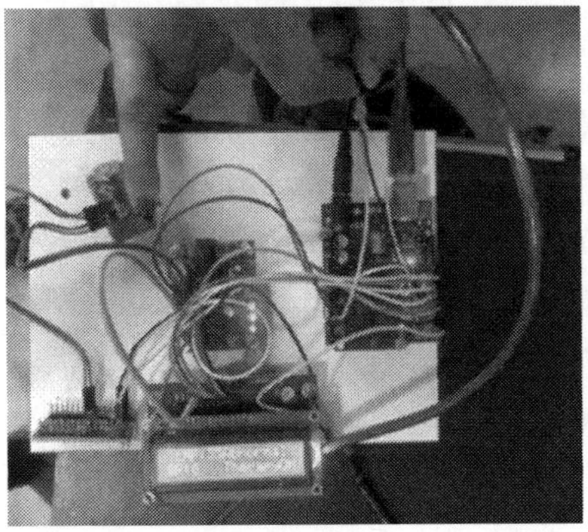

Fig 4: Work set up

Along with this, the output box has other features or plots such as performance plot, training state plot and three other plots to analyse the system. The readings of the vital parameters along with the condition whether normal or abnormal is displayed on a LCD screen. After the program is run, different plots such as performance, training state, confusion, error histogram were analysed.

https://doi.org/10.1051/itmconf/20213701006

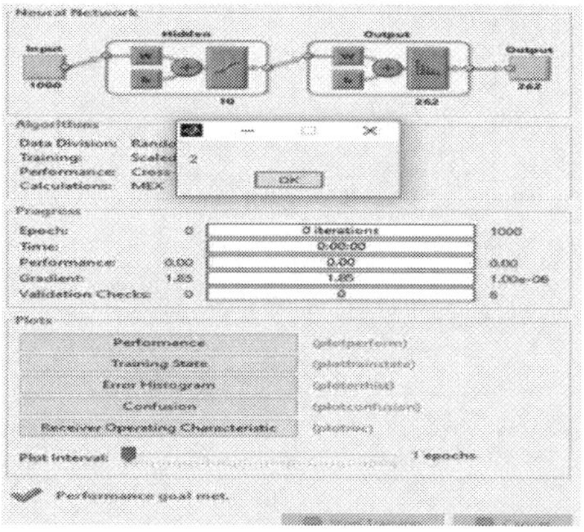

Fig 5. MATLAB output showing conditions

The output is displayed on the hardware through a liquid crystal display. The LCD keeps running a loop by displaying the readings of the vital parameters one after the other. Also, it displays the condition of the infant whether abnormal or normal.

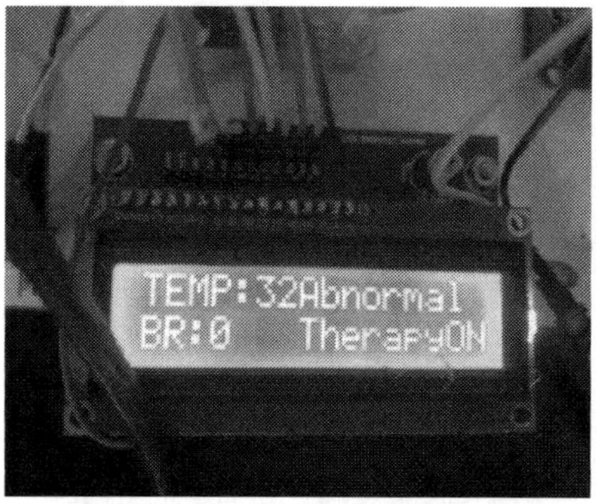

Fig 6:LCD display when the condition is abnormal

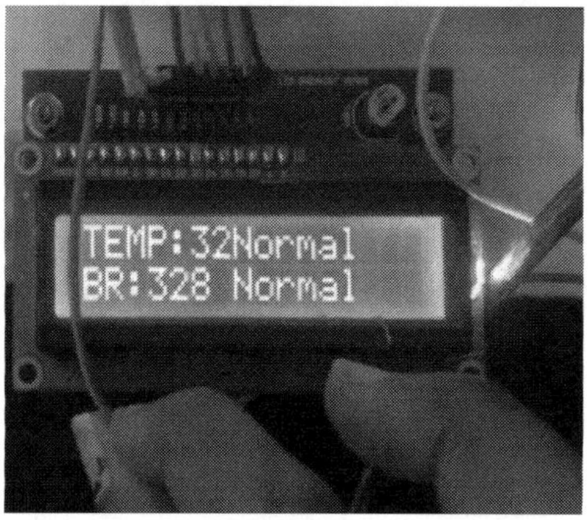

Fig 7:LCD display when the condition is normalised after therapy

Performance plot is used to obtain a plot of training record error values against the number of training epochs. In the default setup, the training stops after six consecutive increases in validation error, and the best performance is taken from the epoch with the lowest validation error.

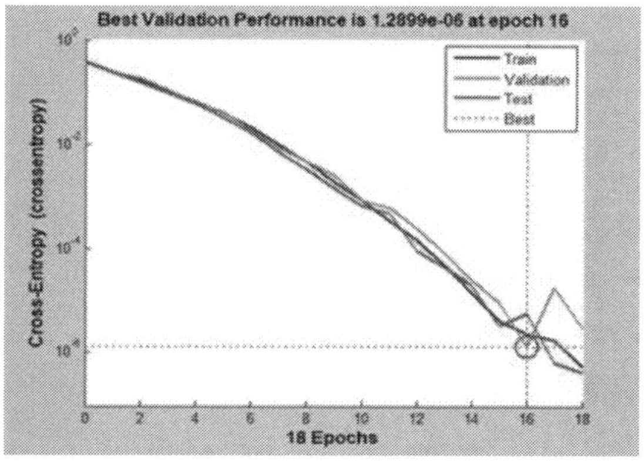

Fig 8 : Performance Plot

The current progress or status of the training at a specific time while training is in progress gives the training state. We have given 6 validation errors,hence the training will stop as soon as 6 validation errors occurs.

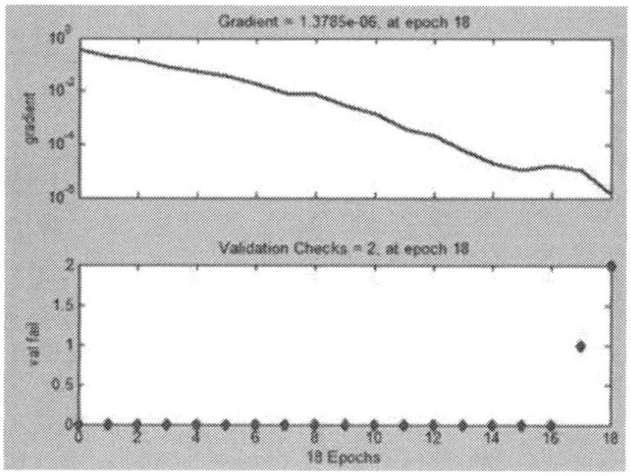

Fig 9 : Training State

The error histogram indicates the distribution of errors during testing instances in a neural network. The figure indicates the errors histogram while running our data.

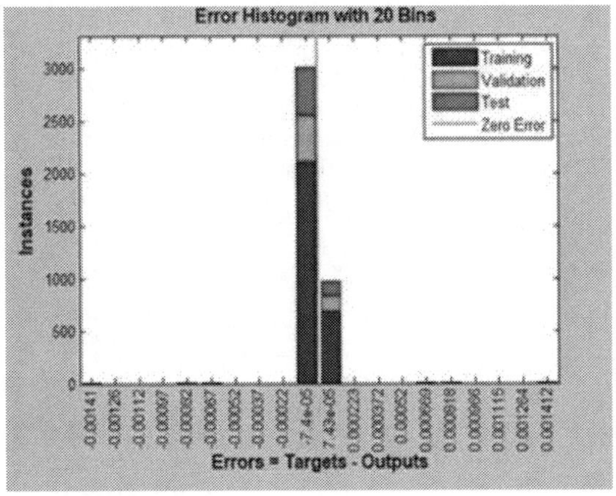

Fig 10 : Error Histogram

The total number of observations in each cell is displayed in a confusion matrix. The actual Yes and No is represented by the rows in the matrix while the column indicates the Predicted Yes or No. The correct and incorrect observations are observed in the diagonal and off-diagonal cells,respectively. Out of the data provided, it is observed that we obtained a predicted YES of 29.

Table 1 : Confusion Matrix for our data

N=135	Predicted YES	Predicted NO
Actual YES	29	2
Actual NO	2	102

The results obtained from running this work showed better results compared to the reviewed papers. Further it is a non- invasive technique which can be highly reliable. Inclusion of Matlab has produced desirable output. The decision model had to make the final decision on the values obtained whethe rabnormal (sleep apnea) or normal. When an abnormality related to breathing rate and oxygen level was noted by the system, it sent a trigger to the flow sensor which was connected to the oxygen supply and the mask. This trigger turned ON the therapy unit and started sending oxygen from the supply to the mask on the infant's face. The performance plot, the training state plot, confusion plot and error histogram were obtained in order to analyse the performance of the device. The therapy unit goes on till the condition of the infant becomes normal and it starts again when an abnormality is detected.

4.Conclusions

An effective solution has been used to detect sleep apnea in preterm infants and also therapy (NCAP) is provided when sleep apnea is detected. The results show successful analysis of the live data obtained from the infant and also shows 98.51 accuracy. The work will help in reducing mortality of preterm infants due to sleep related breathing disorders such as sleep apnea. In the future, this work will pave way for even more advanced way of detecting sleep apnea in infants and also for providing therapy.

References

1. A .Bandyopadhyay, AS Daftary. ClinPediatr (Phila). Jul;**59**(8):752-759(2020).
2. K. Gipson, Lu M, TB Kinane. Pediatr Rev.;**40**(1):3-13(2019).
3. W. Göpel, A. Kribs , A.Ziegler , R Laux, T Hoehn, C Wieg, et al. Lancet; 378(9803):1627-34(2011).
4. D Gozal, HL Tan, L Kheirandish-Gozal. J Clin Med. Mar 24;**9**(3):888(2020).
5. V Guru Anand,., G Hari Krishna,., G Mohandass,., R.J Hemalatha,., S Sundaram,. TISC-2010, 5714621, pp. 122–124, (2010).
6. G Hari Krishnan, R.J.Hemalatha, , G. Umashankar, N Ahmed, S.R Nayak,., Adva in Int Sys and Computing 308 AISC **1**, pp.635-641,(2015).
7. G .Hari Krishnan, R.AnandaNatarajan, A. Nanda, J of E E and T, **9**(6), 2114–2117, (2014).
8. R.J. Hemalatha, G.H. Krishnan, G.Umashankar, S.Abraham, Biosci Biotech Research Asia **11**(2), pp.907-910,(2014).
9. JJ .Ho, P .Subramaniam, PG.Davis Cochrane Database of Systematic Reviews 2015, Issue 7.
10. N. Ilangovan,, G. Hari Krishnan,RJPBCS, **6**(3), pp.568-570, (2015).
11. G.H.Krishnan, A.Nanda, A.Natarajan, BPJ, **7**(1), 221–224, (2014).
12. G.H.Krishnan, A. Natarajan,A.Nanda, BPJ, **7**(1), 379–382, (2014).
13. G.H.Krishnan, R.A.Natarajan, A. Nanda, ICICES 2014, 7034178, (2015).
14. B Lemyre, PG.Davis , A.G.Paoli,H. Kirpalani . Cochrane Database of Systematic Reviews (2017), Issue 2.
15. Marta Marin-Oto, E. Eugenio Vicente, Jose M Marin, Mul Resp Med, **14**(21) (2019).
16. G.Mohandass,R.AnandaNatarajan, G.Hari Krishnan, Biomedical Research (India), **26**(2), pp.273-285, (2015).
17. A. Nagarjuna Reddy,G. Hari Krishnan, D. Raghuram, RJPBCS, **7**(6), pp.570-575, (2016).
18. RM.Patel , S. Kandefer , M.C.Walsh , New England Journal of Medicine;**372**(4):331-40(2015).
19. W.F.Qubty,A. Mrelashvili , S. Kotagal . J Clin Sleep Med. Nov 15;**10**(11):1213-6 (2014).

20. B Lemyre , P.G.Davis A.G.Paoli , H. Kirpalani Cochrane Database of Systematic Reviews, Issue 2.**B,** (2017)
21. R.J. Martin , H.S. Nearman , P.G Katona , M.H.Klaus J of Pediatrics,**90**(6):976-819(010).
22. J.J.Ho , P.Subramaniam , P.G.Davis. Cochrane Database of Systematic Reviews, Issue 7(2015).
23. G.M.Schmölzer ,M.Kumar , G.Pichler , K Aziz , M. O'Reilly , P.Y.Cheung , BMJ; 347:f5980 (2013).
24. A.Telerant, G.L. Dunietz, A. Many, Sci Rep 8, 10768 (2018).
25. K.I.Wheeler ,C. Klingenberg , C.J Morley,P.G Davis . Neonatology; **100**(3):219-27 (2011).
26. Z. Xu , Y. Wu , J.Tai . J Otolaryngol Head Neck Surg.;**49**(1):11(2020).

ITM Web of Conferences **37**, 01007 (2021) https://doi.org/10.1051/itmconf/20213701007
ICITSD-2021

Survey on Template Engines in Java

Jishnu Saurav Mittapalli[1*] and *Menaka* Pushpa Arthur[1]

[1]School of Computer Science and Engineering, Vellore Institute of Technology, Chennai Campus, India 600 127

Abstract - In today's fast paced world every minute is very important. In the corporate world and also otherwise Documentation is very important for various purposes like version control, proofs, consent, copyrights and expectations and outcomes/reports. So, because of these reasons template engines have become very important and extremely necessary for the world. Template engines are basically software that help us create result documents from data models and templates by combining them. This paper presents a survey on the newest development of research work on template engines for java, along with an in-depth outline of its research. To the best of what developments have been achieved till now, this survey paper has been written. Finally, the differences, advantages and disadvantages of the various template engines for java, have been tabulated as a part of the results.

1 INTRODUCTION

A template engine is a software that provides static template files for our application. During runtime, template engines can replace the variables by real values in our file and it helps us to transform the template into a HTML file. This basically supports the developer to make a better HTML webpage. A template language or templating language is the language that the templates are written. a template. Most of the template engines generally have the common features for all high-level programming languages, where features for processing plain text are emphasized [1].

The common features of template engines are as follows:

- Replacement of text.
- Functions and variables
- Loops and conditional statements
- Inclusion of files

In many different contexts, template processing has many interesting uses, also very different and unique ones. One of the main uses is that it depends on the software application that is created for this purpose. Basically, the template engine's flexibility helps us in

- Organization of source code.
- Helps in version control.
- Improves productivity and efficiency.
- Improves teamwork or team collaboration in the projects.

The importance of automatic generation of code is increasing at a tremendous rate. As today in model-driven software applications, flexible code generation is very important. This paper basically discusses the various existing template engines in Java programming language and their pros and cons. Along with this, we recommend an appropriate template engine as per the developers demand and requirement.

many other uncommon ways different from the ones the creators require [2]. The following are the main uses of template engines:

- A software that has modules of all major programming languages and is generally a part of the software development.
- Main functionality of the software - Basically, the source data and the templates are used typically to generate output web pages or parts of the output web pages. Generally, used in web application development.
- Template engines can also be used for generation of source code from very abstract or a little unclear data models for example UML or relational data models. That is, it helps the developer, by simplifying the production process and making his work easier.
- Template engines are also used for documentation purposes, i.e, they help us in generating documents for software products.

The main benefits of using template engines are as follows:

- Saves a lot of time in the software development life cycle.
- Helps in the proper

2 APACHE VELOCITY

One of the commonly used templating engines for Java is Apache Velocity. Apache velocity is open source and it is designed to be a web framework and can be used very efficiently in the MVC (Model-View-Controller) architecture. XML files, SQL, PostScript and many other formats can be generated using Apache Velocity. Reading, parsing, content generation and many other things can be made easily by using this apache velocity. It is one of the cleanest ways of incorporating webpage content that is dynamic using references. We can download velocity from their official website. Building web applications is the most major use of apache

* **Corresponding author:** jishnu.2018a@vitstudent.ac.in

velocity. Servlet-based framework or servlet will be needed by us to develop a web application using apache. Velocity can also be used to generate PDFs from HTML documents. Coming to the price details, apache velocity is a free and open-source templating engine. This can be installed in various IDEs like eclipse, netbeans and also in visual code studio as a plugin and can be used as the template engine. And the free version itself has all the features that we can use in this framework.

The following are the main use-cases of apache velocity:

• Dynamic Web Pages: HTML page structures can be rendered Web applications by Web developers. Dynamic information populates the content. VelocityViewServlet or a number of frameworks that support Velocity process the web pages [3][4]. Various applications that require the generation of automatic email reminders for password, reports automatically sent or account signup. The email that is a text file storing the content can be sent and this can be embedded in the java code itself using velocity [5].

• AWS API Gateway: Use the velocity engine for body mapping templates and the output is the JSON file.

• Conversion of HTML into documentation and vice-versa: This can be used to convert HTML pages into documents and documents into beautiful web pages.

The following is the technology used behind Apache Velocity:

Context objects are used for pulling references from a set and get methods for setting and receiving objects. The values can be directly inserted into the page. Macros can be created and files should be included. Java methods which are arbitrary can be called and this ability is available to be used multiple times making this a strong and still a simple method with which we can develop text files or other web pages dynamically.

MVC is enforced by velocity and this way of creation where the html code is divided with the Java code is very efficient and reusable. Velocity features are not implemented using other functions unlike PHP., Velocity does not implement features with other functions. The greatest strength of velocity as it allows to create web pages with better designs as it uses MVC [6]. The overall architecture of the Velocity is given in Figure 1.

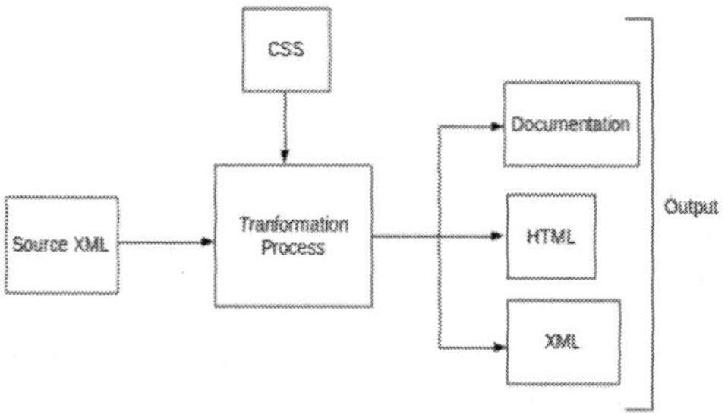

Figure 1. Architecture diagram of Apache Velocity

3 Apache Freemarker

Apache Freemarker is also a free Java based template which was mainly designed for dynamic web page generation based on software architecture such as MVC. The main advantage of using this is that it is a general template engine with no particular dependencies and it is not dependent on specific technologies like HTTP or HTML or servlets, therefore allowing us to use it for all purposes of a template engine like document generation and source code generation.

A few important features of FreeMarker are listed below:

• It is really lightweight and multipurpose, has no dependencies at all and any output format is allowed and can be used as just a plugin in text editors like VSC etc. It also has a lot of configuration options.

• It is aware of what kind of program it is working on and when it is locale sensitive with respect to date and timeline. It has a lot of capabilities like XML processing. It can traverse files of various different types and also process them in a more efficient way than many other engines.

• The data model is a very versatile one [3]

Basically, as shown in Figure 2, the template and the given Java objects act as the input to the Apache freemarker and it parses and combines in order to give the output as the template.

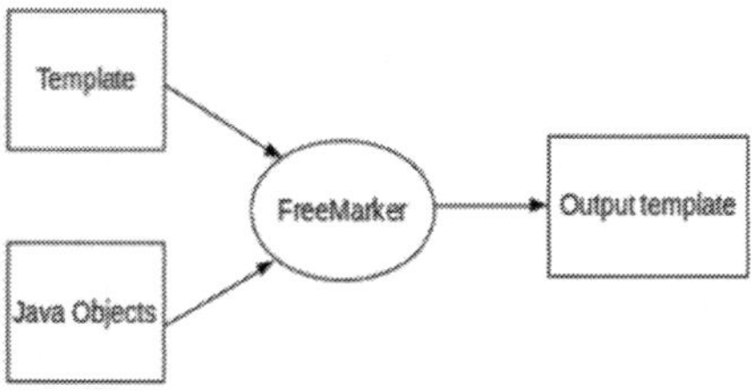

Figure 2. Architecture diagram of Apache Freemarker

4 Thymeleaf

Thymeleaf is basically a template engine for Java that can work on both in non-web as well as servlet-based environments and applications. It is more apt to work with HTML5 for web applications but it can also work well with offline applications. It can basically act as a substitute for JSP, its template files can be opened directly from the browser itself.

The main features are as follows:

• Can be used for full documentation.

• If required it can be used as a template engine framework.

• It can detect the type of document and convert as and when required.

• Very good support

• Reduces the input output to the minimum due to the feature known as parsed template cache which gives a high performance and efficiency.

• It is capable of working in both online and offline environments.

The main advantage of this in comparison to the other engines that we have discussed is that in it templates work as prototypes enabling a natural templating system. The templates open in the browser without any server running i.e., templates are statically displayed. It is basically like a template engine framework.

5 Apache Tiles

Apache Tiles basically is a composition framework for templates. Web application development and User interface development in Java was the main aim for the development of apache tiles, but it is no longer restricted to the JavaEE web environment. Here, page fragments are allowed to be defined. And the complete page can be assembled from the page fragments generated at runtime.

A template is basically the layout part of a webpage. Every page created has a structure with a few gaps called attributes, that have to be filled. There are three types of attributes, string, template and definition. A composition that has to be rendered to the end user and this is called a definition. A definition is completely or partially filled attributes, composed of a template. It can be rendered to the end user if all of its attributes are filled and if all of its attributes are not filled it can be called as an abstract definition and can be used for missing attributes or extended deadlines i.e, it can be used a base definition for it and the attributes can be filled at runtime.

Many of the time definitions need to be prepared and only then can they be rendered. And this is the reason we have something called a view Preparer and it can be used and after which the definition is rendered, so that all the things that are required are rendered correctly in the correct form.

6 Mustache.java

Basically, mustache is a web-based template system with applications of so many different languages like, PHP, Rust, Xquery, Python, Rupy, C++, R, Scala etc. Another important feature is that mustache is described as a logic less templating engine. The main reason for this is the lack of any explicit conditional / looping statements and the evaluation can be achieved using section tags , lists and lambda functions.

The reason Mustache got its name is because of the huge use of curly braces '{ }', which resembles a side 'mustache'. Mustache is used mainly for mobile and web applications. This is mainly used for mobile and web applications. It is available as plugins in many text editors and syntax highlighting is available in many text editors like Textmat,Vi, Visual studio code, Atom, Coda etc.

The templates of mustache support are there by default in many frameworks including web application frameworks like cakePHP. Client Side, ajax frameworks such as jquery, dojo as well as server-side javascript programming libraries are also included and are supported by this.

The following are the important features of mustache:

For the mustache engine many implementations are existing, and all of them have a few common specifications that the results have a common syntax. Most important method is the render method and the most common class is the parser class of this technology.

The development of the mustache template engine inspired many JS template libraries that improved the functionality and efficiency of this template/rendering engine. Lack of logic statements and another feature is that it is compatible with all javascript frameworks. On the client side Jquery, YUI, Dojo and on the server side Node.js are supported by this.

Main advantages of this engine are Code is very understandable, Re-use of code, and easy maintenance. Source code, HTML and config files are the major applications of this.

Hash or objects have the values which are used for expanding the tags, and this is how it works.

A Mustache template have A mustache template containing a string and it can have any number of mustache tags. The double mustaches indicate the tags. The sections's tag value can be undefined, false or null and then only an inverted section's block is rendered. Mustache.compile can be used to create javaScript functions by compiling mustache templates for improving the performance of the rendering process.

Table 1. Features of Template Engines

Name	Functions	Error handling	Natural Templates	Inheritance
Apache velocity	Yes	Yes	No	No
Freemarker	Yes	Yes	Yes	No
Mustache	Yes	Yes	Yes	No
Thymeleaf	Yes	Yes	Yes	No
Groovy	Yes	Yes	Yes	Yes
Tiles	Yes	No	Yes	No

7 Groovy

A language which can work well with the OOPS concepts especially in Java was developed and called groovy. It is supportive of dynamic as well as static language. It has features similar to ruby, python etc [7]. We can use it as a scripting as well as a programming language for java.

Table 2. Comparison of Different Template Engines in Java

Name	Name of developer	Company owned	Year	IDE	Paid/ Free	Online / Offline	Enough documentation	Completed / Ongoing
Apache velocity	Apache Software Foundation	ASF (The Apache Software Foundati on)	2001	NetBeans, Eclipse, BBEdit, Dreaweaver, InteliJ IDEA, JEdit	Free	Both	Yes	Most probably satisfy the requirement of the user
Freemarker	Jonathan Revusky, Attila Szegedi, Dániel Dékány, and others	The Apache Software Foundati on	2000	Eclipse, NetBeans	Free	Both	Yes	Most probably satisfy the requirement of the user
Mustache	Chris Wanstrath and other contributors	Available in github	2009	IntelliJ	Free	Offline	MIT Licensed Documentation	Ongoing
Thymeleaf	Daniel Fernández	ASF(The Apache Software Foundati on)	2018	IntelliJ, Eclipse	Free	Both	Yes	Completed
Groovy	James, Strachan	AFS (The Apache Software Foundati on)	2020	IntelliJ, Netbeans	Free	Both	Yes	ongoing
Tiles	ASF	ASF (The Apache Software Foundati on)	2012	IntelliJ, Selenium, Eclipse, NetBeans	Free	Both	Yes	Ongoing

The syntax with a curly bracket is used by groovy. Expressions in strings, multiline strings and closures are supported by groovy.

Most of the valid Java files that are existing are also out of groovy. Groovy code can be very small and clear which is a great advantage.

It is easy for developers to learn and make themselves familiar with the basic java syntax as it does not does not require all the components that normal java requires. The features that groovy offers are not available anywhere else. Official syntax for lists, polymorphic iterations, maps, helper methods are a few to name. From its second version groovy also supports the jar files etc to improve the modularity of the programs. Still research is going on and people are working to improve the efficiency of this platform.

Native support is provided for many markup languages like HTML, XML etc. Not like Java, groovy code can be executed as a script i.e., without compilation itself

8 Difficulties and Disadvantages of Template Engines

The use of template engines itself has quite a few disadvantages. The following are a few of them:

- Firstly, using these is a skill and it has to be acquired, which will take time. As it is not very easy to learn them.
- There definitely will be some amount of performance overhead as it is something additional and extra processing will be required when we use them.

- Time consuming and add a component into the development lifecycle.

9 Results and Discussion

The various features of the template engines in Java such as Apache velocity, Freemarker, Mustache, Thymeleaf, Groovy, Tiles are tabulated in Table 1. and Table 2. We, have presented the review to the best of our knowledge. From Table 2, the following observations we made from our comparison. Other than Mustache, remaining engines are developed by Apache Software Foundation. It shows the quality and feature richness in the tool. As a developer, we can contribute our work in Mustache since the code is in GitHub. Groovy and Tiles are latest release with current trend features. The IDE IntelliJ is almost compatible with all listed template engines. Each template engine has its own pros and cons, we have to choose the template engine based on our requirements.

10 Recommendation

Since Apache Velocity and FreeMarker are free open-source code, students and research scholars can utilize this engine for their application development. Web and non-web applications can be developed by the Apache Velocity template engine. Groovy's template engine is very simple and intended for limited purposes. Like mail merge in the word processor, Groovy can provide such a functionality for Map [8]. Mustache generates an output for the input with simple logic-less tags. Mostly used for dynamic web-page generation and server-side programming. Thymeleaf can be a template engine for source code development and web-based applications.

11 Conclusion

This paper reviews all template engines that are currently being used in Java. In this paper as we discussed every template engine has its own pros and cons. We have to choose our template engine as and when based on our templating requirements and the type of application we are developing. Though there can be more research and development in this field we still can still make the use of these template engines a little more widespread. These technologies basically are very useful and if used properly as per the requirements we can achieve a greater efficiency in our work and have a more efficient and a faster development lifecycle. So, basically, we in this paper have given the details about the various template engines that are existing in the market to the best of our knowledge and research, for further research and industries to choose the template engine required by them according to their needs.

Many enhancements and different new developments can be made to this field. These technologies can be more widespreadly used and more advancement can be brought in. And these engines can be made more efficient and faster.

References

1. Sturm T., von Voss J., Boger M., Generating Code from UML with Velocity Templates. In: Jézéquel JM., Hussmann H., Cook S. (eds) ≪UML≫ 2002 — The Unified Modeling Language. UML 2002. Lecture Notes in Computer Science, vol 2460, 2002.
2. Michiaki Tatsubori and Toyotaro Suzumura, HTML templates that fly: a template engine approach to automated offloading from server to client. In Proceedings of the 18th international conference on the World wide web (WWW '09). Association for Computing Machinery, New York, NY, USA, 951–960, 2009.
3. Tatsubori M., Sasaki T., Chiba S., Itano K., A Bytecode Translator for Distributed Execution of "Legacy" Java Software. In: Knudsen J.L. (eds) — Object-Oriented Programming. ECOOP 2001. Lecture Notes in Computer Science, vol 2072. Springer, Berlin, Heidelberg. 2001.
4. Barclay, Kenneth; John Savage, Groovy Programming: An Introduction for Java Developers. ISBN 978-0-12-372507-3. Archived from the original on 2010-01-12. Retrieved 2007-07-19, 2007.
5. Tilevich E., Smaragdakis Y., J-Orchestra: Automatic Java Application Partitioning. In: Magnusson B. (eds) ECOOP 2002 — Object-Oriented Programming. ECOOP 2002. Lecture Notes in Computer Science, vol 2374. Springer, Berlin, Heidelberg, 2002.
6. The Apache Velocity Project, https://velocity.apache.org/
7. Kenneth Barclay and John Savage, Chapter 21 - Template engines, Groovy Programming, Morgan Kaufmann, 2007, Pages 297-300
8. Kenneth Barclay and John Savage, Chapter21-Template engines, Groovy Programming, Morgan Kaufmann, 2007, Pages 297-302

Disruptions on India's FinTech landscape: The 5G wave

Mohanasundaram T[1,*], *Sathyanarayana* S[2], and *Rizwana* M[3]

[1, 3] Department of Management Studies, Ramaiah Institute of Technology, Bengaluru, Karnataka, India
[2] M. P. Birla Institute of Management, Bengaluru, Karnataka, India

Abstract. The emergence of FinTech companies disrupts the banking and financial service industry through its technology-driven services. But with 5G, the 5th generation mobile network, the entire FinTech outlook is disrupted. The 5G technology is expected to create a new FinTech ecosystem in India. The 5G features like enhanced mobile broadband (eMBB), Lower latency and robust security aspects are certain to bring transformation in banking and FinTech industries. 5G's application in different domains offers a 'blue ocean' of new prospects to FinTech companies. 5G technology offers necessary features to enable FinTech companies to bring a massive change in the banking and finance field. As 5G is on the brink of being a next-generation mobile network in India, it is necessary to examine its influence on the FinTech landscape. In this study, the authors have analyzed the effect of 5G on evolving bank technologies, with special focus on FinTech. This study also projects the authors' view on the impact of 5G on FinTech firms' future challenges and prospects.
Keywords: 5G, FinTech, disruption, Banking, Financial services, digitalization.

1 Introduction

5G technology enables the new industrial revolution, *industry 4.0*. A huge ongoing automation is in progress across different sectors. Innovative and life-enhancing business models are emerging and creates disruption in the way business functions. The arrival of 5th generation (5G) telecommunication technology is expected to create new ecosystems of Fintech and digital banking by offering new features and tools. 5G technology is creating new business models in the financial industry and expected to change financial landscape, especially in FinTech. There is a growing attention from the international tech firms to make a way into financial industry. 5G features and its potential applications include: (i). Enhanced Mobile Broadband (eMBB) for virtual reality and improve video conferencing with extreme network speed and processing capability. (ii). Highly reliable and low latency in processing a very large volume of data. (iii) Massive Machine to Machine Communication enabling Internet of Things (IoT) etc., and (iv). Robust security aspects leading to high reliability. Innovation is the need of any business which is all about novel thoughts that produce great value. We are living in the age of digitalization and we are constantly moving in that direction by adopting pioneering technology to ensure more effective and prolific way of life. In the area of Finance too, we are consistently upgrading the products, services, methodology and system with new technology.

The financial sector is one of the largest users of digital technologies. Such innovation and upgradation in Finance through technology is known as FinTech (Financial Technology). FinTech as a concept developed only in the 21st century. FinTech is the combination of 'Financial Service' and 'Technology'. The adoption of technology in the field of financial services viz., Banking, Share market, Insurance sector, Mortgage loans etc. with database management, analytical tools, and innovative approach. The prominent operators of FinTech are; Banking and Insurance sector, Financial Institutions, Small and Medium Enterprises (SMEs), Start-ups, and individuals. Since 2016, India is actively engaged with FinTech after the demonetization of high denominated currency notes. From then, Unified Payments Interface (UPI) payment technique is highly engrossed by the Government of India. Many start-ups started using innovative technology to make easy and cashless payments. FinTech will play an important role in achieving the cashless economy. [1] stated that the scope of FinTech is huge, as there were 1,400 FinTech firms existed in 2015 with investment of more than $33 billion.

Digitalization not only minimizes the time and cost but also increases efficiency. The fintech industry has progressed enormously from challenging and team up with traditional banks and has today entered a new era at the front of digital revolution prioritizing technology. The digital revolution has shaped several innovative technologies but none of them had a massive widespread influence than internet and mobile communication technology. These technical revolutions independently have transformed the way we communicate with others, but collectively, they have transformed the entire world. 5G's application

*Corresponding author: tmohansun@gmail.com

in different domains offers a 'blue ocean' of new prospects. [2] referred blue ocean as a rise of an entirely new industry or developments in the prevailing industries that amends the boundaries of competition. The most apparent opportunities from a bank and financial service institution's viewpoint is the sustained progress of digital banking solutions aimed at each distinct customer. These digital devices and Fintech firms endlessly collect data on each customer [3]. 5G being the next-generation technology is expected to change the role of mobile in society. 5G alone is forecasted to add $2.2 trillion to the world economy[2]. Improved connectivity, facilitated by 5G technology, shall be a catalyst for economic growth in the Industry 4.0 with a projected economic value of $13.2 trillion by 2035[3].

The mobile-oriented business restructuring is undergoing in banks and financial service firms by functioning meticulously with Fintech providers for providing better services, retaining their current customers, and attracting new or potential ones. As 5G is on the brink of being a next-generation mobile network in India, it is necessary to analyse its effect on the FinTech landscape. In this study, the authors have analysed the effect of 5G on evolving bank technologies, with special focus on FinTech. This study also projects the authors' views on future challenges and opportunities.

2 Fintech Industry – An Overview

As the name 'FinTech' proposes, it is the blend of financial service and technology. FinTech comprises many tools and technological advancements that are way different from traditional banking. In recent years, a vast number of FinTech firms are coming into the market by offering a variety of new financial solutions and services that are in great demand.

Fig.1. Fin Tech

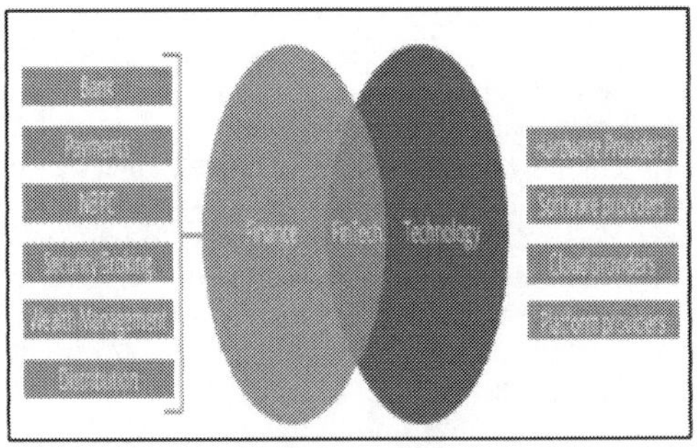

Source: Deloitte International Analysis

FinTech is an application of technology to offer novel and better financial services. Technology has been a crucial facilitator in the digital economy's expansion. Over the time period, banks and financial institutions in India have gradually and carefully adopted technology to expand the reach, service and working efficiency with growing market and advancements in technology.

In today's world, financial institutions have higher significance; and even a minor change makes an enormous impact on the global economy. Although finance as a domain met numerous demanding revolutionary waves over the period, it has succeeded to adopt and endure from those changes. At present, financial institutions and the industry as a whole are witnessing massive change because of FinTech. The consequences of this change are yet to be fully explored and they may bring a huge challenge for different stakeholders viz., policymakers, academicians, service providers, intermediaries, economists and the general public. [4] specified that information technology will inspire the specialized firms which are generating niche markets with superior customized product in order to cater to consumer preferences. Peer-to-Peer (P2P) lending, cryptocurrencies and many other innovations of all segments of an emerging technology-supported personalized financial services. FinTech has come a long way since its evolution and is expected to reach much greater heights in the days to come.

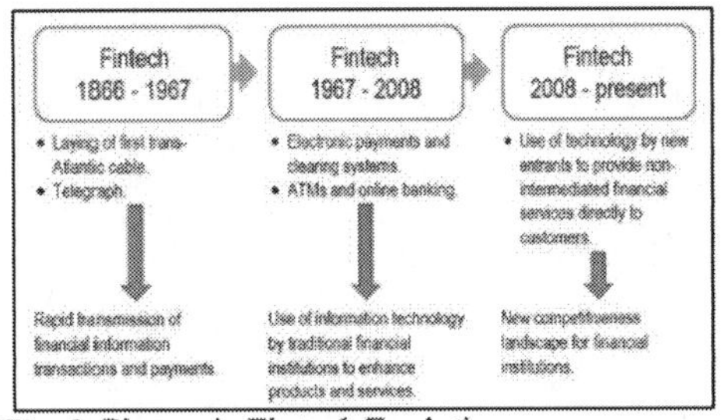

Fig. 2. Phases in Fintech Evolution
Source: Consumers International, 2017

Technology has constantly affected the finance industry with developments thereby altering the way the finance industry functions. For example, the emergence of Automatic Teller Machines (ATM) or Online transactions of money were some of the key innovations in Financial service. It is interesting to know that what makes the current FinTech revolution so special than earlier innovations? First and foremost, the speed at which new innovative financial know-how and technologies are tested is much quicker than earlier. Secondly, the most crucial thing about FinTech revolution is; it is unique, as the change happening is from the external environment (i.e.) new start-up firms and large recognized technology companies are endeavouring to intrude the financial services by presenting

[2] GSMA report 2018 'Study on Socio-Economic Benefits of 5G Services Provided in mmWave Bands'

[3] The impact of 5G: Creating new value across industries and society, World Economic Forum report.

innovative products and technologies thereby offering a significant competition in an innovative manner.

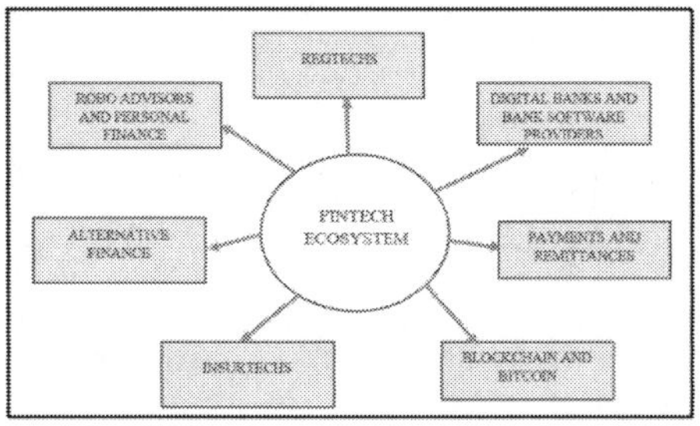

Fig. 3. Fintech Ecosystem

FinTech firms' ecosystem includes various stakeholders, partnering institutions, technology platforms and regulatory compliances. [5] in their book discussed the FinTech revolution to highlight the persistent disruption and to create methods for analyzing flash crashes, hostile high-frequency trading, and other specific facets of the market. According to the IMF, among the low- and middle-income countries, India has grown phenomenally in the number of registered mobile money accounts between 2015 and 2019. India had witnessed 17-fold jump during that period[4]. [6] observed that big strides happened in different fields such as mathematics, big data, computing technology, statistics, econometrics and cryptography have integrated well to make an explosion of FinTech.

3 5G Technology in FinTech

Change in any form is not easy to deal with, particularly when it happens in banking and financial services, as the process involved and the number of stakeholders is very high. In a digital era, the speed and ability of data transfer along with compatibility of data are vital for FinTech to prosper. The 5G technology just offers those features to enable FinTech companies to bring a massive change in the banking industry. Rapid internet speed has become imperative for improved financial transactions. 5G also has an edge at handling heavy load compared to the previous technologies. 5G will certainly help the FinTech players to bridge the gap between the existing reality and customer expectation by offering safer and faster service to end-users. 5G mobile technology services also support FinTech companies to provide affordable loans by refining the lenders' assessment of borrowers based on financial soundness. 5G is expected to disturb information security, machine learning, IoT, data management system etc. which are used extensively by FinTech companies. In the Banking and Financial Services industry (BFSI), 5G will be mainly useful in making communications more secured, processing real-time information and improving customer experience. 5G services have the potential to connect numerous devices, minimize power consumption, reduce latency, offer higher security, and better support for augmented/virtual reality experience. It enables to manage millions of IOT devices. 5G mobile technology will empower the extensive implementation of blockchain, artificial intelligence, machine learning and quantum computing. The Organization for Economic Cooperation and Development (OCED) predicts an upsurge in world GDP, higher employment and digitalization with quick 5G penetration.

Being new and technology-driven company doesn't assure a new player an easy journey especially when it comes to legal compliance. 5G along with RegTech will help the financial institutions to better adhere to the legal compliances, and this is more important for the banks and institutions operating at the global level as they have to comply with numerous legal aspects and norms. India is having a good 4G connectivity which leads to the highest mobile data consumption in the world, offers a great potential for 5G technology implementation. According to Ericsson, India has the world's highest data usage per smartphone at an average of 9.8GB per month and this is expected to double to 18GB by 2024. As the financial institutions seemingly leverage the value of data, the method in which the data is gathered, warehoused and deployed are sprouting to maintain pace with ever-growing customer expectation and legal requirements. The benefits of 5G in banking include; enriched ATM facility with features like virtual reality in order to directly interact with the bank, enables richer customer experience with lower cost, Risk prevention, complying with regulatory norms. 5G technology along with blockchain and other revolutions will affect different areas in finance like capital markets, cross-border payments, automated payments, mobile banking etc. 5G provides an extensive platform for Fintech firms to innovate. Australian telecommunication providers have realized speeds up to 1200 Mbps (1.2 Gbps) in 5G testing[5] which is significantly higher than 20 Mbps to 50mbps in 4G technology. No doubt, if implemented, 5G will immensely improve the user experience and real-time delivery of services.

Startups in Finance are providing computerized automatic guidance, touchless payments, and many other services that brings many social and economic benefits which includes lesser prices and better

[4] Mobile money a/cs: IMF says 17-fpld jump in 4 years in India, Financial express dated 11.11.2020 – retrieved on 06.12.2020.

[5] The Future of Digital banking, KPMG, 2019.

https://doi.org/10.1051/itmconf/20213701008

access to credit. Purely from the perspective of product categorization, fintech's role can be split into those providing credit, payments process, delivering advice, asset management and legal compliance. Fintech being a relatively new category of companies, the business models are based on digital products. In spite of intensive competition from large technology firms, the chief driving force of fintech innovation has come from numerous technology-based startups attracting investments in billions of dollars each year. India is doing fairly well in adopting FinTech compared to most other nations.

The destructive impact of the global financial crisis in 2008 has created every reason to suspect the purpose and working ethics of financial services firms across the globe. Even big financial institutions all across the world suffered the worst from both economic as well as from public trust perspectives. The failure of big firms to meet the expectations of different stakeholders has opened the doors for a small-scale innovative technology-based financial institution. The substitute to existing banks and financial institutions emerged with small, controllable, and transparent features. Therefore, it is judicious to express that the global economic and financial crisis in 2008 signified the landmark to trigger fintech 3.0 phase. As Fintech firms move towards 5G service, the likelihoods of forming truly universal financial services turn out to be a greater reality. International firms view 5G as a disruptive technology of this era. The leading global aeroplane manufacturing firm, Boeing has already created a tech-based ecosystem for its commercial planes [7]. Many international banks like JPMorgan Chase, Bank of America, Citigroup and leading FinTech companies like, Ant Financial, Paypal, Qudian, etc. are actively penetrating 5G technology in their ecosystem to reap the benefit out of it.

Table. 1. FinTech Country Ranking

Global Fintech Rank	Change from Startup Rank	Country	Total Score
1	▶ ±0	United States	31.789
2	▶ ±0	United Kingdom	23.262
3	▲ +18	Singapore	19.176
4	▲ +14	Lithuania	17.343
5	▲ +3	Switzerland	16.018
6	▶ ±0	The Netherlands	14.464
7	▶ ±0	Sweden	14.272
8	▼ -3	Australia	13.555
9	▼ -6	Canada	13.322
10	▲ +3	Estonia	13.303
11	▼ -2	Germany	12.787
12	▼ -8	Israel	12.771
13	▼ -3	Spain	12.372
14	▼ -2	Finland	12.110
15	▲ +2	India	12.024

Source: *www.findexable.com*

FinTech firms such as GoCardless, Transferwire, Currency cloud etc are started disrupting the Forex market by competing aggressively with the bigger banks having larger market shares. [8] suggested that the digital footprints may be used to assess individual default risk. FinTech companies across the globe are attracting huge investments from a different set of investors like investment banks, venture capital firms etc. Fig. 5 will depict the quantum of investments happening in FinTech companies worldwide.

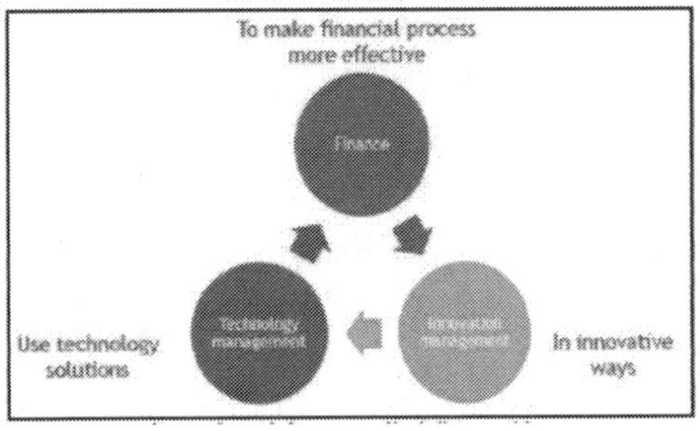

Fig. 4. FinTech – Cross Disciplinary nature

4 India and 5G Services in FinTech

5G primarily functions in three bands– low, mid and high-frequency spectrum – each having their own advantages and limitations mainly on account of speed and coverage. In 2018, India had intended to start 5G technology at the earliest possible. The major telephone players in India have urged the Department of Telecommunication (DoT) to roll out a detailed future plan on 5G frequency bands and spectrum allocation. However, Reliance Jio planned to unveil its own 5G network. The major challenges of 5G in India are; inadequacy of spectrum and TRAI's higher reserve price for spectrum auction. 5G has the ability to augment the day-to-day use of banking products and services. The financial institutions that embrace 5G technology are effectively poised far ahead of its rivals as they enjoy offering data-backed personal services for the customers of Gen Z. Moreover, 5G can overcome the major shortcoming of earlier technologies - inefficient handling of increased financial transactions.

5G potential in Indian banking and FinTech domain is boundless as the arrival of 5G coincides with a boom in Internet of Things (IoT) inventions, including smartphones and wearables. This will help banks and fintech firms to make seamless financial transactions, innovate and to provide secure services to their customers. 5G could also enable high-frequency trading and effective diversification of the bank's portfolio. 5G services may be effective in capturing how sentiment analysis can be used to gain more customer understandings. No doubt, 5G will bring innovative and better services apart from bringing more digital-savvy users for the banking industry.

5 Indian FinTech Scenario

It is well-known that the Financial sector across the world, more specifically in India, is undergoing a digital revolution. As per Fintech Global, Indian FinTech firms have raised over $13 billion through more than 460 deals since 2014. Many researchers have studied the emergence, evolution and effect of FinTech and Technology in Financial services [9 – 12]. The FinTech business has phenomenally grown and no more made up of only new start-up companies, the FinTech industry has many seasoned firms that provide a wide range of financial services and functions at an international level. In Emerging markets, FinTech adoption is much higher than the global average. India and China were leading the FinTech adoption index in 2019.

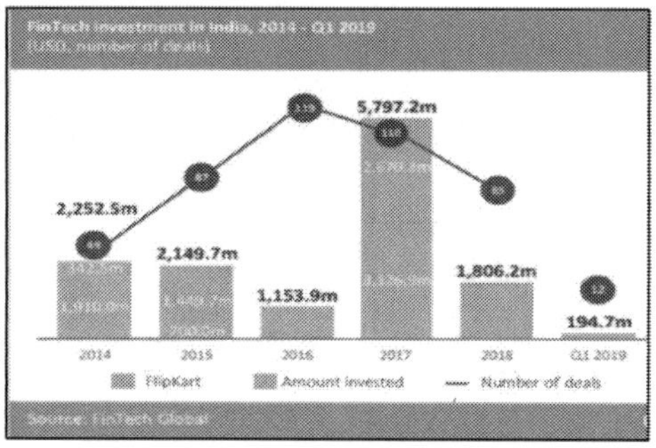

Fig. 5. Investments in World-wide Fintech Companies (in billion US$)

Some of the challenges faced by Indian FinTech firms are; (i). Small and inexperienced management teams who lack the capability, (ii). Fewer financial and economic resources that avert appropriate scaling, (iii). Inadequate credit facilities and lack of startup exposure, (iv.) Non-availability of established business models, (v). Failure to attract talents in analytics domain and skillful personnel, (vi) Lack of full compliance knowledge and (vii). Absence of more reliable, transparent, secured and speedily information processing and customer service.

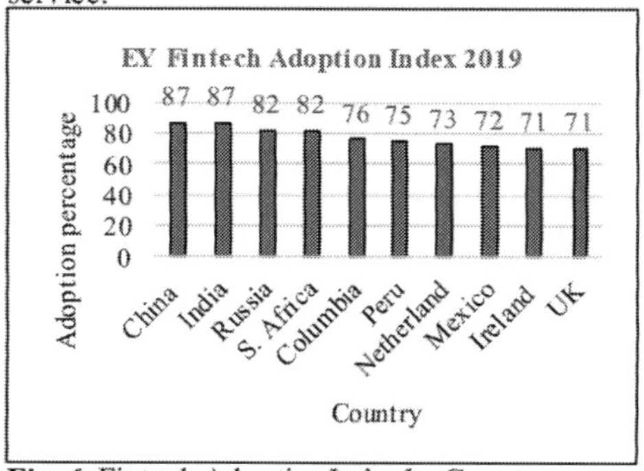

Fig. 6. Fintech Adoption Index by Country
Source: EY FinTech Adoption Index 2019

5G technology will certainly address the challenge of transparent, secured speedily process of data. The higher penetration of smartphone consumers and persistent build-up of the digital infrastructure will enable quicker adoption of 5G in the FinTech industry. India's FinTech sector is thriving and changing the financial services landscape to achieve a secure and comprehensive financial environment. FinTech offers chances to expand financial services to all members of the society, particularly the deprived and financially excluded, through better access to information and lending.

6 5G and FinTech effects on Banking and Financial Sector

Post Financial Crisis, FinTech firms progressed rapidly which enables to raise the question of whether or not FinTech companies are heading to replace traditional banks and financial institutions. However, many believe that banks and financial institutions are going to adopt fintech developments into their core businesses and successfully overcome this transition phase. Similarly, the ongoing 5G implementation along with FinTech revolution, is expected to have a massive change in the way banks are functioning. There is also a view that banking and financial institutions cannot embrace 5G service and enhanced financial technology instantly because of strict regulations. Though regulating guidelines are not a hindrance to innovation and change, the strict guidelines will keep the banks under the pressure of preventive actions to attempt innovative process and actions. Most used FinTech services are for Transfers & Payments, Insurance, Savings and Investments, Borrowing and Financial Planning. Advantages of 5G technology along with FinTech innovations are; More accessible financial markets for ordinary consumers and Easy distribution of financial services.

The challenge in banking and financial services sector is to ensure perfect competition and greater transparency. As the competition increases, there will be a wide range of quality and assortment of banking products. Higher transparency in banking services helps the customers to understand the offerings much better. Fintech businesses with the adoption of 5G technology are predicted to expand the services presently being provided by the traditional banks and technology-driven financial firms. As FinTech firms are entering into a new period of growth, regulating authorities are having a constant eye on them to protect consumers and look for objectionable practices. FinTech firms have mainly created disruption for big banks and age-old financial service firms. One of the common examples of such disruption can be witnessed in a quantum of mobile applications (apps) that provide share trading without levying charges per trade from users. Despite all the efforts, the speed of technology adoption has not as expected compared to its potential and hence there are gaps in the penetration of financial services. Although FinTech is creating the banking experience more instinctive, tailored and empowering, the convergence of growing technologies and financial services holds the key to make a robust digital Indian economy. Equipped with better data management, analytics competencies and nearly zero process cost, FinTech firms are supplementing and in certain areas competing the existing banking and financial institutions worldwide. Be it exploring new financial segments or making an inroad into foreign markets, Indian FinTech start-up firms are chasing several ambitions. As per NASSCOM, India's FinTech software market is predicted to reach US$ 2.4 billion by the end of 2020.

India's age-old cash transaction-based economy has responded superbly to the FinTech prospect which is mainly because of e-commerce boom and smartphone usage growth. The domains that Fintech affects in banking and financial services are: (a) lending, deposit, investment and financing services; (b) remittances, clearing and settlement, and payment services, (c) Trading advisory and investment management, and (d) insurance. The reality is very few banks in India are having a robust financial forte to achieve global competency. FinTech is an ideal path to emerge as a robust institution in digital payments and transactions. According to the Payments Council of India, the digital payments industry has a higher growth rate than before the demonetization period. Many FinTech companies combine automated analysis of retail customers with user-friendly interfaces to offer financial services that are more appropriate at a lesser fee. One suitable example is that 'marketplace lending' platforms have become a new lending forum that entices borrowers because of its easy and simple loan application method. Traditional banks and financial service firms have seen technology as a support system to business propositions, rather than creating new business propositions themselves. But, FinTech firms are changing that function by taking advantage of digital technologies in order to make innovative business propositions and target new market segments. The effect of new technologies on banks and financial firms due to innovative frameworks like blockchains and cryptocurrencies are already disrupted the way industry functions and yet to witness the full impact. However, the likely benefits of these technologies are adequate to induce significant attention from technology-savvy people, entities and governments.

India's financial intermediation cost is high at 3.15 percent compared to 2 percent in advanced nations (Payments Council of India report, 2017). Application of 5G services in FinTech companies may help to reduce this existing high intermediation cost which will benefit the consumers ultimately. The convergence of information technology, financial services and regulatory guidelines will play a significant part in creating a robust Indian digital economy (Digital Money Conference, 2017). Some of the leading fintech companies are effectively conducting their operations in India are Policy Bazar, lending kart, Pay Sense, Mobikwik. Capital Float, Bank Bazar, Paytm, PhonePe, PayU etc. These companies are looking forward to using 5G services to enhance their business and improve end-user satisfaction.

Table. 2. Electronic clearing and Card Payments

Item	Volume (Million)		Value (Rs. Billion)	
	2017-18	2018-19	2017-18	2018-19
Total Retail Electronic Clearing	6,382	12,467	193,113	267,515
Total Card Payments	8,208	10,781	10,607	14,097

Source: RBI Annual Report, 2019

7 5G and FinTech – Conclusion and a Way Forward

The disruptions created by these new digital revolutions has the possibility to reduce the role and significance of the existing banks. The banks started acknowledging that they require to come out of institutional complacency. In the coming days, banking and financial services will be like a layer on which the whole system is built on. It is no more monopoly or oligopoly market in the finance and banking industry. To sustain and grow in the industry and to become more competing, banks have to implement 5G technology and collaborate with FinTech companies. Numerous startup companies are looking forward to speeding up their progress by means of a faster, affordable and wider distribution of services contrary to the old generation financial institutions which are costlier and lagging in service quality. In the forthcoming years, banks are expected to either outsource or start the job of technology management with external firms which allows them to focus on the core business and risk management aspects. All physical operating venues of banks may be closed as most of them are now on the path to get their process digital, and this becomes more possible with 5G technology. The social and economic transformation potential through FinTech is very high. Many view 5G and FinTech as 4th industrial revolution. The emergence of 5G mobile technology along with FinTech firms is just the new innovative trend to disturb the banking sector, it is expected that a lot of innovations are in the pipeline for banking and financial service sector. Despite all the merits, one of the main challenges for FinTech companies is to collaborate with banks is the cultural differences. However, in the recent past, many big banks are transforming into a shared service model in various areas like due-diligence and technology, this is where the FinTech firms can capitalize on the collaborative benefits. As Fintech signs the new era of banking and financial services, it is indeed not the solution for all open problems. It is equally an improving drive and a disruptive force; it requires to be affected thoughtfully.

According to an information services firm Markit's 5G economic impact study, "5G mobile value chain alone can generate revenue up to $3.5 trillion by 2035 and support up to 22 million jobs". A report on The guide to capturing the 5G industry digitalization business potential, suggests operators could grow revenues between 12-36% (revenue of $204-$619 Billion) to their current forecast service revenue of $1.7 Trillion in 2026. This would mean a 13.6% annual growth against current operator service revenue growth targeted at 1.5%. Technology is not just a change enabler, but, a driver of innovation as we already seeing the emergence of high impact technologies in banking and financial services industry, especially in FinTech. [13] analyzed the likely impact of 5G on the Society. Along with 5G, technologies such as Artificial intelligence, IoT, Cloud computing, Quantum computing and Blockchain are expected to change the nature of services and how the way the financial services are offered to the customers. Overall, these technologies will combine to redefine the bank-customer relationship forever. Despite getting huge customers in recent time, FinTech and neo banks are still facing a lack of trust among many people, most of them even suspect the existence of these institutions after a year or two. The government's strong signal to support and nurture these institutions will go a long way in creating trust. There is also a possible gender split, with FinTech adoptability by females may be very meagre. Even among the users, many do not use FinTech and digital banks as a primary account. By implementing a robust legal and regulatory framework. The technologies like 5G, AI and blockchain would be highly beneficial in various fields for the world to adapt to. It will certainly improve efficiency and remove the biases that sneak into the financial system. [14] found that the normal cost of intermediation is held firm at about 2% of the transaction value. 5G technology can benefit to minimize operational and administrative overheads, potentially to the advantage of both FinTech companies and their customers.

References

1. S. Srinivasan, Big data and financial fraud using fintech methods (Working paper). Houston, TX: Texas Southern University. (2016).
2. W. Kim, R. Mauborgne, Blue ocean strategy. Harvard Business Review. **82**, 10 (2004).
3. S. Newell, M. Marabelli, Strategic opportunities (and challenges) of algorithmic decision making: A call for action on the long-term societal effects of datafication. Journal of Strategic Information Systems, **24**, 1 (2015).
4. A. V. Thakor, Information technology and financial services consolidation, Journal of Banking & Finance, **23**, (1999).
5. I. Aldridge, S. Krawciw, Real-time risk: What investors should know about fintech, high-frequency trading and flash crashes, Hoboken, NJ: Wiley (2017).
6. S. R. Das, The future of FinTech, Financial Management, **48**, (2019).
7. R. Nolan, Ubiquitous IT: The case of the Boeing 787 and implications for strategic IT research, Journal of Strategic Information Systems, **21**, 2 (2012).
8. T. Berg, V. Burg, A. Gombović, M. Puri, On the rise of fintechs: Credit scoring using digital footprints, The Review of Financial Studies, **33**, 7 (2019).
9. O. Acar, Y. E. Citak, FinTech Integration process for Banks. 3rd World Conference on Technology, Innovation and Entrepreneurship, Procedia Computer science (2019).
10. K. Leong, A. Sung, FinTech: What is it and How to Use Technologies to create business value in FinTech way?, International Journal of Innovation, Management and Technology, **9**, 2 (2018).
11. M. Kaur, Digital Currency and its implication for India, The Management Accountant, **54**, 11 (2019).
12. A. Fuster, M. Plosser, P. Schnabl, J. Vickery, The role of technology in mortgage lending, The Review of Financial Studies, **32**, 5 (2019).
13. Abel Nyamapfene, The Impending 5G Era and its Likely impact on Society. Journal of independent Studies and Research, **14**, 1 (2016).
14. T. Philippon, The fintech opportunity. (Working paper). Manhattan, NY: New York University (2016).

CHARTER AUTHENTICATED ANTI-THEFT SECURE MOTOR SYSTEM

Sharmila P[1], Brindha Dev[1], Akshaya 1, Dhanvarsha D[1], Pavai A[1]

[1] *Department of Information Technology, Sri Sairam Institute Of Technology, Chennai.*

Abstract: The prominent mode of transportation being used by the public is roadways. Increased usage of roadways may lead to many accidents and thefts. The major reason is violation of rules drafted for driving and causing accidents mainly by the youngsters who does not have a proper knowledge and license. So this proposed system has some restrictions to follow the rules drafted for driving and for higher safe and security, the License is scanned by means of NFC tag. When it is read by the NFC reader, Node MCU provides two-way authentication facility by sending an SMS to both the owner for authorization and the authorized current user would be able to control our vehicle by means of switching on and off the engine. The owner authorizes the user for a particular time period by using real time clock module. Without authorizing the license, the engine cannot have turned on. So, in this case license become mandatory to drive the vehicle. The features of the application to control the vehicle through the attached device. This will help a person to control the vehicle from anywhere. If a robber tries to snip the vehicle, immediately the Micro-Electromechanical System sends a signal to the controller and will be notified by a text message. Any violation to unlock the vehicle, emergency alert will send to owner as well as the current user and alarm blows out to seek the attention. This device can setup both in two-wheeler and four-wheeler.

1. Introduction

User friendly and safety are important in today's world. Now-a-days automobile burglary cases and road accidents are increasing day by day. Securing the vehicles are the prime concern. Initially, the License is scanned by means of NFC tag and secondly an android application which sends SMS to both the owner for authorization and the current user would be able to control our vehicle by means of switching on and off the engine. Without authorizing the license, the engine cannot have turned on. So, in this case license become mandatory to drive the vehicle. Once the license is Scanned the application will send the message to the owner, with the help of this system a person can find the vehicle's current location, stop the vehicle and raise an alarm. If a suspicious person tries to snip or ride the vehicle will alert an owner via text messages. Any violation to unlock the vehicle, emergency alert will send to owner as well as the current user and alarm blows out to seek the attention. User Authentication plays an important role in Security Systems, authentication allows a person to access the system based on the identity verification.

In this proposed motor System, two type of Authentication is used. Initially, by the owner for authorizing the user and the authorized user for further access of the vehicle. This provides higher safety of our vehicle that being theft. Anti-Theft Security is a complete and intuitive application. Antitheft Security has many features to increase chances to not only recover our vehicle but also identify the theft.

Corrosponding author:sharmila.it@sairamit.edu.in

2. Literature Survey

An Anti-Theft System for Vehicles using Fingerprint Sensor for preventing the vehicle from probable theft. This vehicle tracking system is distributed, to track the position of vehicle and provide security to vehicle [1]. Anti-Theft and security System for two Wheelers uses GSM or GPS to find the current location of the vehicle. The direction and the distance are transmitted to GSM through Microcontroller An Anti-Theft Security System for Vehicles gives an alarm when someone tries to access the vehicle. This is a security technique for our vehicle which avoids trespassers to mishandle it [3][4].

2.1 Existing System

Vehicle tracking system is becoming more popular because the count of vehicle theft is increasing enormously, Vehicle theft is happening on parking and sometimes driving in unknown places. The existing system contains embedded device integrated with global system for mobile (GSM) and global positioning system (GPS) along with a microcontroller installed to the vehicle. The use of GSM and GPS technologies allows the user to track the vehicle location details then and there. Moreover, fingerprint verification is used in the implemented system to ensure the correct person is driving the vehicles. But the use of sensor is difficult in Emergency situation and there may be a possibility of locking the vehicle. This issue is resolved in our proposed system. Few of the important features supported by our system are alerting owner by SMS about the theft attempt, tracking the location of vehicle using GPS technology. Our system has further enhancement of controlling the vehicle remotely by SMS and authorizing the user by the owner for providing higher security of the vehicle [7][8]

3. Proposed System

Road accident is one of the major problem in current situation. In order to bring out the safer driving, this following system is proposed. It consists of three modules. First, the user's license is scanned by means of NFC tag and when it is read, the microcontroller sends a message to the owner for authorization. The authorized user can access the vehicle for an authorized period of time. [1]

Second, if the license is lost and finds any illegal activity, the user is allowed to lock the vehicle by using mobile application to prevent from theft. [2]

Third, if anyone tries to unlock the vehicle, the micro-electromechanical system sends a signal to microcontroller, immediately an emergency alert message sends to both the owner and the current user, simultaneously alarm blows out to have a surrounding attention.

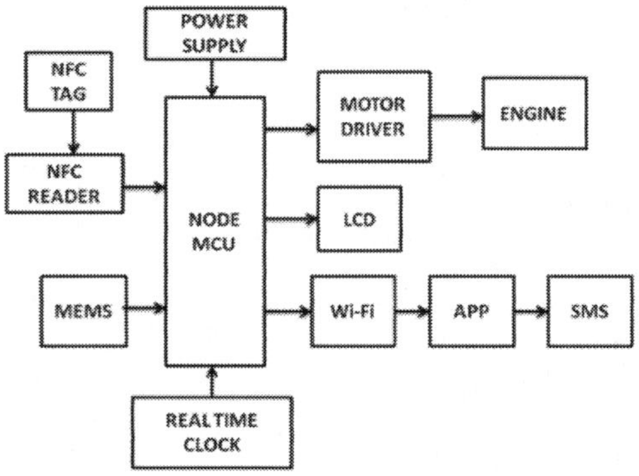

Fig.1. Data flow diagram represents the flow of data and control in the system

4.1. Block diagram description

4.1.1. Power supply

The supply of 5 volts is given as an input to the micro controller.

4.1.2. NFC Tag and Reader

A near field communication tag (NFC) is a small microchip in which it stores the user license information and that can be read by the NFC reader. The stored information can be used for authorization. With NFC capabilities, the tag has the ability to send data to other system. This module can operate without power supply.

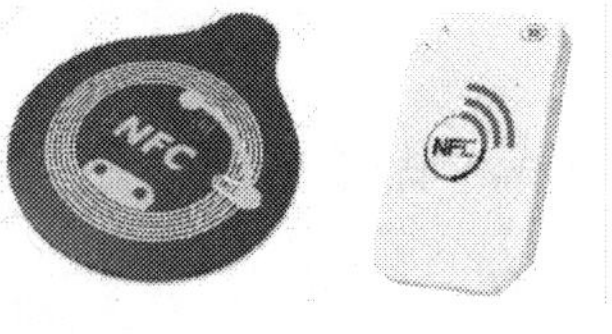

Fig.2. Near Field Communication Tag

4.1.3. Node MCU

Node MCU (Node Microcontroller) is an open source firmware to develop IOT applications. It is used in the proposed system to manage and control the entire system.

Fig.3.. Node MCU

4.1.4. Micro-Electromechanical System

A MEMS is a small device that has both electronic and mechanical components, it is used to sense, control and actuate and generate. It response the slight changes that made in the vehicle and transmit the signal to the microcontroller. Then that signal can be changed to create a stable output signal and it is transmitted to the microcontroller for further actions [5]. This result in sending the emergency alert message to the user and the owner.

4.1.5. Motor Driver

Motor Driver is an amplifier which accept the signal from the controller (Node MCU) and helps to drive the motor depending upon the condition given by the current user through mobile application. Depending upon the signal given by the microcontroller, a motor driver controls the speed of the vehicle and switching ON/OFF of an engine motor[13].

Fig.4.. Motor Driver

4.1.6. Liquid Crystal Display(LCD)

LCD used to display the current status of the system, it uses active matrix display for more clarity [14]. It works on the principle of blocking light rather than emitting light.

Fig.5. Liquid Crystal Display

4.1.7. Mobile application

For increasing accessibility, the entire system can be connected to the mobile application. Through mobile application, with the help of Internet, one can control the system by clicking on the respective function. The major advantage of having a mobile application is faster to access and authorization is done by the owner through this application.

Fig.6. Mobile Application

4.1.8. Real time clock(RTC)

RTC is used to provide an authorized access for a particular time. Often this device used in personal computers, servers and embedded system. This module takes a low power consumption for performing a time-critical tasks[3].

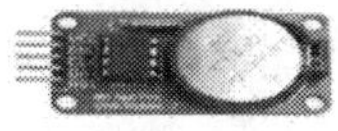

Fig.7. Real time clock

5. Working

The proposed system "LICENSE AUTHENTICATED MOTOR SYSTEM IN ROADWAYS" have developed with the help of NFC tag, NFC reader and a Node MCU. In this technique the scanning of license is introduced. The information of user's license is stored in NFC tag. The user can access the motor with the help of this tag. Once the tag is read by the reader, the SMS sends to the owner with the help of GSM module which was controlled by Node MCU. The owner authorizes the user for a particular time period by using real time clock module [6]. The authorized user can access the motor for a specified time. Once the time expires, automatically the engine cut-off. For further continuation of driving, the user should repeat the process again. When the tag is lost, the owner may not aware of authorizing who is going to drive the vehicle. In this circumstance, the user who lost his license can lock the engine through SMS. When the thief tries to unlock the vehicle, the Micro-Electromechanical System sends a signal to the controller. As soon as the alert SMS will send to owner as well as the current user. Simultaneously, the alarm blows out to seek the attention of surrounding people. This device can setup both in two-wheeler and four-wheeler [4].

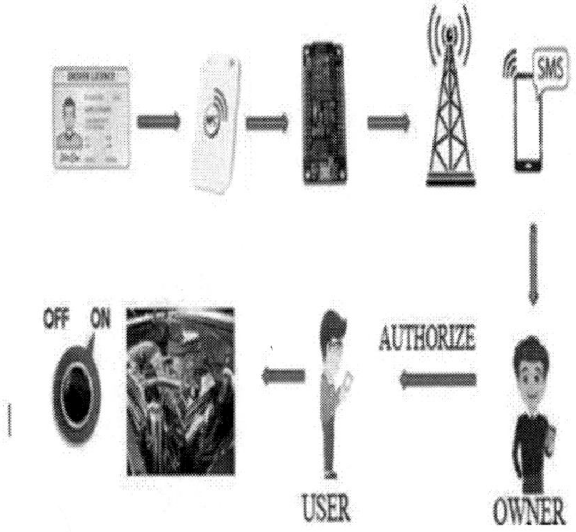

Fig.8. Authentication of an owner and access of an authorized user

Authentication of a Motor System: To make a license to be a mandatory one, we have proposed a system in which license information are stored by means of NFC tag [10]. Once the tag read by the NFC Reader, the information is transformed to the microcontroller (Node MCU). It will send a SMS to the owner to authorize the user for a particular time period. The authorized user is allowed to use the motor for a given period. If the requested user is not authorized by the owner, then the user access is denied [15]. Locking The System by Using Mobile Application: In case the stolen license of the user is used to operate the vehicle, the microcontroller will send the alert message to the owner. Without the knowledge of the theft, the owner authorizes the user. But the user can lock the vehicle by using mobile application [11].

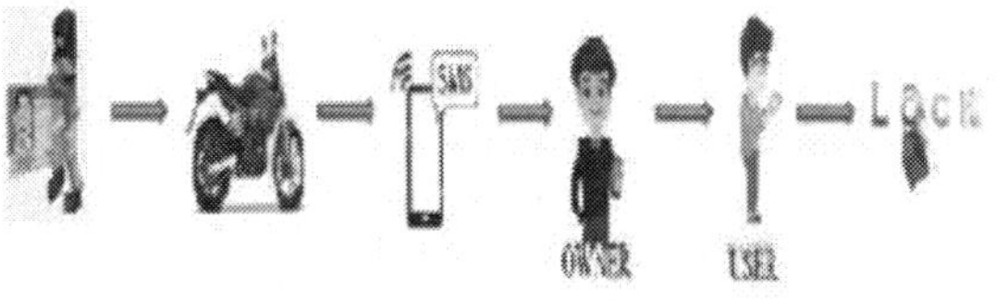

Fig.9. Locking the entire system by using mobile application

Security Alert System: If there is any illegal activity like vehicle theft, the alert message will send to the owner as well as the current user. Simultaneously, the alarm blows out to seek the surrounding attention.

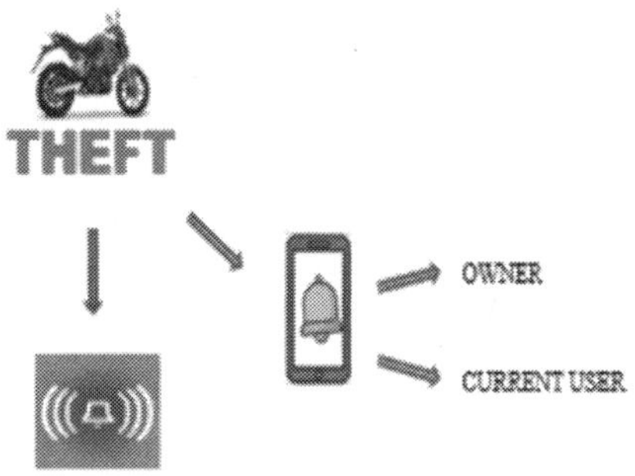

Fig.10. Emergency alert to have higher security

6.Experimental Setup and Result

This system has developed with the help of NFC tag, NFC reader and a Node MCU. In this technique the scanning of license is introduced. The information of user's license is stored in NFC tag. The user can access the motor with the help of this tag. Once the tag is read by the reader, the SMS sends to the owner with the help of GSM module which was controlled by Node MCU.

The owner authorizes the user for a particular time period by using real time clock module. The authorized user can access the motor for a specified time. Once the time expires, automatically the engine cut-off. For further continuation of driving, the user should repeat the process again. When the tag is lost, the owner may not aware of authorizing who is going to drive the vehicle. In this circumstance, the user who lost his license can lock the engine through SMS.

When the thief tries to unlock the vehicle, the Micro-Electromechanical System sends a signal to the controller. As soon as the alert SMS will send to owner as well as the current user. Simultaneously, the alarm blows out to seek the attention of surrounding people. This device can setup both in two-wheeler and four-wheeler.

Fig.11. Hardware set up

Fig.12. shows the alert message sent to the user from

Charter Authentication System.

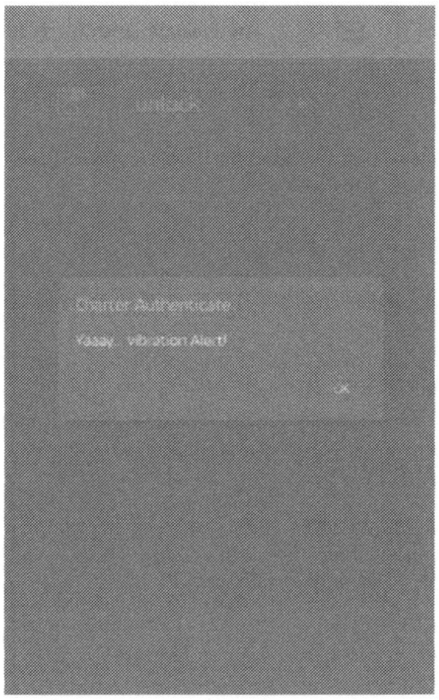

Fig.12. Application showing alert message.

7.Performance Evaluation

The reliability test result shows the system performance was good when the distance is minimum, to turn engine on/off for the distance of 1 meter the system takes 0.5 seconds, for maximum of 20-meter distance the system takes 4 seconds. License scanning message to reach the users application takes

1 second, for maximum of 20-meter distance the system takes 5 seconds.

ACT	Distance in meter	Time in seconds	Response
MOTOR ON /OFF	1-4	0.5	Good
MOTOR ON /OFF	5-10	1	Good
MOTORON /OFF	11-15	2	Good
MOTOR ON /OFF	15-20	4	Good
MOTOR ON /OFF	21-25	-	Connection Lost
Licence Scaning message	1-4	1	Good
Licence Scaning message	5-10	1.5	Good
Licence Scaning message	11-15	3	Good
Licence Scaning message	15-20	5	Good
Licence Scaning message	21-25	-	Connection Lost

Table1. System performance

8. Conclusion and Future work

In this work provides a safety and security mechanisms for motor vehicles. This system consists of a license scanner methodology as well as authentication methodology. This device can be implemented in all motor vehicle to increase the security system and also the safety of the riders, and this system will effectively have applied in the future and also the result will be better. In this method, License plays a key role in accessing the vehicle [9]. Thus it reduces a rate of accidents in day-to-day life. Furthermore, the proposed method has a strong advantage in that it provides two-way authentication system. Initially, by the owner for authorizing the requested user. Secondly, the authorized user can unlock the vehicle by using mobile application in order to access the vehicle. In case of theft of license, the authorized user can lock the motor in a fraction of second by using mobile application. This provides more security to the vehicle and prevent from illegal activities. In future, the motor vehicle manufacturing industry can inbuilt this device in their production in order to provide higher safe and security for both the human and the vehicle. Unlike motor vehicle industry, we can also implement for four wheeler and heavy vehicle to provide safe environment for the users.

References

1.JoelSachin,Kiran Rana Gill,Anti-Theft System For Vehicles Using Fingerprint sensor,IJSER,**7**,1436-1441(2016).

2. Shanmuganathan J , B.C.Kavitha, Tracking and Theft Prevention System for Two Wheeler Using Android,IJETT,**21**,355-359(2015).

3. Samir Rana, Ritu Mewari, LataNautiyal ,Anti-theft Security System for Vehicles,IJET,**7**,42-46(2018).

4. Prakash Kumar , Pradeep Kumar, Arduino Based Wireless Intrusion Detection Using IR Sensor and GSM,IJCSMC,**2**,417-424(2013).

5.Jayanta Kumar Pandey, R.N. Das choudhary,Embedded Automobile Engine locking System, using GSM technology,IJICA,**1**,154-158(2011).

6. Pritpal Singh, Tanjot Sethi, Sujit Kumar Pattanayak ,a smart anti theft system for vehicle security,IJMMM, **3**, 22-43(2015).

7. Nurul Hutha.S and ArunKumar.B, Vehicle Monitoring and Theft Prevention System Using ARM Cortex, IJSETR, **2**, 992-997(2013).

8. Pritpal Singh, Tanjot Sethi, Bunil Kumar Balabantaray, Bibhuti Bhushan Biswal,Advanced Vehicle Security System" IEEE Xplore(2015).

9. L. K. Hema, Achal Mercy Banra, Raju Kumar and Ranu Rakesh Singh,Int. J. Chem.,An Intelligent Automobile Anti-Theft System ,IJCS,**14**, 863-868 (2016).

10. B. R. Prudhvi and Yuvapreethi Ganesh, Gravity Lock: Next Generation Auto Theft Prevention System, International, ICAEE,175-178 (2013).

11. S. Zhou, Q. Shan, F. Fei, W. J. Li, C. P. Kwong, C. K. Wu et al., "Gesture Recognition for Interactive Controllers using MEMS Motion Sensors", IEEE Xplore, 935- 940 (2009).

12. Geeth Jayendra, Sisil Kumarawadu and Lasantha Meegahapola ,RFID-Based Anti-theft Auto Security System with an immobilizer" ICIIS ,8 – 11 (2007).

13. Robert Bosch GmbH, 2001, "Device for protecting a motor vehicle against theft ", United States Patent 6856044. Nov 7(2001).

14. M. A. A. Khedher, Hybrid GPS-GSM localization of automobile tracking system,IJCSIT. **3**, 75-85, (2011).

15. P. P. Wankhade and S. O. Dahad, ,Real time vehicle locking and tracking system using GSM and GPS technology-an anti-theft system,IJTE, **2**,272-275, (2011).

AIR 100 UAV FOR LAW AND ENFORCEMENT APPLICATIONS

Jegatha.R[1,*], *Sharmila.P*[1], *Ganesh U*[1], and *Dinesh*.S[1].

[1] *Department of Information Technology, Sri Sai Ram Institute of Technology, Chennai-600044, Tamilnadu, India.*

Abstract: In India the crime increased with a percentage of 1.3% as per the NCRB data, in number, there are around50.74 lakh crimes under the Indian Penal Code and around 20 lakh crimes under local laws. There are many ways to detect where crimes are committed but those locations needed frequent monitoring. During an emergency call foran incident quicker response is required. The cops may/may not reach the location in time because of road transportation.But with UAV, the reaction for the call can be made better .UAV make an incredible time in reaching the exact location through the air medium. So the people in Law and enforcement sometimes use COTS (Commercial off the self) drones. They are not purposely made for it .The purpose of the paper is to make a custom UAV with the perfect specification to handle various critical situations.

1. Introduction

In India, cops use Commercial off the shelf Drones for surveillance, search operations, spying, etc.but those drones are not made for law and order purposes and those cost multiple lakh and replaceable parts are also pretty high at the cost. So for this problem, we can make a custom drone particularly made to serve law and enforcement purposes especially for maintaining large crowds, riots, responding even faster to an emergency call. By letting the UAV reach the destination before time and performing previous-analysis, live data transmission, and responding accurately to GS (Ground Station) Command. Similarly, we can add person detection with the help of machine learning features to identify whether the particular person is innocent or criminal. By this, we can stop more crimes happening in India by faster response through Drones. This custom drone can even be used for search operations, fire departments to do previous analytic of the situation. In some typical scenarios, if a building is set to fire, firefighters arrive at the spot with a piece of limited information about the extent of the fire and damage caused in that area and they begin their operation. Instead, with a UAV, firefighters can begin with an aerial assessment of the scene and know the extent of the fire before starting their rescue operation using a UAV. Deploying the UAV first into the tight spaces since they are nimble and agile, we can avoid putting humans in unsafe situations.

2. Literature survey

India holds a very high population rate and vast landmass. The cops and the people at Law and enforcement cannot reach a particular destination in time when it is in case of a remote area but on another hand the Indian government is using military-grade UAV for monitoring over a particular facility area which comes under government maintenance. By the same concept with slight custom modifications, we can implement the concept for protecting civilians, monitoring high crime rated locations, chasing in over speeding vehicles, monitoring wildlife,previous analytical working on an emergency by firefighters, search operations, maintaining large crowds using megaphones on drones. By the government supporting emerging engineering students, firms, companies can make this concept real as many people get benefited and the crime rate can be reduced overall at a greater rate.

2.1 The research on unmanned aerial vehicle remote sensing and its applications

Remote sensing with UAV (unmanned aerial vehicle) contains unique advantages like high resolution, real time, and cost efficient spatial date in dangerous and critical without the involvement of the flight crews. For both categories of space borne and airborne remote sensing, the implementation of the UAV is a powerful potential and bright prospect. The applications such as flood disasters, land supervision, land use, meteorology and , geological disasters, surveying, mapping and forest fire disasters are summarized as key technologies of Uunmanned Aaerial Vehicle remote sensing systems are presented in this paper.

2.2 Aerial surveillance system using UAV

In today's world, maintaining of the decorum at a place,people's safety and security surveillance is required.In this regard,the implementation of aerial surveillance system is feasible.In this paper, using an unmanned aerial vehicle,How can an aerial surveillance system can be builtis described

*Corresponding author: jegatha.it@sairamit.edu.in

 https://doi.org/10.1051/itmconf/20213701010

2.3 Internet of Drones

The Internet of Drones (IOD) is a layered network control architecture designed mainly for coordinating the access of UAV to the controlled airspace and by providing the navigation services of UAV between the desired locations.The IOD provides a generic service for various drone applications, such as package delivery, traffic surveillance, search and rescue. How such architecture of a conceptual model can be organized and how the features of an Internet of drone's system architecture of our concept are implemented is presented in this paper.

2.4 Self-Organization Drone-Based Unmanned Aerial Vehicles (UAV) Networks

The rapid drone network deployment can offer the access to high difficulty areas.In this research paper the deployment of multiplea-hop drone based UAV networks with a focus on the self-organization aspect. When rescue drones carry out their rescue operations which may fly far away from the gateway,To maintain the connectivity with the Ground Station.Relay drones are Aautonomously deployed for the missions. How Each rescue drone is connected to the gateway via Themultiplerelay drones through dedicated connections is presented in this paper.

3. Unmanned Aerial Vehicle role in law and Enforcement

3.1 Responding to SOS

The custom-built UAV can respond faster when compared to people in law and enforcement. Multi- rotors and VTOLS can effectively used against these emergencies. Multi rotors and VTOL have a high rate of Authority over all Tait–Bryan angles are also referred as nautical angles; Cardan angles; elevation, heading, pitch, yaw, and roll. Occasionally, both categories of sequences are called "Euler angles" as compared to fixed-wing aircraft. But multi rotor requires high Capacity Battery source to maintain its posture in the air for a long period. With the various external payloads, it is achievable.

3.2 Example

For instance, if an emergency call is made the cops want to take road transport which has multiple factors to increase in Time of cops for reaching the destination and in most of the cases they route they are taking is not a straight they want take multiples to turn and multiple signals to reach the destination. But for a drone, they are no such restrictions in air transport. They can take a straight route from the halting position to the Destination Point.

4. UAV anatomy

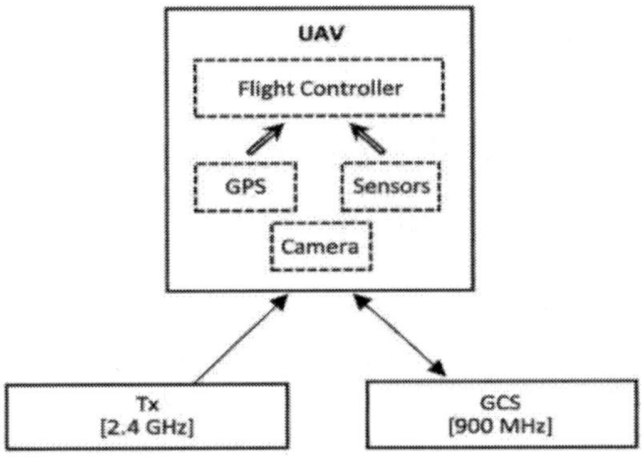

Fig.1. Simplified UAV System representation.

4.1 Surveillance

What is surveillance **UAV**? **Surveillance UAV**are also Drones used to gather images, live video transmission of targets such as vehicles, people or specific areas.UAVcan be integrated with a various types of **monitoring payload** equipment system that can fetch high definition videos, images during both day and night time. Generally drones are be equipped with on board computational systems allowing them to determine/calculateit own GPS locations, tracking the target,restrict cell phone calls.Aerialvehicles (UAV) can be deployed in multiple altitudes for 24/7 to monitor/for security of the user/customers assets,live stocks,etc.

4.2 Way Point Navigation:

A waypoint is a place or point a stopping point on a route or line of travel at which the navigation course is changed. In modern terms, it most often refers to the coordinates which specifies one's position on globe at end of each stage of an air flight. The generation and checking which are generally done computationally.

5. AIR 100 UAV Frame work

An autonomous UAV with an on board computer system and GPS for performing its mission commands with a LTE video transmitter and LTE router, carrying a payload of micro UAV which can also transmit video through the WiFi router .18650 Lithium -ion cells serve asPower source for both the WiFi Router and AIR 100 Lithium polymer power source for micro drone.This fundamental framework design can be utilized for making multiple variants of UAV.This framework is more efficient on multi rotors and VTOL .

The WiFi video transmission can be achieved by an esp32 cam module which has inbuilt WiFi module for video transmission to the server .The AIR 100 UAV FW is the introduction of the Dual Drone Configuration Methodology with WIFI Based Video Transmission.

5.1 Dual Drone Configuration

-Parent/Master (Primary - UAS).
-Child/Micro Quad rotor (Secondary - MAV)
MAV - micro aerial vehicle.

5.1.1 Primary UAV

The framework is more efficient for the following categories of UAS.

-Applicable UAS
-Muti rotors.
-VTOL.

5.1.2 Secondary MAV

The secondary drone should be strictly a quad rotor.It offers high rate of stability compared to all categories of UAV.

5.1.3 Size Ratio

The ratio in size of the MAV to the UAV determines how efficient the UAV in the real word scenarios and also determines how many MAV units can be accommodated in UAV.

Efficiency of the UAV ∝ (Size of MAV: Size of UAV)

The above proportionality is applicable only when Number of MAV in UAV = 1 and range of the ratio should be in range of 1:4 to 1:7.
For ratio greater than 1:7 the no of MAV can be increased .The MAV can be increased in multiples of 2 for maintaining the CG of the UAV
Example:

For 1:8 - 2 * MAV of 1:4 ratio
For 1:16 - 4*MAV of 1:4 ratio

5.1.4 Weight Ratio

The weight ratio of MAV to UAV is related to thrust generated by UAV.The Thrust of the UAV should be double the time of the combined weight of the UAV and the MAV.The MAV/child drone strictly should be in Micro category i.e the weight of the MAV should be less than 250 grams. Since it is in micro category it is addressed as MAV throughout the framework.

Required Thrust:

Assume that Thrust of the UAV=1
Weight of the UAV=1

For Hovering

$$Thrust\ (UAV) = weight\ (UAV) \quad (1)$$

$$Combined\ weight\ (UAV) = weight\ (UAV) + weight\ (MAV) \quad (2)$$

Here,

$$Weight\ (UAV) = combined\ weight\ (UAV) \quad (3)$$

But, now the Framework consists of MAV where it has own weight

.

Worst case:

$$Weight\ of\ the\ MAV = 250\ gms \quad (4)$$

Applying equation (4) in equation (2)

$$Combined\ weight\ (UAV) = weight\ (UAV + 250\ gms) \quad (5)$$

By equation (3), applying equation (5) in equation (1)

$$Thrust\ of\ the\ UAV = weight\ of\ the\ UAV + 250\ gms \quad (6)$$

In the above the equation the 250 gms is constant value Converting the constant to our assumption values.
=250/1000
=0.25

$$Thrust\ of\ the\ UAV = weight\ of\ the\ UAV + 0.25 \quad (7)$$

From the above we can get clear

$$Thrust\ of\ the\ UAV = 1 \quad (8)$$
$$Weight\ of\ the\ UAV = 1 \quad (9)$$
$$Weight\ of\ the\ MAV = 0.25 \quad (10)$$

By this we can conclude the minimum weight ratio of the MAV:

$$UAV\ is\ 1:4 \quad (11)$$

We know that from equation (1)
For Hovering,

$$Thrust\ of\ the\ UAV = weight\ of\ the\ UAV$$

But now we have additional weight of 0.25 proportionality. Since 1:1 ratio is acceptable but not efficient the ratio has to be increased to 1:2 for better performance and 1:3 for best performance and efficiency. So,

$$Required\ thrust = \{2*weight\ (UAV)\} + \{2*weight\ MAV\})) \quad (12)$$

Applying our assumed values to equation (12)
Required Thrust=2*(1+0.25)
=2.50
Hence for AIR 100 UAV FW the required Threshold thrust generated should be 2.50 times of the combined weight of the UAV and MAV.

 https://doi.org/10.1051/itmconf/20213701010

6. WiFi based video transmission

6.1 General Video Transmission Procedure

6.1.1 On-Board

The common practice of video transmission in COTS is using a camera connected to video transmitter with presence of OSD(on screen display),power module for the camera and OSD ,Gimbal mount , Gimbal controlling circuit,additional power supply for Gimbal and camera.

6.1.2 Ground Station

Video reception setup with digital screen to view the live video footage and the real time statistics of the drone parameters.

6.2 AIR 100 UAV FW video transmission methodology

An IOT development board based WiFi video transmission to a private server.
Required Components:
-ESP32 CAM development board
-USB to UART TTL serial adapter module
-WiFi dongle (portable)
- Micro USB to USB cable.

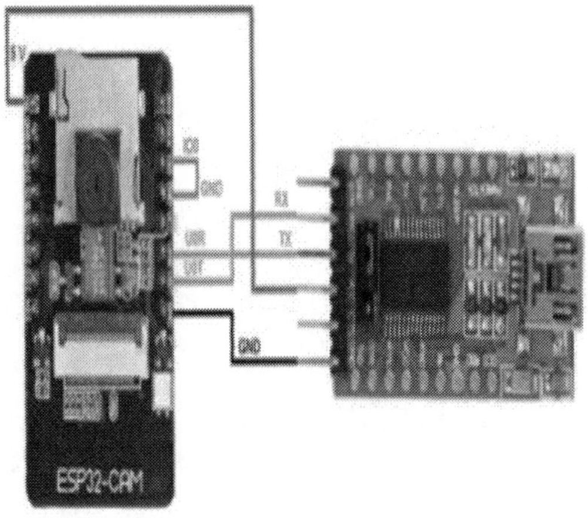

Fig.2.esp32 WiFi cam module with UART serial adapt

7. Initial Prototype

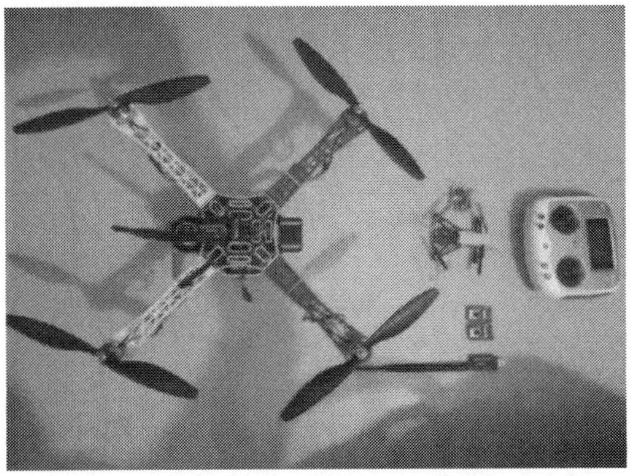

Fig.3.initial prototype model UAV, MAV (on progress).

8. Technical Challenges/Issues

To overcome challenges likesafety standards of air/road traffic conditions, DGCA regulations, local laws drones should be licensed to follow the above mentioned protocols.

These challenges include:

- Simulation will help us to optimize certify whole computational systems, individual hardware parts, etc.
- Develop, test and certify functionally safe and secure software.
- Design, test and certify sensors, radars and cameras capable of observing various environment scenarios, the above test on drone will make the drone to withstand when they encounter.
- Adverse weather conditions.
- Carrying payload by improving aerodynamics, increasing travel distances by reducing weight and propulsion system performance by increasing battery power.
- Current drone technology has limitations. An expert is required to typically supervise the flight mission, even most of the system in automated In General consideration, the drone technology isn't completely capable to manage complex and unpredictable situations
- Handling adverse weather conditions.
- Flying without permissions or restrictions in populated areas Certify, optimize, and verify radar,lidar, GPS,camera and other external/additional sensor performance.
- Verifying and certifying computational software logic's and Artificial Intelligence systems.

- Connection initialization/maintenance and Testing of the link between sensors, external modules, software, virtual worlds.
- Engagement of the functional safety assessments/data links with ground stations/servers, software's, AI and sensors.
- Utilization of Lightweight components for build.
- Optimization of the battery source, motors and electric propulsion system performance.

9. Policy Requirements (In India)

Except the MAV (micro aerial vehicle) nano category drones others must be registered and issued with a UIN (Unique Identification Number) .For commercial drone operations a permit is required (the flight altitude should be below 50 feet for nano category and below 200 feet for the micro category drones).While operating a drone the drone pilot should maintain the direct visual line of sight throughout the entire flight duration. According to DGCA protocol's Drones should be flown below 400 feet vertically.Drones/UAV should not be operated in locations like airports, international borders, strategic locations, military installations and similar places because these areas are specified as "No Fly Zones". By completing the filing process of a flight plan and obtaining a unique Air Defense Clearance (ADC)/Flight Information Center (FIC) number from aviation authority after that only we have permission to fly in controlled airspace.

10. Requirements

10.1 Functional requirements

The UAV comprises of an GPS receiver, flight controller, data link module, two digital cameras, IMU, microcontroller The UAV must be entirely autonomous. During all times of flight manual override must be available for emergency purposes.. The UAS system must be able to calibrate its flight path to avoid both statics and dynamic obstacles. The UAS system shall generate its own waypoints to imaginary data of the entire user defined area. The survey over a vast area can be done in less than hour (say 2- 3kms)

10.2 Camera Requirements

For obstacle detection one front facing camera of lower resolution is required. To capture the near infrared pictures a ground facing camera with high resolution is required. To capture a high resolution image a ground facing camera with a 16 megapixel is required.GPS information must be tagged with all images captured using the ground facing.

11. Future works

11.1 Swarm Drone Technology

A new advanced breed of unmanned aerial vehicles (UAV) integrated with network has it own vastly greater power and advantages than their predecessors, has been developed.Artificial intelligence and machine learning is been harnessing by the aerospace industry. The advanced ability of the drones/UAV to operate independently as well as in coordination with multiple UAV/Drones is obtained by the utilization and implementation of artificial swarm intelligence technology and networking which reflects the collaborative behaviour which is exhibited by birds ,animals and insects.By applying the concepts of this framework civilian accessible Drone RPA can be implemented in warehouse management, Efficient Crowd Control.

12. Conclusion

With the Assistant of custom-built UAV with help of artificial intelligence and machine learning algorithms, the Law and Enforcement Department can reduce the rate of crime by reacting with greater reflexes to critical situations during an emergency call or during a Search operation during disaster management.

Reference

1. M. L. Smith, "Regulating law enforcement's use of drones: The need for state legislation", Harvard J. Legislation, vol. **52, pp. 423**, Sep. (2015).

2. S. Hayat, E. Yanmaz and R. Muzaffar, "Survey on unmanned aerial vehicle networks for civil applications: A communications viewpoint", IEEE Commun. Surveys Tuts. vol. **18, pp. 2624-2661**, 4th Quart. (2016).

3. L. Gupta, R. Jain and G. Vaszkun, "Survey of important issues in UAV communication networks", IEEE Commun. Surveys Tuts. vol. **18,no. 2, pp. 1123-1152**, 2nd Quart. (2016).

4. N. H. Motlagh, T. Taleb and O. Arouk, "Low-altitude unmanned aerial vehicles-based Internet of Things services: Comprehensive survey and future perspectives", IEEE Internet Things J., vol. 3, **no. 6, pp. 899-922**, Dec. (2016).

5. A. Kumbhar, F. Koohifar, I. Güvenç and B. Mueller, "A survey on legacy and emerging technologies for public safety communications", IEEE Commun. Surveys Tuts. vol. **19, no. 1, pp. 97-124**, 1st Quart.(2017).

6. Y. B. Sebbane, Smart Autonomous Aircraft: Flight Control and Planning for UAV, Boca Raton, FL, USA: CRC Press, (2015).

7. A. G. Korchenko and O. S. Illyash, "The generalized classification of unmanned air vehicles", Proc. IEEE 2nd ICAPAVD. (APUAVD), pp. **28-34**, Oct. 2013.

8. A. Valcarce et al., "Airborne base stations for emergency and temporary events", Proc. ICPSS, pp. **13-25**, (2013).

9. M. Silvagni, A. Tonoli, E. Zenerino and M. Chiaberge, "Multipurpose UAV for search and rescue operations in mountain avalanche events", Geomatics Natural Hazards Risk, vol. **8,** no. 1, pp. 18-33, (2017).

10. J. Scherer et al., "An autonomous multi-UAV system for search and rescue", Proc. 1st Workshop Micro Aerial Vehicle Netw. Syst. Appl. Civilian Use, pp. **33-38**, (2015).

11. M. A. R. Estrada, "How unmanned aerial vehicles–UAV's–(or Drones) can help in case of natural disasters response and humanitarian relief aid?", (2017).

12. D. C. Macke, "Systems and image database resources for UAV search and rescue applications", (2013).

13. J. Sun, B. Li, Y. Jiang and C.-Y. Wen, "A camera-based target detection and positioning UAV system for search and rescue (SAR) purposes", Sensors, vol. **16, no. 11**, pp. 1778, (2016).

14. M. Quaritsch, K. Kruggl, D. Wischounig-Strucl, S. Bhattacharya, M. Shah and B. Rinner, "Networked UAVs as aerial sensor network for disaster management applications", EIEI, vol. **127**, no. 3, pp. 56-63, (2010).

15. R. Austin, Unmanned Aircraft Systems: UAVS Design Development and Deployment, Hoboken, NJ, USA: Wiley, vol. **54**,(2011).

16. T. F. Villa, F. Gonzalez, B. Miljievic, Z. D. Ristovski and L. Morawska, "An overview of small unmanned aerial vehicles for air quality measurements: Present applications and future prospectives", Sensors, vol. **16**, no. 7, pp. 1072,(2016).

17. M. Madden, T. Jordan, D. Cotten, N. O'Hare, A. Pasqua and S. Bernardes, "The future of unmanned aerial systems (UAS) for monitoring natural and cultural resources", Proc. Photogramm. Week, vol. **15,** pp. 369-384,(2015).

Blockchain based mechanism to eliminate frauds and tampering of land records

Devi D[1]*, *Sai Rohith* G, *Shri Hari* S, and *Sri Ramachandar* K
[1]Department of Electronics and Communication Engineering, Sri Krishna College of Engineering and Technology, India

Abstract: Data tampering and fraud in land records have increased drastically in the modern world. A data storage model using Blockchain and Interplanetary File System (IPFS) is proposed in this work. Land records and the farmer's information are stored inside the Interplanetary file system. To avoid data faking, the hash address of the respective data generated by IPFS is stored in the blockchain. This proposed system when deployed on a large scale can outperform the existing methods of securing user data. One of the latest technological advancements in the software industry is the innovation of Blockchain Technology. This new technology has opened up a new business relationship platform that delivers feasibility, protection, and cheap rates. It provides a new foundation of trust for transactions that can facilitate a very streamlined workflow and a faster economy.

1 Introduction

With the world around us moving at a faster pace, the security of our possessions has become a major concern. Illegal acquisition of agricultural land and misuse of land records had been increased. However, these problems can be solved with the power of new technologies like blockchain. A blockchain is a decentralized, distributed, and made up of several records called blocks that are connected through encryption. It is a peer-to-peer network that is handled anonymously by a mass collaboration of people (nodes) with self-interests. Blockchain manages a limited number of blockchain operating systems, including the blockchain cryptographic hash that preceded them.

A blockchain-like protocol was first proposed by Cryptographer David Chaum in 1982. Merkle trees were incorporated in blockchain by Haber, Stornetta, and Dave Bayer in 1992 which drastically improved efficiency. However, blockchain was first conceptualized and devised as bitcoin, by Satoshi Nakamoto in 2009 [1]. This has been a breakthrough in electronic cash systems and cryptocurrencies. Blockchain was primarily created to monitor digital currency transactions, however, its robustness and immutable nature have proved useful in many domains. Since it is an append-only ledger, and as the number of blocks generated increases, it becomes very slow and highly ineffective for storing real-time data [2]. So blockchain technology is not applied directly. In this paper, blockchain is combined with an Interplanetary file system (IPFS) to gain security and reliability. The contents of this work are as shown, the consecutive section provides the related work and Section 3 gives an overview of the architecture of our proposed model. Section 4 explains the working analysis of our model in detail, the work outcomes, and the conclusion is given in the final section.

2 Related Work

An overview of related work that leverages access control methods and blockchain-based data storage is proposed in this section.

2.1 Data storage solutions:

Blockchain provides us with an unchangeable database that allows us to collect transactions and never delete or modify them. This enhances security, however, the real-time costs of storing a massive amount of data inside the blockchain is not feasible [3]. For instance, according to etherscan an average size of a block in July 2020 is 40kb.1kb costs around 0.032 ETH or 828 INR when using Ethereum. Hence the developer should decide which data to keep off-chain and which data to keep on-chain.

*Corresponding author: devi@skcet.ac.in

2.2 Data access approach:

Traditional access control mechanisms rely on a centralized database system to store and access data. However, under a centralized system, an organization maintains the identities and access rights of the user. Some large media file-sharing systems, such as BitTorrent, KaZaA, and Napster, are introduced to store massive data. Alam et al presented the Interplanetary Wayback as a permanent Web archive to distribute data files into the IPFS network [4]. Header and payload are split for every response record, then they are disseminated into IPFS. The average indexing rate can be boosted by this method. However, various studies show that blockchain can be used as an access control manager for distributed systems. We have proposed an idea, where the data collected in the decentralized database in which a third-party application has control over it. This is achieved by combining blockchain and IPFS with an off-chain solution to track user identity and corresponding document identity which is described by a tuple.

The key advantage of using blockchain is that the user identity data cannot tamper inside this third-party application. The smart contract manages and provides access and control [5].

2.3. Privacy:

The key concept behind privacy in the blockchain is public and private key cryptography. These systems use asymmetric cryptography to secure transactions [6]. Private keys are used to protect the user data through digital signature, hence adding a layer of authentication. Ethereum uses the Elliptic Curve Digital Signature Algorithm (a version of the Digital Signature Algorithm) which uses elliptic curve cryptography, to generate key-value pairs. The proof of work function used in Ethereum is Ethash (256-bit), belonging to the keccak family (same family as SHA-3) [7].

3 Architecture Overview

This section provides an introduction to the data storage model and how smart contracts control the flow of data and add a layer of security.

3.1 System Overview:

A highly secure and reliable system has been designed to store farmer details and their land details using blockchain as the layer for monitoring transactions, Interplanetary File System for data storage, and a local database management system for tracking public user details and transaction hash codes as shown in Figure 3.1. Currently, all the records are manually maintained by the department leading to difficulty in access and retrieve the records, also increasing vulnerability. The user details and land information are obtained through the government servers or any offline methods devised by the government. The data is processed and stored by the data model architecture.

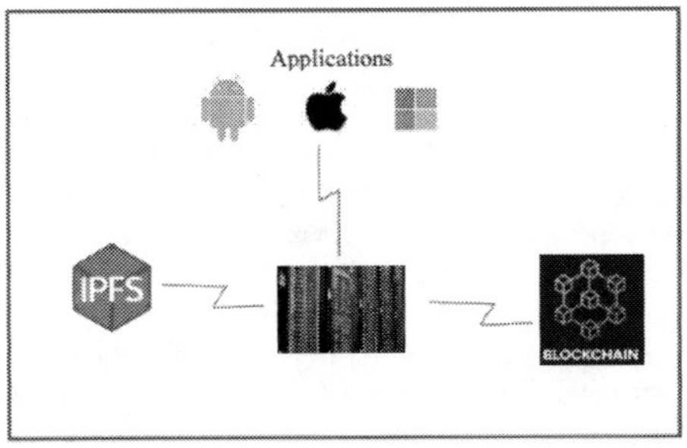

Fig. 3.1 System Overview

3.2 Interplanetary File System (IPFS):

The state-of-the-art file storage system which is decentralized in terms of architecture and a robust system called The Interplanetary File System (IPFS) is being implemented. IPFS provides a block storage model that is content-addressed hyperlinks and has high throughput [8]. Hence a generalized Merkle directed acyclic graph (Merkle DAG) is formed, a data architecture onto versioned blockchains, file systems and permanent web are built. IPFS combines an induced exchange of blocks, a distributed hash table, and a self-certifying name box. There is no single point of failure in IPFS since there is no need for two nodes to trust each other.

IPFS envisions to develop a distributed web, making it immune to central server attacks. In IPFS the system relies on content-based addressing rather than location-based addressing followed in HTTP protocol. IPFS uses a Distributed Hash Table or DHT to store data. This system leverages physical proximity. Upon receipt, a network of friends with content belonging to hash is called and the content is obtained straight via the nodes holding the necessary information instead of connecting to a standalone server. Data transfer between nodes in a network is as same as to the mechanism in a BitTorrent.

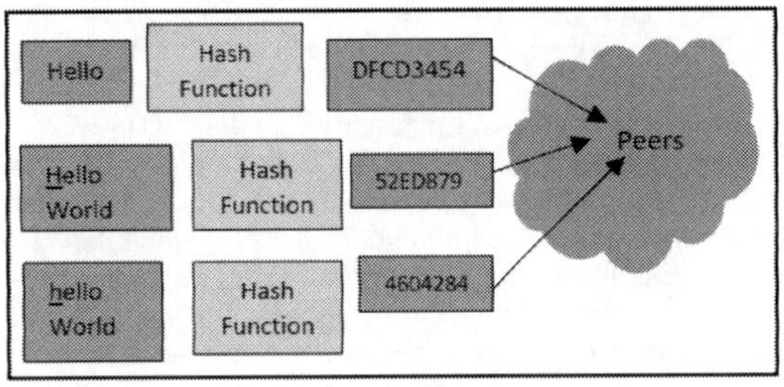

Fig. 3.2 Working of hashing in IPFS

Since IPFS uses hashing to store data as shown in Figure 3.2, data duplication is avoided, hence increasing efficiency. Files greater than 256 kilo-bytes are split into

IPFS objects and their respective hash codes are generated and further linked into another IPFS object.

3.3 Ethereum and Smart Contract:

Ethereum is one of the most popular blockchain operations. Its custom currencies called tokens and native cryptocurrency ether are used to make transactions [9]. Its use of smart contracts and extensive integration make it highly secure and reliable. It can be used to create an alternative protocol to develop decentralized applications, with particular emphasis on scenarios where security for rarely used and small-scale applications, fast development time, and the ability of applications to interact very efficiently, are crucial. Ethereum has an intellectual basic layer: a built-in theme-based programming language, allowing developers to write contracts and decentralized apps where they can define their own configuration rules for state transition tasks states, forms of trade and property. The Ethereum protocols follow the principles, i.e., modularity, agility, simplicity, universality, non-discrimination, and non-censorship. No entity has control over the execution in a decentralized network. Firstly, since the business logic present in the contract is agreed upon by consensus the resultant from the execution of code can be trusted. Secondly, due to proof of work, consensus, and append-only ledger in Ethereum protocol, code tampering or changing is not possible.

Solidity with compiler version 0.4.22 is used for writing and compiling the contract and storing the access control list. Solidity is typed statically and has various variables, functions, and structures. It avails msg.sender which is implicitly available and holds the path of the user who is initiating the transaction, thus ensuring authenticity [10]. The architecture named OwnerDetails is depicted in the contract containing two variables of type address, called PO_hash and data_hash, a process to map called _approve, and a variable array of paths named user. Variable sender is mapped to an example of the structure OwnerDetails, called map.

Only some functions are accessible by the user. The validation is done using if- statements. If the resultant of the if-statement is false then the contract does not update the transaction, hence the state of the contract is not altered. Those functions which do not make any changes to the smart contract are called constant functions, and they do not require any costs to execute the transaction (since they do not alter the state of the contract).

addIpfsHash: Cryptographic hashes help us identify ipfs files. A hash value of bytes32 type is accepted as an only argument in this function. It checks whether the hash represents some data that already exists in the network or whether the file is empty. If either of these checks fails, the transaction gets declined. Otherwise, the hash is stored in the OwnerDetails structure and as a key in map mapping, and it's the key holds for the owner which routes to msg. sender. The contract now commits this obtained data to the database.

retriveFileAccess: This function accepts a PO_hash and a data_hash representing an IPFS hash as an argument. The transaction fails if the entered hash doesn't match the mappings. Otherwise, the relevant data is retrieved utilizing the hash variables as the key from the storage. The msg.sender maps the recorded data of ownerDetails at an instance and approached to the destined path via a map.

In our proposed system a feature has been added in the smart contract, in which the hash address of the previous owner along with the data from whom the land was bought or inherited is stored as shown in Figure 3.3. This allows us to track the hierarchy and retrieve several useful information about the land.

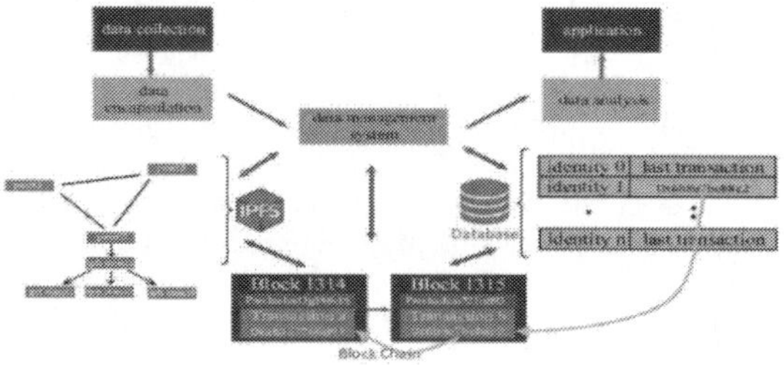

Fig. 3.3 A proposed model overview

3.4 Data Model Architecture:

A highly secure system is designed to store farmer details and their land details. Currently, all the records are manually maintained by the department leading to difficulty in accessing and retrieving the records, also increasing vulnerability. Firstly, the user details and land information are obtained through the government offices, encapsulate the data, process the above data, and store them into IPFS as shown in Figure 3.3. The user can verify that the correct information is uploaded by encrypting it with his private key and various offline methods, which can be devised by the government. The authenticity of the data obtained in the system is ensured by storing the hash generated by the IPFS in the blockchain. The transaction hash generated by the blockchain after completing a successful transaction along with the public details uniquely identifies a person, which is then stored in a database management system to track the hash codes generated during this process.

When extracting user information, the system uses the transaction hash to query transactions and obtain the IPFS hash address. With this hash, the data from IPFS can be retrieved. The data analysis module analyses the data taken from the database management system and returns it to the application layer.

4 Working Principle

This model uses DAG in IPFS to manage the user data. The PropertyOwner object stores the related details of the farmer and his owned property [11]. Firstly, the smart contract is deployed in ropsten test network and injected web3 environment. In our testing, the contract was deployed in the block 8950103.

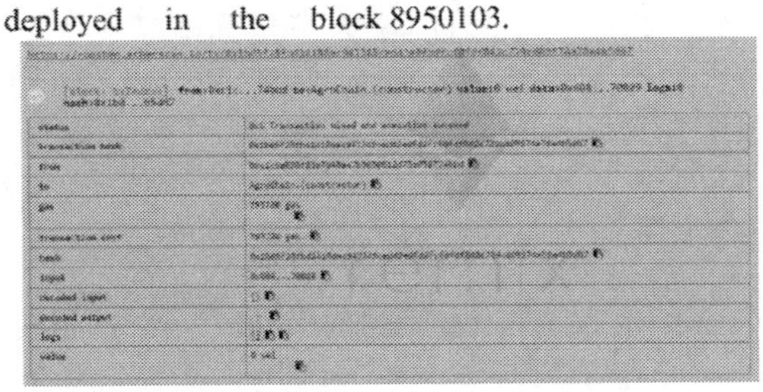

Fig. 4.1 Transaction log after deploying a smart contract

The data attributes are used to store objects and related information. Links property is used to store the previous owner of the particular record along with the current record hash address of the data in IPFS

{
"data": {"type": "PropertyOwner", "user_id ": "0x567b"},
"links": [
{"hash":"QmQZzTMN2X54SxC2jMuAug6Qcz1KYS5ZB12i3gGusvn", "name": "record_11010 ", "size":"7987"},
{"hash":"QmUgeQaCjhZ8V42DKBLtTCvzxojUt-LpaX6QmXX9rTLNnxF", "name": "record_0850", "size":"5765"}
]
}

The data obtained during data collections after processing is then stored in the IPFS as shown in Figure 4.1.

Algorithm: Writing data to IPFS
Input: data, hash of the previous owner. Output: data IPFS hash, PO IPFS hash
1: get old PO_data → get old property owner data with old PropertyOwner hash (old PO_hash);
2: data_hash → store the data into IPFS and obtain the data hash;
3: generate new PO_hash → link the old PO_hash with new PO_data and generate a new PO_hash;
4: return data_hash, PO_hash;

There is a risk of storing the PropertyOwner IPFS hash address directly into the blockchain. The tamper only needs to modify the latest blockchain transaction hash and links the contents of the transaction to the spoofed data IPFS hash address to achieve the purpose of tampering with the data. To deal with this problem, the hash of the newly stored data is uploaded beside the IPFS hash of PropertyOwner. The hash address of the provenance data can be matched with the PropertyOwner while querying. Each time, when the data is collected in IPFS, the updated PropertyOwner hash address and the hash path of the corresponding data block are reported to the management system. The management system encapsulates these addresses into a transaction and write it into the blockchain. The transaction is as follows: Encapsulated as blockchain transaction data:

{
"jsonrpc": "2.0", "method": eth_sendTransaction", "id": 1, "params": [{
"from":"0xb60e8dd655152be8058bb8eb970870f07233155",
"to":"0xd456odd67c5d32be8058bb8eb970870f07244567", "gas": "0x76c0",
"gasPrice": "0x9184e-72a000", "value": "0x9184e72a",
"data":"0x38363037313930323339393538313833b33323b3235313b3130323435313b302e32353b3134383237393936333353b333935372e35373231343b4e3b31313632302e39393030353b45"
}]
}

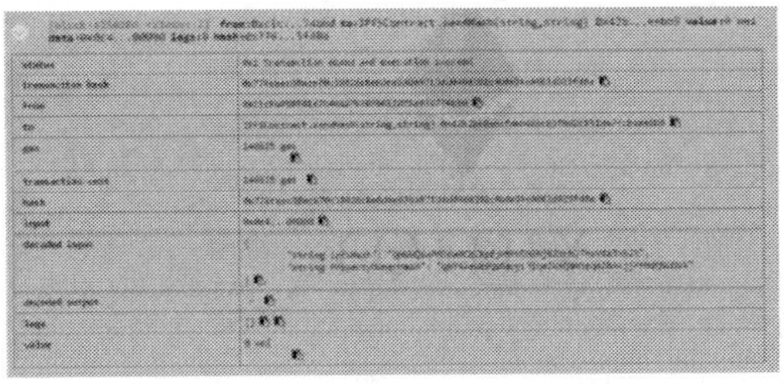

Fig. 4.2 Transaction log after storing the IPFS and PO_hash

The transaction hash is queried from the storage and the IPFS data hash is accessed via the blockchain, which can be utilized to retrieve the original data as demonstrated in Figure 4.2 and Figure 4.3.

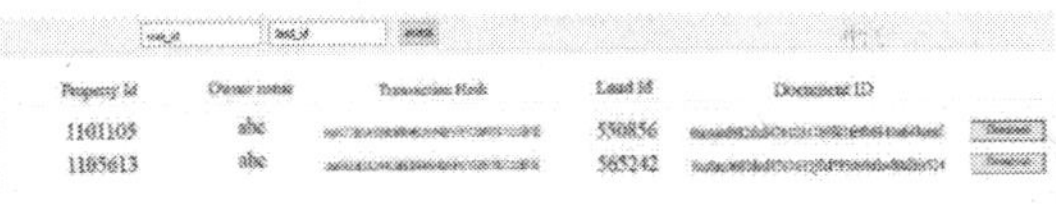

Figure 4.3 Data analysis after querying owner information from database.

Algorithm: Data query algorithm Input: user_id, public details (other parameters)
Output: result (success or fail)
1: Tx_hash → Query transaction hash from the database by user_Id and other parameters;
2: old PO_hash → Get old PropertyOwner IPFS hash from blockchain by Tx_hash;
3: if verification (PropertyOwner hash) == success
Return data, success;
Else
Return fail;

5 Experimental Analysis

Ethereum v1.9 is used as blockchain and IPFS v0.5.0 to store data. The database for tracking the user information is Oracle v18.4.0.0.0. This model was deployed in 6 machines, with each machine having a minimum configuration of 1.8GHz Intel Processor with 8GB ram and windows OS.

5.1 Case Study

The comparative study of IPFS storage and model based on Blockchain is being researched in detail determining their efficiency on real-time scenarios. The time taken to process the uploaded data is shown in horizontal axis also determining the amount of data i.e., the records that are uploaded per minute in kilobytes is represented in vertical axis.

The proposed model is designed to store records efficiently and secure transactions using blockchain. The system was tested by uploading data every two minutes. The processing efficiency of our system increases exponentially as the size of data uploaded increases.

The experimental observations are plotted in Figure 5.1. We expect that IPFS consumes less time however in real-time scenarios when a single person uploads a small file, we observe that both the speeds are more or less similar in both Blockchain and IPFS but as of enterprise-level, IPFS storage must be preferred where the number of transactions is exponentially increased. The upload transaction is depicted in Figure 5.2.

During the download speed testing, we observed deflections in both IPFS and Blockchain, and in contrast, IPFS performance is significantly higher. As the size of the files scales up, we note that the download speed of Blockchain becomes more stable while IPFS even facing deflections serves being a better performant as shown in Figure 5.3. The comparison of upload and download speed between IPFS and Blockchain is depicted in Table 5.1 to show the variations as the file size scales up.

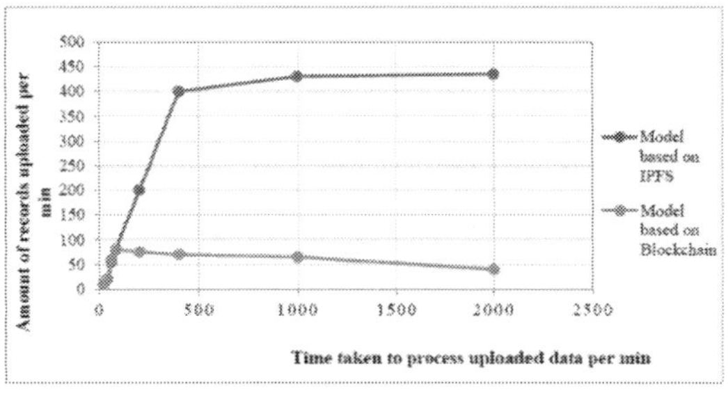

Fig. 5.1. Efficiency of IPFS in comparison with Blockchain

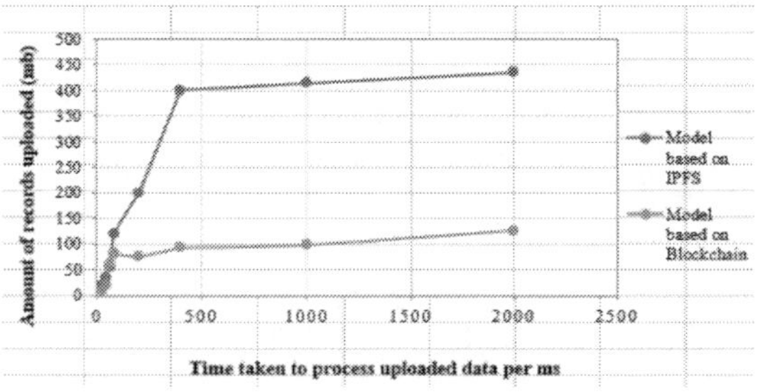

Fig. 5.2. Comparison of upload transactions between IPFS and Blockchain

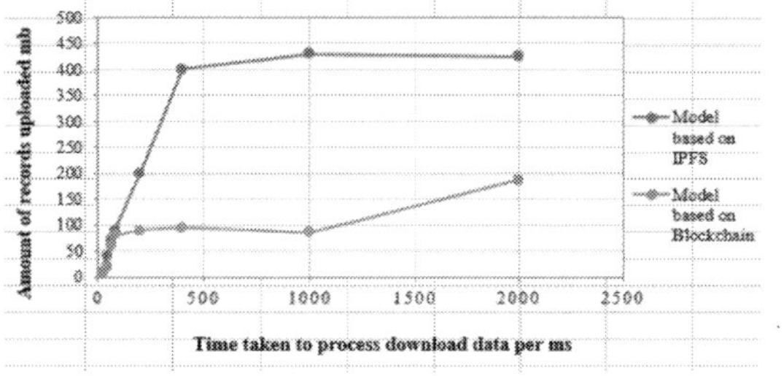

Fig. 5.3. Comparison of download transactions between IPFS and Blockchain

Blockchain (ms)	IPFS (ms)
Upload	
10	20
20	35
60	55
80	120
75	200
93	400
98	415
125	435
Download	
10	10
20	40
60	70
80	90
90	200
95	400
85	430
187	425

Table 5.1. Comparison of upload and download speed between IPFS and Blockchain

6 Results

We have designed an application to maintain the records and protect against frauds and data tampering. After manipulating the data query algorithm, the user could access secure data in the blockchain [12]. This can also help in monitoring the hierarchy of the property by keeping track of the previous transactions in the chain.

Our proposed system with IPFS and Blockchain has shown improved speeds compared to standalone Blockchain use cases. This system when deployed on large scale, would help to attain good real-time speed and reliable security via Blockchain. As we intend to use a centralized database to monitor the transactions, it becomes highly feasible to query any transaction done in past. Hence, it provides a reliable approach to deliver the power of a centralized database alongside the certainty of IPFS and Blockchain.

7 Conclusions

This paper has addressed the requirement to store the data of the user without the risk of data tampering using blockchain technology preventing forging of records and overcoming the hacking and manipulation of centralized databases. As discussed above, storing files directly in blockchain is not feasible. For this purpose, the hash address of the data is stored in the blockchain, along with a third-party database management system to monitor and query the transactions. The delay observed during this process is primarily due to the engagement with blockchain, which may hinder scalability and convenience in real-time. This mode of storing the records in Blockchain with the help IPFS makes a reliable and trustworthy system rather than the existing centralized mode of storing these records. The significant access and transparency are the key to correspond the proposed model.

8 References

1. S. Nakamoto, Bitcoin: A peer-to-peer electronic cash system, https://bitcoin.org/en/bitcoin-paper (2008)
2. Z. Zheng, S. Xie, H. Dai, X. Chen, and H. Wang, *An Overview of Blockchain Technology: Architecture, Consensus, and Future Trends,* IEEE, 6th International Congress on Big Data, 557-564 (2017)
3. M. Steichen, B. Fiz, R. Norvill, R. State, and W. Shbair, *Blockchain-Based, Decentralized Access Control for IPFS*, Conference on Internet of Things, Green Computing and Communications, Cyber, Physical and Social Computing, Smart Data, Blockchain, Computer and Information Technology, Congress on Cybermatics, IEEE, 1499-1506 (2018)
4. S. Alam, M. Kelly, and M.L. Nelson, *InterPlanetary Wayback: The Permanent Web Archive*, Proceedings of the 16th ACM/IEEE-CS on Joint Conference on Digital Libraries, JCDL, 273-274 June (2016)
5. SmartContracts, https://docs.soliditylang.org/en/v0.4.24
6. Privacy and Blockchain: https://en.wikipedia.org/wiki/Privacy_and_blockchain
7. J. Benet, IPFS - Content Addressed, Versioned, P2P File System (DRAFT3), https://ipfs.io/ipfs/QmR7GSQM93Cx5eAg6a6yRzNde1FQv7uL6X1o4k7zrJa3LX/ipfs.draft3.pdf, (2014)
8. H. Gilbert and H. Handschuh, *Security Analysis of SHA-256 and Sisters*, International Workshop on Selected Areas in Cryptography, Springer, **3006**, 175-193 (2003)
9. Ethereum Documentation, https://ethereum.org/en/whitepaper/
10. Solidity 0.7.4 documentation, https://docs.soliditylang.org/en/v0.7.4/
11. J.T. Hao, Y. Sun and H. Luo, *A Safe and Efficient Storage Scheme Based on Blockchain and IPFS for Agricultural Products Tracking*, Journal of Computers, **29**, 158-167 (2018)
12. Transaction made in ropsten test network, https://ropsten.etherscan.io/tx/0x1b05f28fbd14186ac947345ced43e0fddfcf0fdf842c729c809574a78a4b5d67

Modelling a Behavioral Scoring System for Lending Loans using Twitter

Suthanthiradevi P[1*], *Srividhyasaradha K*[2], *Karthika S*[3]
[1,2,3] Department of Information Technology, Sri Sivasubramaniya Nadar College of Engineering, Chennai, Tamil Nadu, India

Abstract. Traditionally, banks follow a risk assessment model in sanctioning loans. Risk assessment is performed by computing a credit score considering certain financial factors. This work proposes a behavioral score that can be computed from social media data. Social media covers almost all aspects of a person's life. Integrating the credit score with the behavioral score of a person lowers the risk that comes with traditional assessment models. The behavioral score is measured by the profile score, financial attitude, and twit score. A general profile score is computed for the data fetched from Twitter. The twit score of a person is calculated by considering multiple parameters like relevance, usage, and authenticity. Additionally, to strengthen the model, a novel multi-level voting ensemble is implemented with 84% accuracy to scrutinize the financial attitude of the individuals. Pair wise comparison is used to reveal the importance of the various criteria analyzed. The behavioral score is derived by aggregating the three scores accordingly. This research work proposes fusing social media details as an added risk evaluation feature in granting loans.

1. Introduction

There has been a major change in the banking paradigms to cater to Digital India. The emergence of a cashless economy doubled by the social and digital technologies of today have put even more pressure on banks to scale down their structures and focus on their most profitable core business segments. To keep pace with these changes, Indian banks are increasingly coming up with technology innovation using mobile, analytics, and social media [11] to suit their customer needs and serve them more efficiently. Banks can expand their horizons from KYC Forms to be aware of a customer. They should tap into the digital identities associated with the customer to perform various analyses and forecasts [10]. One major use case of this would be in the Loans sector.

Loans and lending have evolved into a major banking sub-sector in the 21st century. When an individual applies for a loan, the lenders traditionally use proprietary mathematical algorithms to compute a credit score based on the financial circumstances and account history. It is generally computed considering an individual's capacity, collateral and capital. Despite this, there has been a steady increase in the rate of defaulters' i.e. unpaid loans. An increased number of such fraudulent cases can even lead to the closure of banks in the longer run. This throws light on the fact that potential risks are notexclusive to previous financial transactions but rather behavior as well. In a growing digital era, the best way to assess the behavioral score of a person would be to tap into his or her digital footprint on social media. This also reflects the characteristics of their behavior. Their behavior tends to throw light on their approach towards loan payments. Because of its' popularity and ease of acquiring data, Twitter [9] has been chosen as the source of social media data. Tweet users behavior can be shaped based on optimistic or pessimistic value of the future market [12]. The authors of [14] has developed a systematic risk model for predicting the financial market prices from financial tweets.

Twitter is a micro-blogging and social networking service in which users post and interact with messages known as "tweets". Twitter users can post, like, and re-tweet tweets and it also allows videos and pictures to be shared in the tweets. A person's online presence on a microblogging platform like Twitter would serve as a litmus test in evaluating their characteristics and personality. Their tweets on stock markets [13] or the latest financial initiatives can elucidate their financial attitude.

This paper proposes a behavior scoring model to classify customers based on their presence on Twitter. It aims in generating behavior scores which would be used as portfolio sanctioning decision [15] in conjunction with an existing credit score. The profile score and financial attitude are evaluated by developing a multi-level voting classifier. A multilinear approach is defined in considering these parameters and Bradley Terry mathematical model is subsequently implemented to aggregate the scores.

*Corresponding author: devisaran2004@gmail.com

2. Related Works

2.1. Behavioural Score

Rao, et al. [1] have developed a social score known as the Klout score which assigns an influence rating to users based on their profile across various social networking platforms. They have utilized supervised models to determine the weights for features, and the final Klout Score is obtained by hierarchically combining information from multiple communities and networks. Another approach [2] would be to use deep learning along with network science to mold the social influence and predict human response across various events. They have discussed the correlation between social influence and various activities by analyzing and inferring human real-life behavior through online social networks. This throws light on the fact that a person's social presence reflects their behavior. Jie Tang, et. al.[5] have described the different methods and algorithms for calculating social influence related measures based on statistical media. Similarly, Yanhao Wei, et al. [7] studied social behavior as a model with the exogenous network through tie formation and analyzed the effect of judiciously constructed networks in attaining higher scores. They have investigated how the accuracy of social network-based scores changes when consumers can strategically construct their social networks to attain higher scores. It also throws light on how strategic social networks can be used to improve customer scores by analyzing the strength of ties from friendship formation. In the context of companies, the authors [8] use sentiment analysis to compute the social media score of companies and perform classification using Support Vector Machine. Authors of [16] proposed a graphical Gaussian model, it calculates systemic risk based on the financial tweets. TweetCred, a semi supervised ranking model assigns a credibility score to the tweets based on their credibility [17].

2.2. Credit Score

Yuejin Zhanga, et al. [3] have addressed the issue of loan default following the FICO score and found the significant factors that contribute to the same. They also merge certain social media aspects while computing the criteria. They have used a decision tree to implement an online peer to peer channel for credit scoring. This reports on the importance of assigning proper weightage to the attributes involved. Similarly, upon performing a comparative analysis of the credit classification models, the authors [4] have presented a case study on the 2006 Taiwan credit flow crises. They have considered the model with and without imputation and concluded that an artificial neural network (ANN) is a better alternative than a decision tree and logistic regression when data availability is high in the dataset.

2.3. Gap Analysis

These papers have served as a support system for us to understand the existing features and helped us in developing a new assessment model. The major conclusion that can be drawn is that relevance of the people on social media i.e. their presence among the friends' circle[2][7] plays a vital role in determining their rating. Apart from this, it can be established that the behavioral score should be developed, considering both profile factors as well as analyzing the tweets [8] of the person. Concerning the techniques, it has been observed that Artificial Neural Network[6], Logistic Regression, Decision Tree, and Support Vector Machine[8] are the popularly applied techniques. SVM yields good results as it is used for bi-class classification. This elucidates that these routines can be combined to develop a multi-level voting ensemble classifier that yields better results. About credit score, the factors that lead to a loan default can be identified [3] and weightage assigned accordingly. To combine it with the behavioral score, Gu"l, Sait, etal.[8] rely mainly on performing a sentiment analysis using SVM to define the credit rating of the companies. Aggregation operators are defined to find final rating ofcompanies with credibility distributions. The approach used for companies cannot be replicated as such for individuals. While a company could be primarily assessed on financial standing and tweets, an individual's profile needs to be further analyzed to reflect his or her attitude and habits. The financial score has been specifically developed to segregate the financial tweets of the user. This is used to gain aperspective on his financial views. Unlike the former, sentiment analysis has been performed by developing a novel multi-level voting classifier to scrutinize the tweets and boost accuracy. Further to this, the Twit Score has been developed to find the quality of the Twitter user.

3. System Design

The following, as illustrated in Fig. 1, describes the system workflow to compute the novel score which is defined as an aggregate of behavioral and credit score. The behavioral score is summed up by analyzing the Twitter profile and tweets of a person. The credit score is based on the traditional loan application.

3.1 Data Collection

The details of the profile and tweets posted by the users are accessed using Tweepy. Tweepy is a Python library for accessing Twitter API. It is an open-source library, hosted on Git-Hub and enables Python to communicate with the Twitter Platform. The users whose data is to be collected need to have certain relevant financial tweets. To achieve this, the popular hash-tags used by various banks and their handles are collected from the tweets. The id of these users is fetched from the response JSON object. These users are again passed as objects to the following model. These are then aggregated to form the dataset of the users. These users are again filtered on the range of their financial tweets. Using Tweepy, the various details and tweets of the users are retrieved.

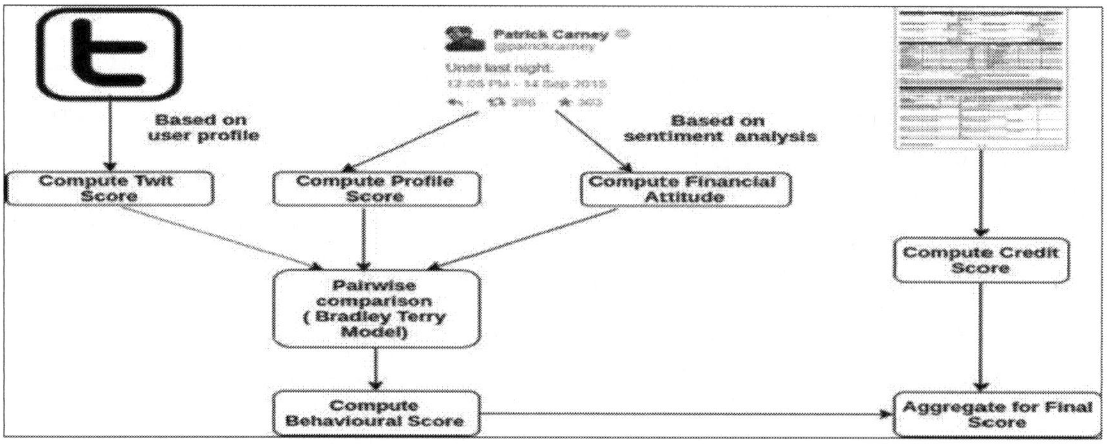

Fig. 1. System Architecture

3.2. Computing the Profile, Financial, Behavioural Score

The Twit Profile, Financial attitude, and Profile Scoreis computed as stated in Table 1.

Table 1. Score Computation

Score	Definition
Twit Score	The user profile of the person forms the basis of computing the twit score. It is an attempt to rate the quality of Twitter user by various metrics available through the API. A Twitter user with low twit score is more likely to be a sign of a spam account or a less safe user.
Profile Score	The tweets of the user form the basis of this score. They are pre processed by tokenization, lemmatization, stop-word removal etc. Subsequently, sentiment analysis is performed and the percentage of positive tweets is chosen to be the profile score.
Financial Score	The financial tweets of the user form the basis of this score. The financial tweets are identified by checking against a corpus of such terms. A threshold is set on the minimum number of financial tweets that need to be present for the analysis. To classify the tweets as positive or negative, a multi-level voting classifier is implemented and the number of positive tweets is chosen as the financial attitude.

3.3. Bradly Terry Model

Pair wise comparison is pivotal to reveal the importance of the three criteria. Pair wise comparison is utilized to rank the three parameters. The three scores have a priority weight attached to them and are combined based on it. The scores need to be prioritized to reflect their relative importance. Bradley Terry is probabilistic mathematical model that has been chosen to derive the priority weights and the number of times a factor is preferred to others becomes the weight assigned.

3.4. Credit Score

The credit score is based on several factors like payment history, credit exposure, and age of the credit. These parameters, in turn, depend on multiple factors, and weights are assigned for all based on the traditional loan application.

4. Methodology

In this section, the implementation of the scoring system is elucidated upon. The following describes the various methods and modules carried out for computing the Twit, Profile, and Financial Attitude score and executing the Bradley Terry Model. The modules have been carried out in the Python framework along with the necessary external libraries.

4.1. Modelling the Twit Score

The Twit Score is an attempt to rate the quality of Twitter users by various metrics such as friend/follower ratio, profile completeness, and other factors that are all available through the Twitter API. The Twit Score is developed giving equal weightage to elements like FF ratio, relevance, usage, and authenticity. A person with an FF ratio around 4 throws light on the fact that the user is more likely to be followed by all the person he or she is following. This indicates that the user is probably more relevant among his or her friend's circle and has acquired this ratio as a result of people liking what he or she talks about. This factor is further supported by the value of relevance score. A person with FF ratio greater than 10 indicates the person does not have enough amount of followers compared to the number of people he is friends with. This is a reflection of a less socially relevant user and complemented by the usage score. The twit score computation factors are further described in Table 2. The Twit score is initially calculated for a total of 20 and it is later evaluated on a scale of 100.

 https://doi.org/10.1051/itmconf/20213701012

Table 2. Twit Score Computation

Sl.No .	Score	Description	Credit	Weightage
I	FF Ratio / Friend Follow Ratio	A person's follower count compared to thenumber of people they follow is a good measure of an interesting or integral Twitter user.	5	25% of Twit
II	Relevance Score	This is to verify how influential the useris among his or her followers i.e. the relevance of the user in his circle. This throws light on the impact factor of an user.	5	25% of Twit
II. a.	Listed Ratio	Listed Count / Followers Count. This is calculated as the ratio of followers who have listed them to their total followers. This throws light on how many followers find them impactful.	5	40% of Relevance
II. b.	Re-tweet	Average number of re-tweets shows the impact of an user.	5	60% of Relevance
III	Usage Score	Too much time on social media is not a great quality. In this computation, highly active users have a lower usage score and less active users have higher score.	5 (5+3+ 1 +1)	25% of Twit
III.a.	Tweets Frequency	Scored based on average time diff in terms of day, hour and minutes basis.	5	50% of Usage
III.b.	Media	Amount of status posted.	3	30% of Usage
III.c.	Twitter Bio	Having a bio is seen as a positive factor.	1	10% of Usage
III.d .	Profile Picture	Having a Profile picture is also seen as a positive factor i.e. confidence.	1	10% of Usage
IV	Authenticity Score	This score is used to track the legitimacyof the user and how sound the profile is for consideration.	5	25% of Twit
IV. a.	Duration	Duration is the best way to verify the legitimacy of a person's twitter profile.	3	60% of Twit
IV.b.	FollowersCount	The no. of people they follow can play quite a large part in calculating Twitter handle interest scores. A person who follows a large number of people may not be a perfectly ideal person.	2	40% of Twit
	Twit Score		**20**	100%

4.2. Modelling the Profile and Financial Score

Firstly, the raw tweets are given as an input for applying standard NLP pre-processing technique to extract contextual features. This is fed to multi-level voting ensemble to train the model. The financial tweets are segregated from the list of tweets based on the predefined list of handles and hashtags used.

4.2.1. Feature Extraction

Data Pre-processing. Twitter data may be incomplete, inconsistent and noisy which can produce misleading mining results. Data preprocessing is a proven method of resolving such issues. It transforms raw data into a cleaner and understandable format. It also filters out useless data. Various steps involved in it are data cleaning, integration, feature reduction, and transformation. The list of pre-processing rules applied is shown in Table 3.

Table 3. Score Computation

Rule	Rule Description
Tokenization	Split a Tweet text into smaller words.
RT removal	Remove RT (retweet).
URL removal	Remove URL links.
Lemmatization	Convert words to base form.
Hashtag removal	Removal of '#' words
Convert to lowercase	Transform tweet into small letters.
Stripping whitespaces	Remove additional whitespaces.
Stopword removal	Remove stopwords like a, an provided by the standard NLTK package
Smiley removal	Remove emoticons from tweets.
Replace user mentions	Replace '@' mentions in a tweet
Punctuation removal	Remove punctuation marks and non-english character.

 https://doi.org/10.1051/itmconf/20213701012

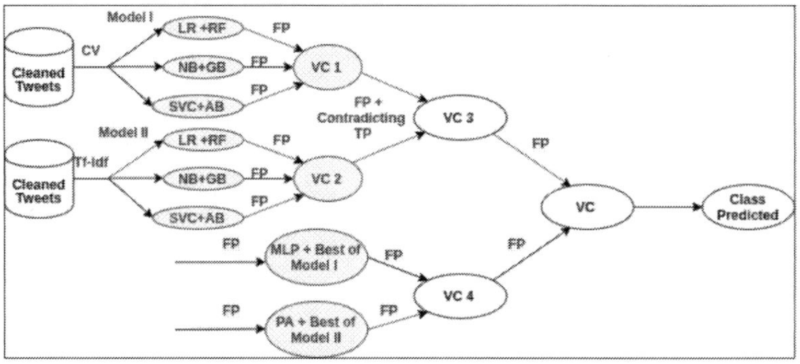

Fig. 2. System Architecture

Count Vector (CV). The vector is a measure of how frequently the word occurs in a document. Count Vector converts a collection of text documents to a matrix of token counts. Consider a Corpus C of D documents {d1,d2.....d_D} (rows) and N unique tokens (columns) extracted out of the corpus C. The size of the Count Vector matrix M will be given by D X N. Example of Count Vector Tokenization is given by

D1: He is a lazy boy. She is also lazy
D2: Neeraj is lazy.

Table 4. Calculation of CV

	He	She	lazy	boy	Neeraj
D1	1	1	2	1	0
D2	0	0	1	0	1

Tf-Idf. This represents Term frequency document frequency. This weight is a statistical measure used to evaluate how important a word is to a document in a collection or corpus. In 'i' is a weight in the document 'j' is calculated as follows

$$TF(i,j) = \frac{Term_i\ frequency\ in\ document_j}{Total\ words\ in\ document_j} \quad (1)$$

$$IDF(i) = \frac{Total\ documents}{Documents\ with\ term_i} \quad (2)$$

4.2.2. Proposed Multi-level Voting Model

Classification Algorithms. The processed dataset retrieved (after preprocessing and feature extraction phase) is fed to the classification phase for the identification of sentiment of the tweets as positive or negative. The machine learning algorithms used are Logistic Regression (LR), Naive Bayes (NB), Support Vector Machine (SVC) along Multilayer Perceptron (MLP). It is built involving a variety of bagging and algorithms such as Random Forest (RF), Gradient Boosting (GB), Adaptive Boosting (AB), and Passive-Aggressive Classifier (PA).

Multi-level Voting Model. In this module, an ensemble classifier is developed and it is implemented as a multi-level voting model that involves various classifiers. The financial corpus built is limited to 925 tweets. Multi-level voting classifiers harnesses the diversity of the individual base models and provide better performance compared to standalone classifiers because they solve the problem posed by a limited volume of data. They are repeatedly combined on the basis of minimum FP ratio and hence overcome the weakness attached with the existing base learning models as illustrated in the results section. Majority voting is implemented by initializing a higher weight to the better performing classifier. Fig.2. illustrates the architecture of the model.

Model I. The inputs are transformed into a CV matrix. The linear machine learning algorithm is combined with bagging or boosting model to form the three classifiers Logistic Regression with Random Forest, Naive Bayes with Gradient Boosting, and Support Vector Machine with Adaptive Boosting. By fusing a bagging or boosting algorithm with a machine learning model, it results in the creation of a stronger classifier.

Model II. The inputs are transformed into a Tf-Idf matrix. The linear machine learning algorithm is combined with bagging or boosting model to form the three classifiers Logistic Regression with Random Forest, Naive Bayes with Gradient Boosting, and Support Vector Machine with Adaptive Boosting. By fusing a bagging or boosting algorithm with a machine learning model, it results in the creation of a stronger classifier.

Voting- Classifier 1. Based on the performance of the set of classifiers in Model I, they are combined with hard voting based on the FP rate. The lower the FP rate, the better the classifier performs, or a greater weight is attached to that classifier.

Voting -Classifier 2. Based on the performance of the set of classifiers in Model II, they are combined with hard voting based on the FP rate. The lower the FP rate, the better the classifier performs or a greater weight is attached to that classifier.

Voting- Classifier 3. This combines the results of both

the feature extraction phases. The third level of the voting classifier is retrieved by merging Voting Classifier 1 and 2 based on their FP rate. This classifier combines the FP values of I and II along with the contradicting TP results of the same for testing. This overcomes the weakness attached with the individual learning models of the feature extraction phases and presents a better result.

Voting- Classifier 4. This combines MLP with the best performing model of VC1 and PA with the best performing model of VC 2 based on their FP rate. The best performing model is chosen as the one with the highest accuracy. The Voting Classifier 4 is constructed by retrieving these classifiers based on their minimum FP rate. This strategic combination of these classifiers can reduce the total error rate.

Final Voting- Classifier. The final voting classifier is derived by merging VC 3 and VC 4. Subsequently, the final predicted class is marked on the basis of voting. Hard voting ensures the majority class label predicted at each level is allocated to the tweet. Thus, the classifier is optimized to achieve the highest accuracy of prediction.

4.3. Pairwise Comparison

The Bradley–Terry model [18] is a probability model that can predict the outcome of a paired comparison. Given a pair of individuals i and j drawn from some population, estimates the probability that the pair wise comparison P(i>j) is turns out true if i > j is defined by

$$P(i > j) = \frac{P_i}{P_i + P_j} \tag{3}$$

If the competitions are assumed to be mutually independent as in our model, then the probability is found to satisfy the logit model. The three scores Twit, Profile, and Financial Attitude score are compared as such and the probabilities allocated. The competitions are calculated by using p_{ij} is derived by

$$\frac{p_{ij}}{1 - p_{ij}} = \oint i - \oint j \tag{4}$$

4.4. Computing the Behavioral Score

The behavioral score of the user is to be computed as follows.

$$Score = (PS \times W1) + (FS \times W2) + (TS\ W3) \tag{5}$$

Where PS, FS, TS stand for Profile Score, Financial Score and Twit Score respectively and weights of W1, W2, W3 represents

$$W1 = \frac{(P\,(PS + TS) + P(PS + FS))}{2} \tag{6}$$

$$W2 = \frac{(P\,(FS + TS) + P(PS + FS))}{2} \tag{7}$$

$$W3 = \frac{(P\,(PS + TS) + P(TS + FS))}{2} \tag{8}$$

All the formulas implemented are original except the default working scores of the mathematical model.

5. Experimental Results

The various experimental results that were recorded during the development of this research work have been presented in the following section. The classification algorithms discussed have been applied to the two different datasets (training and testing) and their results have also been given.

5.1. Data Collection

The tweets and user details required for analysis are not readily available. The required tweet dataset has to be constructed by collecting the tweets, as mentioned, from Twitter API (using Tweepy). Dataset was collected from February and March in 2020. All the pre-processing rules described in Table 3 have been applied and the cleaned tweet is displayed against the original tweet in Table 5.

Table 5. Tweets and their pre-processed text

Tweet	Preprocessed Text
#CanaraBank Shorted 110 PE today for Zero Target, keep tracking. 26th March, 2020 it can be zero. CMP 1.3/-...	today zero target track marchzero
@canarabank i tweet because i cant use mynet banking app due to this problem i m working somewhere i m not taking h...https://t.co/9ivsvJvzDm	tweet cant use net bank app due problem work somwhere take

5.2. Profile and Financial Score

5.2.1 Evaluation of metrics of the Multilevel Voting Classifier

The evaluation metrics such as true positive, false positive, true negative, false negative can be found. It has been observed that VC 2 (Tf-Idf) is found to perform better than VC1 (CV) as the features are unigram. Tf-Idf is generally found to cover unigram features better.

Table 6. Evaluation of metrics of the classifier

Model	Accuracy	Precision	Recall
Logistic Regression, Random Forest (CV)	76	72	76
Naive Bayes , Gradient Boosting (CV)	70	76	71
SVC, AdaBoost (CV)	71	71	71
Voting Classifier I (CV Combined)	70	100	71
Logistic Regression, Random Forest (Tf-Idf)	73	75	73
Naive Bayes , Gradient Boosting (Tf-Idf)	68	68	68
SVC, AdaBoost (Tf-Idf)	71	72	71
Voting Classifier II(Tf-Idf Combined)	80	100	80
Voting Classifier III (VC I = VC II)	89	96	89
MLP + Best of Model VC I	89	96	89
PA + Best of Model VC II	63	95	63
Voting Classifier IV (MLP + PA)	89	95	63
Voting Classifier Final (III + IV)	84	100	84

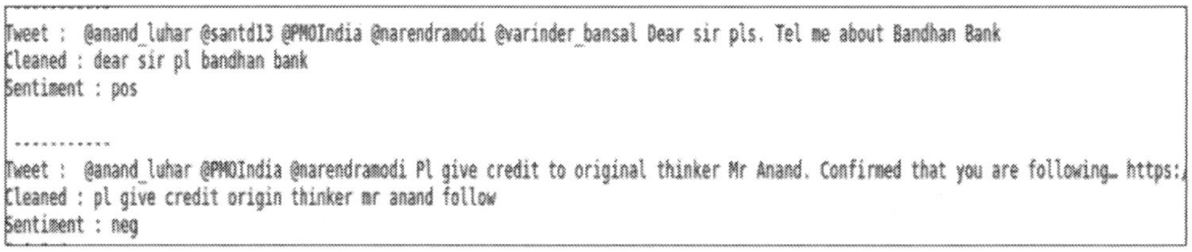

Fig. 3. Class for tweets predicted by classifier

This is because adding extra features in the CV may lead to over-fitting. Multi-layer perceptron is found to give the best result when combined with the best performing classifier of model 1. The final voting classifier achieves an accuracy of 84 %. Table 6. illustrates the tweets and their corresponding prediction classification by ensemble. Fig 3 depicts prediction of tweet along with their corresponding class.

5.3. Behavioral Score

Following the method prescribed in the previous sections, the behavioral score of the user is computed as illustrated in Table 7. Pair-wise comparison is implemented on the basis of Bradley-Terry model. The weights have been derived as the average of these probabilistic measures. The threshold value of the behavioral score is aggregated from the Pair-wise comparison. The algorithm has computed this score for 51 relevant users. As mentioned previously, users with null value are a reflection of a negative Profile and Financial Attitude. Hence their score has been 0. The ones whose values are found to exceed the fixed threshold of 170 have a good behavioral score. Further analysis of the scores have been presented as case studies.The behavioral score computation for tweet users is shown in Table 8.

Table 7. Pairwise Comparison

Probability	Weight
P (Profile Score > Twit)	0.13
P (Profile Score < Twit)	0.87
P (Profile Score > Bank)	0.84
P (Profile Score < Bank)	0.16
P (Twit > Bank)	0.44
P (Twit < Bank)	0.56
Average Behavioural Score (for 51 users)	**170**

Table 8. Behavioral Score Computed

Twitter Handle	Profile Score	Twit Score	Financial Attitude	Behavioral Score
@reach4smiles	88.88	55	100	177.66
@saitejapanems	77.14	75.75	88.88	156.83
@pennewstweet	70	60.5	100	171.24
@Sinha024Sinha	64.28	72.5	100	205.37
@ViraragavanV	0	65	0	0

Table 9. Loans Criteria

Attribute	Factor	Young	Middle-Aged	Senior-Citizen
Other Factors	Educational Qualification	5	2	2
	Nature of Employment	5	3	3
	Experience	2	7	5
	Income	5	8	5
	No. of. dependents	3	5	5
Financial	Loans if any	3	3	3
	Liabilities	2	2	2
	Current Account Balance	6	3	5
	Average credit balance of last 6 months	10	5	10
	Assets purchased on credit if any	4	2	7
Assets	Current Deposit	5	5	5
	Cash Investments	5	5	4
	Movable Assets	3	5	4
	Immovable Assets	2	5	8
Security	Collateral	5	5	7
	Value of collateral	3	3	5
	Guarantor's Net worth	2	2	10
Credit Exposure	Credit Limit	15	15	10
	Credit card usage	15	10	10
	Total Score	**100**	**100**	**100**

5.4. Evaluation of Credit Score

Credit score can be defined in simple terms as the chances of how likely a person is to pay back the debt. Typical credit scores when summarized over several factors range from 300 to 900. There are generally credit bureaus that calculate these scores. This score not only affects a person's chances of getting a loan sanctioned but even plays a crucial role in deciding the interest rates. The weights are fixed to the following credit score based on the age group. The factors have been based on a traditional bank application. The weights have been allocated on a scale of 100. On a scale of 900, the threshold is fixed as 500 i.e. a value above that was traditionally granted a loan. Based on that, the threshold is set as 55 for the same. The following weights are assigned as in Table 9.

5.5. Case Studies

The threshold value for behavioural score is fixed as 170 i.e. users having behavioural score greater than the specified limit are considered. Similarly for credit score, on a scale of 100, 55 is the threshold. Due to the non-availability of details or attributes required to compute the credit score, the values presented here are based on assumptions. The total score is computed on a scale of 400 and the values above 226 (thresholds combined) can be sanctioned. The case studies are represented in the following Table 10.

6. Conclusion

Introduction of behavioural score in the banking sector is an advancement that replaces the traditional credit score computation methods and factors. This paper proposes a scoring system that considers both the financial aspects and the social attitude of a person. The first major contribution of this paper is three measures that are used to calculate the behavioural score. The profile score and financial attitude throw light on the personality of the user and calculated by performing sentiment analysis. While traditional sentiment analysis models rely on individual classification algorithms, this system applies a novel multi-level voting classifier. This enhances the accuracy of the results and overcomes the weakness attached with individual prediction models. It can be observed that the model composed of logistic regression and random forest in Level I and along with multi-layer perceptron in Level II reports the best performance. The classifier achieves an overall accuracy of 84%.

Apart from these two scores, there is an emergence of a third unique twit score. The twit score performs a thorough analysis of the profile and scales for factors like authenticity and influence. The second major contribution of this paper is implementation of pairwise comparison (Bradley-Terry Model) which ensures proper aggregation of three scores to compute the behavioural score. A major incentive of this approach is precisely disseminating the magnitude of the three scores while assigning the weights. The behavioural score is summed up with the credit score and the values are evaluated.

Table 10. Case Studies

Sl. No.	Type of User	Profile Score	Financial Score	Twit Score	Behavioral Score	Credit Score	Total Score	Inference
1	Low behavioral score user	50	72.5	76	170	65	235	The user was found to have an average Profile score. A better trend of Twit score and Financial score was observed. This is an example of low risk neutral profile. On analysis of Twit Score, the person has a genuine account but the user's frequency score does not seem convincing. Apart from it, the user enjoys a decent influence in his social circle and has a sound financial attitude. The fact that he has 50 as his profile score serve as litmus to a neutral presence. Since the total score is above the threshold 226, the user is sanctioned a loan
2	Sound behavioral score user	85	90	50	180	70	250	The user was found to have a sound Profile and Financial score. The Twit score was found to be a relatively average value. This case is very similar to the case study 1. In earlier case, the authenticity score was found to be 0. The fact that the person has an authenticity score 4.5 on 5 indicates a strong genuine profile even if slightly pulled down by the FF ratio. A decent influence and frequency score ensure the person is accepted. The combined Behavioral score exceeds the threshold and since the person has a good credit score, the loan can be sanctioned.
3	Good behavioral core user	90.9	87.5	67	221	63	284	It can be reported that user has a very good behavioral score. The person seems to have a good financial attitude as indicated by type of tweets. The Twit score was found to have been a relatively decent value. A strong FF ratio and authenticity score verify the credibility of a person. Though a slightly hampered frequency score, A remarkable relevance score indicates a certain amount of influence in the circle of the user. The computed Behavioral score exceeds the threshold and since the person has a good credit score, the loan can be sanctioned.

This is evaluated against the threshold for loan sanction. This paper assumes the value of credit score due to non-availability of data sources for the same. The third major and main contribution of this paper is that this neoteric multi-linear approach could be incorporated by banks to sanction loans in real time. This would also lower the risk that comes attached with traditional assessment models.

References

[1] A.S Rao, N Pasojevic, Z Li, T Dsouza,. 'Klout score: Measuring influence across multiple social networks', 2015 IEEE International Conference on Big Data (Big Data),(2015)

[2] L. Luceri, T.Braun, S.Giordano, 'Analyzing and inferring human real-life behaviorthrough online social networks with social influence deep learning', Applied Network Scince, 4(1) ,(2019)

[3] Y. Zhang, H. Jia, Y. Diao, M. Hai, Li H,'Research on Credit Scoring by Fusing SocialMedia Information in Online Peer-to-Peer Lending', Procedia Computer Science, 91, pp.168–174, (2016)

[4] S.J.Imtiaz,'A Better Comparison Summary of Credit Scoring Classification.' International Journal of Advanced Computer Science and Applications, 8(7),(2017)

[5] J.Sun, J.Tang, 'A Survey of Models and Algorithms for Social Influence Analysis.' Social Network Data Analytics, pp.177–214, (2011)

[6] A.Bagherjeiran, and R.Parekh, 'Combining Behavioral and Social Network Data for Online Advertising',2008

https://doi.org/10.1051/itmconf/20213701012

IEEE International Conference on Data Mining Workshops,(2008)

[7] Y.Wei, P.Yildirim, C.V. Bulte, C.Dellarocas, 'Credit Scoring with Social Network Data.' SSRN Electronic Journal, pp 234-258,(2014)

[8] S.Gül,Ö. Kabak, I. Topcu, 'A multiple criteria credit rating approach utilizing social media data.'Data & Knowledge Engineering, 116, pp.80–99, (2018)

[9] E.Yasa, B.Mucan, "How Turkish Banks benefit from Social Media: Analyzing Banks Formal Links", International Journal of Strategic Innovative Marketing, Vol. 1, pp. 120-129,(2014)

[10] F.M. Masuabi, L.D. Erasmus, 'Social media service innovation in South African retail banking: A case study', 2016 Portland International Conference on Management of Engineering and Technology (PICMET), (2016)

[11] X.Cheng, J.Wu, S.S. Liao, 'A study ofcontagion in the financial system from theperspective of network analytics', *Neurocomputing*, 264, pp.42–49,(2017)

[12] J.Bollen, H.Mao,'Twitter Mood as a Stock Market Predictor Computer', 44(10), pp.91– 94,(2011)

[13] S.Rönnqvist, P.Sarlin,'Bank distress in the news: Describing events through deep learning', *Neurocomputing*, 264, pp.57–70,(2017)

[14] P.Cerchiello,P.Giudici,G.Nicola, 'Twitter data models for bank risk contagion', *Neurocomputing*, 264,pp.50–56,(2017)

[15] V.Rao, L. Kupper, "Ties in Paired Comparison Experiments: A Generalization of the Bradley-Terry Model", *Journal of the American Statistical Association*, vol. 62, no. 317, pp. 194–204, (1967)

[16] Paola Cerchiello, Paolo Giudici, Giancarlo Nicola, "Twitter data models for bank risk contagion",Neurocomputing, Volume 264,Pages 50-56,ISSN 0925-2312, https://doi.org/10.1016/j.neucom.2016.10.101 , (2017)

[17] A.Gupta,P.Kumaraguru,C.Castillo,P.Meier, 'TweetCred: Real-Time Credibility Assessment of Content on Twitter', In: Aiello L.M., McFarland D.(eds) Social Informatics,SocInfo 2014, Lecture Notes in ComputerScience, vol 8851, (2014)

[18] Taylor Francis, 'Extending The Bradley–Terry Model For Paired Comparisons To Accommodate Weights', (2020)

Securing Pharmaceutical Supply Chain using Blockchain Technology

Vishwesh Lingayat[1,*], *Isha* Pardikar[1], *Shubham* Yewalekar[1], *Shyamal* Khachane[1] and *Sachin* Pande[1]

[1]Information Technology, SCTR's Pune Institute of Computer Technology (PICT), Pune, Maharashtra, India

Abstract. The production and distribution of counterfeit drugs is an urgent and increasingly critical worldwide issue, especially in pandemics. The imperfect supply chain system in the pharmaceutical industry is one of the reasons for drug counterfeiting. Drugs ownership changes from manufacturers to wholesaler, distributor, and then pharmacist before it reaches the customer thus making it difficult to keep track of it. In this paper, we have compared the existing proposed architectures of blockchain and IoT based supply chain management systems. The system implemented using hyper ledger fabric ensures sharing, storing, transparency, and traceability of data in each link of the supply chain. On the other hand, Ethereum architecture utilized the features of smart contracts to manage the interactions between sender and receiver. Finally, the study mainly focuses on increasing the safety of pharmaceutical products and reducing the manual operation of the supply chain with the most efficient architecture.

1 Introduction

The supply chain passes through all the various checkpoints associated with assembling and dispersion of the products. These days, a supply chain can include hundreds of stages and numerous geological areas. This makes it hard to follow a fool-proof structure in a supply chain as there are data misfortunes and obstructions in each progression. Additionally, if the supply chain of the pharmaceuticals' sector is taken into consideration, the security of the process flow becomes extremely important. There's a possibility that a counterfeit medication would be loaded up with something more harmful just like floor wax or boric corrosive. Sometimes, it is discovered to be rodent poison. These can make hypersensitive disorders which might cause genuine injury to death.

To improve the supply chain's traceability and security, Blockchain technology has been producing interesting research zones due to its inventive attributes that give productive answers for the current loopholes identified. In simple terms, Blockchain can be defined as a conveyed information base, which is shared among and concurred upon by a distributed organization, also known as a peer-to-peer network. When a component is annexed to the blockchain, it can't be altered, making a blockchain into a changeless record of past movement, thus helping to improve overall performance and security. To conclude, the features that makes Blockchain an ideal solution are a) Immutability, b) Transparency, c) Verification of information, d) Secured by cryptography.

Blockchain technology mainly offers two types of services: a) Ethereum, b) Hyperledger. These give a base to fathom the responsibility issue. At whatever point products will trade hands, the transaction will be recorded and put away on blockchain for credibility and confirmation. Currently, the trust will be founded on assembling organizations. Crude materials provided to them by the provider will be tried in the laboratory and subsequent testing, as well as the manufacturing cycle, will begin. The main aim of this paper is to share the detailed implementable architecture of the supply chain powered by both the services offered by blockchain. Both the methodologies will be compared based upon various parameters that are vital in the complete cycle.

2 Literature survey

Jianfeng Shi et Al.[1] proposed a Blockchain and IoT based pharma supply chain to ensure the authenticity of data sources, IoT devices, sensors, locators, and QR codes. For this, distributed Ledger was implemented for sharing and storage of data thus maintaining the transparency of data and increasing the safety of products.

Yen-Chih Liao, Bin Chong, Jen-Hung Tseng, and Shih-Wei Liao,[2] put forward a method named Gcoin Blockchain as the base of the drug data flow. It is a combination of a decentralized autonomous organization regulation model and open government which ensures transparent drug transaction of data. It is also proven to be a system that can handle millions of transactions per day unlike Ethereum or hyper ledger blockchain.

Haya Hasana, Esra AlHadhramia, Alia Al Dhaheri, Khaled Salaha, and Raja Jayaraman,[3] proposed a blockchain-based solution which utilized the functionalities of smart contracts in Ethereum blockchain. It manages the transactions between the sender and receiver. IoT sensors are used to monitor and track previously defined shipping conditions such as humidity, temperature, pressure, geographic location, broken seal, light exposure, etc.

*Corresponding author: vishwesh7231199@gmail.com

Sunny, J., Undralla, N., Madhusudanan Pillai, V [4] explained the significance of transparency and traceability in any supply chain management system. They also gave an overview of traceability solutions using blockchain in the food, pharmaceutical, consumer electronics, and automobile supply chains. The Proof of Concept was provided by the implementation of solidity smart contracts in Microsoft Azure Blockchain Workbench.

T. Bocek, B. B. Rodrigues, T. Strasser, and B. Stiller, [5] provided the solution developed by modum.io, which uses IoT and Blockchain Technology for ensuring quality control during the shipping of medicines. Measuring the temperature and humidity of medicine containers while shipping using the IoT sensors, this data is stored on a public blockchain architecture provided by Ethereum.

Akhtar, Mohd & Rizvi, Danish,[6] described two approaches for using blockchain framework in the pharma chain - Ethereum Public Blockchain and HyperLedger Blockchain Framework. They concluded that HyperLedger Blockchain Framework provided better scalability, an identity management system, better TPS, and accountability.

B.M.A.L. Basnayake and C. Rajapakse,[7] proposed an Ethereum based public blockchain for organic food supply chain. Each deployment of product returns an address used to create a Quick Response code which can identify the actual product. A token system was used to indicate the farmer's reputation with their respective products. Solidity based smart transaction model was developed to validate products. IoT devices were used to track product locations.

Ijazul Haq and Olivier Muselemu Esuka,[8] proposed an ethereum based permissioned blockchain to prevent counterfeit drugs in the pharmaceutical industry. Drugs are contemplated as digital assets where every unit has a unique key. The key in the form of QR code will be attached to the product to track it. Any extra information about the drug is stored off chain, it's hash-digest is generated and linked on chain. The effect of the drug is recorded which will be further used by the doctor to suggest a dose to a patient without hampering the patient's privacy.

Krishna Mohan Botcha et. al,[9] proposed usage of IoT edge devices and blockchain to enhance traceability in the pharmaceutical supply chain . The ecosystem of the pharmaceutical industry is elaborated to explain the need of such a system. Edge devices are used to aggregate data from various sensors which help in automation of data collection, processing and validation.The data which is stored on edge devices is further synced to the cloud. Also, feedback is taken from physicians and other beneficiaries for further renovation.

Youness Tribis, Abdelali El Bouchti, Houssine Bouayad,[10] put forward a systematic mapping study of the related parameters in the system architecture of the supply chain based on blockchain. A bubble plot is designed to systematically map the topics against research categories. Topics such as traceability, security, management, performance measurements, quality management were plotted against categories like validation, evaluation, solution, etc. and it was found that most of the articles were based on solutions such as Blockchain ledger and ethereum.

Gregor Blossey, Jannick Eisenhardt, Gerd Hahn,[11] proposed to use Blockchain technology to secure the supply chain. As many as 53 blockchain applications were analyzed in the field of management of supply chain. Predominant use cases cover provenance following of resources just as robotization of supply chain tasks while parts of supply chain account are just upheld once in a while in blockchain applications. Applications of blockchain exclusively center around catching the advantages of executing a novel broadly useful innovation. These issues are identified with the administration model of information possession and ordinarily low information quality in supply chain settings. This paper mostly centered around the issues of assembling and coordination inside SCM.

3 Proposed methodology

Blockchain technology provided the base for the peer-to-peer digital currency and led to the bitcoin platform. But the applications of blockchain technology are far more than crypto-currencies, and it can provide an excellent solution for tracing drugs in the pharma chain. So, let's understand the working of the blockchain network and how it can solve the problem of drug-counterfeiting

Blockchain is nothing but an unchangeable distributed database shared by a group of nodes over a p2p network. Each activity on this network is called a transaction. A group of such transactions forms a block. A block is associated with a timestamp, hash value, and hash of the prior block. This chaining of blocks together makes the ledger tamper-proof and immutable.

The blocks are appended to the network after verification by the Proof-of-Work consensus or any other custom consensus algorithm, which provides accountability and integrity to the blockchain network. The ability to store the transactions based on their sequence makes blockchain an ideal candidate for solving traceability problems in the supply chain management system.

https://doi.org/10.1051/itmconf/20213701013

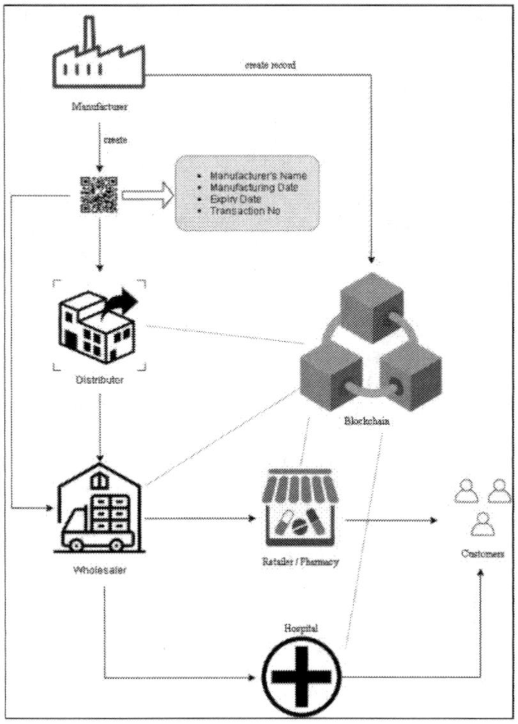

Fig. 1. System Architecture

Initially, the manufacturer will create a QR code containing the manufacturer's name, manufacturing date, expiry date, and transaction number. The transaction number will help in identifying the medicine on the blockchain. Then the distributor will scan the QR code and check for validity. When the shipment leaves the distributor's storage, the distributor should add his signature details in the medicine's record. The wholesalers, retailers, and pharmacies will carry out the same tasks at their end. By recording the transactions at each stage of the supply chain, tracking and traceability can be made possible in the drug supply chain.

Each square has a hash, timestamp, and the hash of the past square. Past hash keeps from adjusting blocks in the middle. Along these lines, the organization is altered, straightforward, and permanent. Blockchain in short is a disseminated permanent record that keeps up the uprightness of the organization by accomplishing agreement through calculations like Proof of Work or Evidence of Authority or own custom agreement calculations.
Furthermore, there are two kinds of blockchain - Public and Private which have some drawbacks. These all give a base to address the responsibility issue. Presently, at whatever point products will trade hands, it will be recorded and put away on blockchain for genuineness and confirmation. Presently, our base trust will be founded on assembling organizations. Crude materials provided to them by the provider will be tried in the lab and subsequent testing manufac- turing interaction will begin. The principal cycle of these assembling organizations like Sun Pharma, Abbott, Pfizer, Cipla, and so on is to create a remarkable QR code name that will contain the name of the producer, expiry date, and exchange number or square number on the medi- cine pack itself. At the point when the medication is produced, they will be put away and their QR code will be filtered, and it will go about as the confirmation of the medication from their side. Presently when this medication will leave the production line, the merchant should filter the medication to realize that it is coming from the assembling industrial facility plant and not from other off-putting places. They will at that point add his particular subtleties and further send it to various wholesalers, retailers, and so forth With this procedure we will accomplish FTP (Production line to Pharmacy), and all the exchanges will be recorded on the blockchain furthermore, the client can basically check this by filtering the QR code at the nearby phar- macy store or in medical clinics. All the stages will be recorded on the blockchain and it will tackle the issue for fake prescriptions.

Blockchain is worked for certain advantages. Be that as it may, out of many, no one but few can be acquired at this purpose of time, due to being in early stages of advancement of blockchain. Consequently, execution component for responsibility in our writing study will be in view of the accompanying:

Decentralization or centralization?

Frankly, why you will not confide in your Government or Ministry of Health to keep your information Blockchain, in principle, states to decentralize establishments by giving no focal position. Be that as it may, we are not inspired by the decentralization particular. On the off chance that we are utilizing any stages like, we will basically let hubs and excavators do the work without getting profound into centralization versus decentralization. So if a solution is incorporated or decentralized, both will be considered as equivalent.

Transparency

Are exchanges straightforward for anybody to confirm on the blockchain. This will be a significant factor.

Security and Privacy

Indeed, the framework will be gotten consistently, and we can likewise follow back the stock chain to recognize who traded right medications with the fake ones.

Scalability

Who can take part, and are there any limitations? What will be the size of the blockchain after a particular time period?

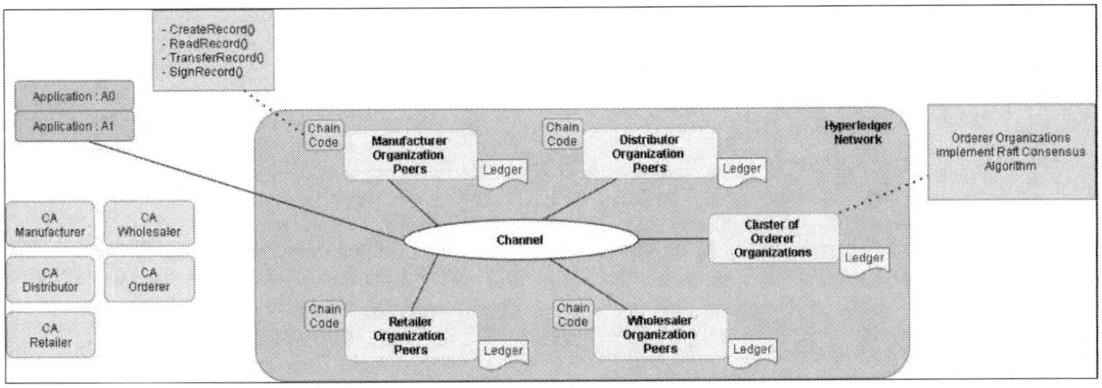

Fig. 2. Hyperledger Network

The Raft consensus algorithm -
Raft is a calculation for dealing with a duplicated log of the structure. Raft executes agreement by first choosing a recognized leader, at that point giving the leader total obligation regarding dealing with the recreated log. The leader acknowledges log sections from customers, repeats them on different workers,
also, advises workers when it is protected to apply log sections to their state machines. Having a leader improves on the administration of the imitated log. For instance, the leader can choose where to put new sections in the log without speaking with different workers, and information streams in a straightforward design from the leader to different workers. A leader can fizzle or get separated from different workers, in which case another leader is chosen.

Raft basics -
A Raft group contains a few workers; five is a common number, which permits the framework to endure two disappointments. At some random time every worker is in one of three states: leader, adherent, or up-and-comer. In ordinary activity there is actually one leader and the entirety of different workers are supporters. Devotees are detached: they issue no solicitations on their own yet essentially react to demands from leaders also, applicants. The leader handles all customer demands (if a customer contacts an adherent, the supporter diverts it to the leader). The third state, applicant, is utilized to choose another leader. Raft separates time into terms of self-assertive length. Terms are numbered with continuous whole numbers. Each term starts with a political decision, in which one or on the other hand more up-and-comers endeavor to become leader. In the event that an applicant wins the political race, it fills in as leader for the remainder of the term. In certain circumstances a political race will bring about a split vote. For this situation the term will end with no leader.

Leader election -
Raft utilizes a heartbeat component to trigger leader political races. At the point when workers fire up, they start as supporters. A worker stays in an adherent state as long as it gets substantial 5 RPCs from a leader or
up-and-comer. leaders send intermittent pulses (AppendEntries RPCs that convey no log passages) to all supporters to keep up their power. On the off chance that a supporter gets no correspondence throughout some undefined time frame called the political race break, at that point it expects there is no feasible leader and starts
a political race to pick another leader. To start a political race, an adherent additions its current term and changes to competitor state. It at that point votes in favor of itself and issues RequestVote RPCs corresponding to each of different workers in the group. An applicant
proceeds in this state until one of three things occurs: (a) it wins the political race, (b) another worker builds up itself as leader, or (c) a timeframe passes by with no victor. These results are talked about independently in the passages underneath. An up-and-comer wins a political decision in the event that it gets votes from a larger part of the workers in the full group for the equivalent term. Every worker will decide in favor of all things considered one applicant in a given term, on a first-start things out served premise. The larger part guarantees that at most one competitor can win the political race for a specific term (the Election Safety Property). When a competitor wins a political race, it becomes a leader. It at that point sends heartbeat messages to all of different workers to build up its position and forestall new decisions. While hanging tight for votes, an up-and-comer may get an AppendEntries RPC from another worker professing to be leader. On the off chance that the leader's term (remembered for its RPC) is at any rate as extensive as the applicant's present term, at that point the up-and-comer perceives the leader as authentic and gets back to the supporter state.
On the off chance that the term in the RPC is more modest than the competitor's current term, at that point the up-and-comer dismisses the RPC and proceeds in competitor state. The third conceivable result is that a competitor not one or the other wins or loses the political decision: if numerous adherents become

applicants simultaneously, votes could be part so that no applicant acquires a larger part. At the point when this occurs, each applicant will break and begin another political decision by augmenting its term and starting another round of RequestVote RPCs.

Log replication -
When a leader has been chosen, it starts adjusting customer demands. Every customer demand contains an order to be executed by the duplicated state machines. The leader affixes the order to its log as another section, at that point issues AppendEntries RPCs in corresponding to every one of the other workers to duplicate the section. At the point when the section has been securely duplicated, the leader applies the section to its state machine and returns the aftereffect of that execution to the customer. On the off chance that supporters crash or run gradually, or then again if network bundles are lost, the leader retries AppendEntries RPCs uncertainly (even after it has reacted to the customer) until all devotees at last store all log passages. Each log passage stores a state machine order alongside the term number when the passage was gotten by the leader. The term numbers in log passages are utilized to identify irregularities among logs and to guarantee a portion of the properties. Each log section additionally has a number record distinguishing its situation in the log. The leader chooses when it is protected to apply a log passage to the state machines; such a section is called dedicated. Raft ensures that submitted passages are sturdy furthermore, will in the long run be executed by the entirety of the accessible state machines. A log passage is submitted once the leader that made the passage has duplicated it on a greater part of the workers. This additionally submits all previous sections in the leader's log, including passages made by past leaders. It talks about a few nuances while applying this standard after leader changes, furthermore, it additionally shows that this meaning of responsibility is safe. The leader monitors the most noteworthy record it knows to be submitted, and it remembers that list for future AppendEntries RPCs (counting pulses) with the goal that the different workers at last discover. When a devotee learns that a log section is submitted, it applies the passage to its nearby state machine (in log request).

4 Results

Blockchain platforms like Ethereum and Gcoin scale well but they fail in providing an identity management system. Moreover, they make use of proof of work consensus technique which is a computationally expensive process. Hyperledger Fabric provides a scalable blockchain platform which not only has an identity management system but also makes use of an unique approach for consensus based on the algorithm pBFT - Byzantine Fault Tolerance. Hyperledger is better suited for building a decentralized platform for tracking and traceability in the pharmaceutical supply chain. Thus, we have compared the existing proposed architectures of blockchain and IoT based supply chain management systems. The system implemented using hyperledger fabric ensures sharing, storing, transparency, and traceability of data in each link of the supply chain. On the other hand, Ethereum architecture utilized the features of smart contracts to manage the interactions between sender and receiver. Basically, the study mainly focuses on increasing the safety of pharmaceutical products and reducing the manual operation of the supply chain with the most efficient architecture.

5 Conclusion

There exists no proper traceability system for the pharmaceutical industry supply chain and hence counterfeit drugs are increasing and it is hard to track them. This paper focused on the need of a blockchain equipped supply chain system in the pharmaceutical industry. This application of blockchain leads to transparency by guaranteeing both tracing and tracking thus preventing the supply chain from counterfeit drugs and also reaching the consumer. Unique ID was generated for each product which was used to create a Quick response(QR) code. QR code was identified as a practical and cost-effective method. Also, consumer feedback was considered which was used for further analysis and rating the actors of the supply chain.

References

1. J. Shi, D.Yi, J.Kuang, Pharmaceutical Supply Chain Management System with Integration of IoT and Blockchain Technology, Springer Nature Switzerland AG 2019 M. Qiu (Ed.): SmartBlock, LNCS (2019)
2. J. Tseng, Y. Liao,B. Chong,S. Liao, Governance on the Drug Supply Chain via Gcoin Blockchain, Int J Environ Res Public Health (2018)
3. H. Hasan, E. AlHadhrami, A. AlDhaheri, K. Salah, R. Jayaraman, Smart contract-based approach for efficient shipment management, Computers & Industrial Engineering Elsevier,ISSN (2019)
4. J. Sunny, N. Undralla, V. M. Pillai, Supply Chain Transparency through Blockchain-Based Traceability: An Overview with Demonstration, Computers & Industrial Engineering, Computers & Industrial Engineering Elsevier (2020)
5. T. Bocek, B. B. Rodrigues, T. Strasser, B. Stiller, Blockchains everywhere - a use-case of blockchains in the pharma supply-chain, IFIP/IEEE Symposium on Integrated Network and Service Management (2017)
6. M. M. Akhtar, D. R. Rizvi, Traceability and Detection of Counterfeit Medicines in Pharmaceutical Supply Chain Using Blockchain-Based Architectures, Sustainable and Energy Efficient Computing Paradigms for

Society, Springer Innovations in Communication and Computing (2020)

7. B. M. A. L. Basnayake, C. Rajapakse, A Blockchain-based decentralized system to ensure the transparency of organic food supply chain, International Research Conference on Smart Computing and Systems Engineering (SCSE), (2019)
8. I. Haq, O. M. Esuka, Blockchain Technology in Pharmaceutical Industry to Prevent Counterfeit Drugs, International Journal of Computer Applications (2018)
9. K. M. Botcha, V. V. Chakravarthy, Anurag, Enhancing Traceability in Pharmaceutical Supply Chain using Internet of Things (IoT) and Blockchain IEEE International Conference on Intelligent Systems and Green Technology (2019)
10. Y. Tribis, A. E. Bouchti, H. Bouayad, Supply Chain Management based on Blockchain: A Systematic Mapping Study, MATEC Web Conf. 200 00020 (2018)
11. G. Blossey, J. Eisenhardt, G. Hahn, Blockchain Technology in Supply Chain Management:An Application Perspective, Proceedings of the 52nd Hawaii International Conference on System Sciences (2019)
12. Saurabh Verma, R. K. (2014). The business of counterfeit drugs in India: A critical evaluation. International Journal of Management and International Business Studies, **4(2)**, 141–148. ISSN 2277-3177
13. Nakamoto, S. (2008). Bitcoin: A peer-to-peer electronic cash system – whitepaper. Retrieved from https://bitcoin.org/bitcoin.pdf.
14. Vitalik Buterin, D. G. (2013). Ethereum: A next-generation smart contract and decen-tralized application platform. Retrieved from https://github.com/ethereum/wiki/White-Paper
15. Asad Ali Siyal, A. Z. (2019). Applications of blockchain technology in medicine and health-care: Challenges and future per-spectives. MDPI Journal Under Cryptography **3(1)**, 3. https:// doi.org/10.3390/cryptography3010003
16. Soundarya, K., & Pandey, P. (2018). Counterfeit solution for pharma supply chain. EAI Endorsed Transactions on Cloud Systems, 3, 154550.
17. Garankina, R. Y., & Zakharochkina, E. R. (2018). Blockchain technology and its use in the area of circulation of pharmaceuticals. Journal of Pharmaceutical Sciences and Research, JPSR, **10**, 2715–2717
18. Haq, I., et.al. (2018). Blockchain Technology in Pharmaceutical Industry to Prevent Counterfeit Drugs. International Journal of Computer Applications, **180**, 8–12. https://doi.org/10.5120/ijca2018916579
19. Patrick Sylim, F. L. (2018). Blockchain technology for detecting falsified and substandard drugs in distribution: Pharmaceutical supply chain intervention. Journal of Medical Internet Research JMIR Publications, **9(9)**. https://doi.org/10.2196/10163
20. Balkrishnan, Y. V. (2019). Blockchain for a robust and efficient supply chain whitepaper, Retrieved from https://www.tes.com/blockchain-robust-supplychain.
21. Hang, F. J. (2019). Novel medical blockchain model for drug supply chain integrity management in a smart hospital. Retrieved from ResearchGate Publication. https://doi.org/10.3390/ electronics8050505
22. Madisetti, A. B (2017). Blockchain Applications: A Hands-On Approach (Ist. ed.). VPT.
23. Szabo, N. (1997). Formalizing and securing relationships on public networks. First Monday, **2**, 9-1. https://doi.org/10.5210/fm.v2i9.548
24. J. S. Arlbjørn, H. de Haas, and K. B. Munksgaard, "Exploring supply chain innovation," Logistics Research, vol. **3**, no. 1, pp. 3–18, 2011.
25. H. Kagermann, "Change through digitization: Value creation in the age of Industry 4.0," in Management of Permanent Change (H. Albach, H. Meffert, and A. Pinkwart, eds.), pp. 23–45, Wiesbaden: Springer Fachmedien, 2015.
26. M. E. Porter and J. E. Heppelmann, "How smart, connected products are transforming competition," Harvard Business Review, vol. **92**, no. 11, pp. 64–88, 2014.
27. M. Crosby, Nachiappan, P. Pattanayak, S. Verma, and V. Kalyanaraman, "Blockchain technology: Beyond bitcoin," Applied Innovation Review, no. 2, pp. 6–19, 2016.
28. K. Siba, Tarun, and A. Prakash, "Block-chain: An evolving technology," Global Journal of Enterprise Information System, vol. **8**, no. 4, pp. 29–35, 2017.
29. H. M. Kim and M. Laskowski, "Toward an ontology-driven blockchain design for supply chain provenance," Intelligent Systems in Accounting, Finance and Management, vol. **25**, no. 1, pp. 18–27, 2018.
30. N. Szabo, "Formalizing and securing relationships on public networks," First Monday, vol. **2**, no. 9, 1997.
31. E. Androulaki, Y. Manevich, S. Muralidharan, C. Murthy, B. Nguyen, M. Sethi, G. Singh, K. Smith, A. Sorniotti, C. Stathakopoulou, M. Vukolic, A. Barger, ´ S. W. Cocco, J. Yellick, V. Bortnikov, C. Cachin, K. Christidis, A. de Caro, D. Enyeart, C. Ferris, and G. Laventman, "Hyperledger Fabric," in Proceedings of the Thirteenth EuroSys Conference, 2018.
32. H. Wu, Z. Li, B. King, Z. Ben Miled, J. Wassick, and J. Tazelaar, "A distributed ledger for supply chain physical distribution visibility," Information, vol. **8**, no. 4, p. 137, 2017.
33. B. Carson, G. Romanelli, P. Walsh, and A. Zhumaev, "Blockchain beyond the hype: What is

the strategic business value?." McKinsey & Company, 2018.

34. B. Notheisen, J. B. Cholewa, and A. P. Shanmugam, "Trading real-world assets on blockchain," Business & Information Systems Engineering, vol. **59**, no. 6, pp. 425–440, 2017.
35. V. Babich and G. Hilary, "Distributed ledgers and operations: What operations management researchers should know about blockchain technology," SSRN Electronic Journal, 2018.
36. S. Davidson, P. de Filippi, and J. Potts, "Blockchains and the economic institutions of capitalism," Journal of Institutional Economics, vol. **14**, no. 4, pp. 639–658, 2018.
37. O. E. Williamson, The Economic Institutions of Capitalism: Firms, Markets, Relational Contracting. New York, NY: Free Press, 1985.
38. V. Schlatt, A. Schweizer, N. Nils Urbach, and G. Fridgen, "Blockchain: Grundlagen, Anwendungen und Potentiale." Fraunhofer-Institut fur Angewandte ¨ Informationstechnik FIT, Bayreuth, 2016.
39. A. Halldorsson, H. Kotzab, J. H. Mikkola, and T. Skjøtt-Larsen, "Complementary theories to supply chain management," Supply Chain Management: An International Journal, vol. **12**, no. 4, pp. 284–296, 2007.
40. G. Baker, R. Gibbons, and K. J. Murphy, "Relational contracts and the theory of the firm," The Quarterly Journal of Economics, vol. **117**, no. 1, pp. 39–84, 2002.

Effective prediction on music therapy using hybrid SVM-ANN approach

K Devendran[1,*], S K Thangarasu[2],P Keerthika[1] R Manjula Devi[1] B K Ponnarasee[1]

[1]*Computer Science and Engineering Department, Kongu Engineering college, Perundurai, Erode.*
[2]*Mechatronics Engineering Department Kongu Engineering college, Perundurai, Erode*

Abstract. In this world, people are moving with lightning speed. Stress has become a usual thing we experience in our day to day routine. Some factors like work tension, emotional obstacles, brutality, etc lead to stress. Many health issues like headaches, heart problems, depression, etc and psychological issues arise in human beings due to stress. Music therapy gives qualitative results in balancing the physical and psychological issues. Music therapy is an expressive type of art therapy. There are many beneficial effects achieved through music therapy like relaxation, maintain blood pressure level, cure on medical disorders, stability in mood, and improve memory and sleep. Here we aimed to establish the main predictive factors of music listening's relaxation and the prediction of music for music therapy using various machine learning algorithms such as Decision tree, Random Forest, Artificial Neural Network (ANN), Support Vector Machine (SVM) and hybrid of SVM - ANN algorithm. The accuracy of these different methods is critically examined with the help of the accuracy performance metric. Various factors like age, gender, education level, music choice, visual analog scale score before and after listening to music for both individual and therapist suggestions on music are considered for prediction. Our study revealed that SVM-ANN hybrid classifier performance is much better than other machine learning algorithms.

1. Introduction

There are several practices to balance psychological health of an individual like speech counselling, music therapy, yoga, exercise etc. But Music play a vital role in day to day life. One can hear music at any place at any time they feel. Music has the power to reduce one's mood swings and make them stable [7]. Some researchers made research in the importance of music listening. Many health issues can be addressed through music therapy. The patients those who undergo regular medication experience pain both physically and psychologically.

Faus and Matas research study is concentrated on how much ability does music helps one to regain after a stressful stimulus. Research has concluded that 20.84% of students to suffer anxiety when facing exams. usually, a person who experiences stress is subjected to many mood issues such as anxiety, lack of motivation, overwhelm, irritability, sadness [16,17]. The research applied to students who experience stress with concern to exams. These items were arranged based on two factors such as positive affect and negative affect. They infer that listening to music should be considered as part of a complex music-based psychological intervention for anxiety and also indicate that listening to music imitating nature than any other imitating features improves a positive mood in person and reduces their negative mood after a stressful academic experience. [1,2]. Experimental study has been made in the field of listening to self chosen music after stress exposure improves mental balance[3,4]. Cancer patients may result in extensive emotional and physical suffering and music can ease symptoms and treatment side effects [13]. This is an comparative and accessibility study on effects of music therapy and music medicine interventions for psychological and physical outcomes in people with cancer. The results are obtained that music has beneficial effect to handle emotions and also have a less effect on heart rate, respiratory rate and blood level [14, 15]. Music genres are considered as the top level descriptors used to organize music files [9].

**Corresponding author email:devacse19@gmail.com*

The researcher Raglio made a study on predicting the effectiveness of music therapy. Here the effectiveness of music is predicted based on the positive effect, negative effect and neutral effect on people after listening to music and also the predictive factors that are associated with music is also determined. Two approaches such as individual choice of music and music based on therapeutical aims [5]. All these studies are concerned with music effects in physical and psychological beneficial aspects of human. But there is any beneficiary prediction on predicting the music for music therapy that associated with personal factors too. Maximum results prove that music has the ability to improve mental health but the music prediction is undetermined. So in our study we predict the music for music therapy on people which associated various personal factors. It helps people and therapist to determine the music and proceed with the practice and get better improvement. Various machine algorithms are implemented for prediction and a comparative analyse is carried and the best performing algorithm is chosen for the prediction.

Our study mainly concentrates on predicting the choice of music for music therapy with concern to personal factors and therapeutically aims and also make a comparative study on various machine learning algorithm such as decision tree, random forest, Support vector machine, artificial neural network, and hybrid SVM – ANN. We introduce a hybrid model to improve its accuracy in prediction whereas the disadvantage of individual model could be overcome through ensemble. From this music prediction, therapist and people can listening to music which gives better outcomes. Here personal factors like age, gender, education level, music interest, self choice of music, therapist choice of music and therapeutical aims such as Visual Analog Scale [VAS] Score before and after listening to music are considered as the attributes. Classification and prediction on music choice is processed in this study.

2. Aim and research hypotheses

The main aim of the present study was to use hybrid SVM - ANN classifier to investigate how personal factors like age, gender, education level, music practice, choice of music, and visual analog scale score before and after listening music can predict and classify the music choice for music therapy. And we also aim to analyze the accuracy performance of various machine learning techniques like decision tree, random forest, Artificial Neural Network, Support Vector Machine, and hybrid SVM-ANN in this prediction.

3. Methodology

The figure1 illustrates the proposed model of the system. The music choice for music therapy is predicted concerning various personal factors and therapeutic results and also we analyze the performance of various machine learning techniques used in our approach. The necessary attributes are identified for the research.

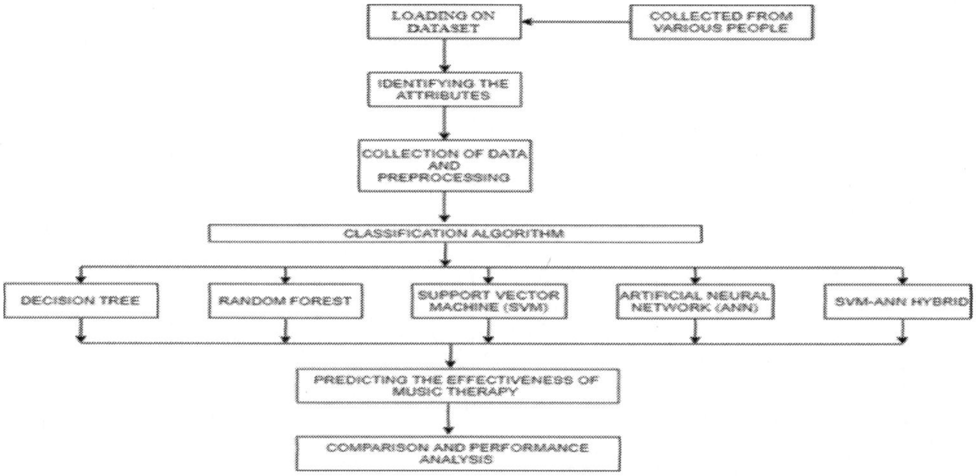

Figure 1. Proposed System

	age	gender	education	musicinterest	selfmusic	ivb	iva	tvb	tva	therapymusic
0	53	Male	non graduate	no	classic	9	3	7	3	classic
1	17	Male	non graduate	no	hip hop	7	5	6	4	pop
2	28	Male	graduate	no	pop	6	2	5	3	pop
3	34	Female	non graduate	yes	pop	7	1	8	2	pop
4	64	Female	graduate	no	classic	8	5	6	2	classic

Figure 2. Sample Instances of the Dataset

The figure 2 illustrates the first five instances from our dataset. For preparing a dataset we collected the information about cause and effects on music therapy from the people. We determined the attributes through the feedback from the people on what are the factors that takes determined place in music prediction. Through collection we received instances of about 2500 instances. There are ten attributes involved in the prediction they are age, gender, education level, music interest, self choice of music, therapist choice of music, visual analog scale score before and after listening to music on both preference of music. The age is an integer, gender, education, music interest, self music are categorical with possible value male, female and others, non-graduate and graduate, no or yes, music classification under class, hiphop and pop respectively. In Figure 3, the feature importance score of the features is depicted. The age attribute contributed more than any other attribute. This feature importance illustrates the important factors that are involved in the classification and prediction of music dataset. This importance feature differentiated through relative score on which feature may be most applicable to the target attribute and the vice versa on least applicable. This would be beneficial for those gather data on specific area.

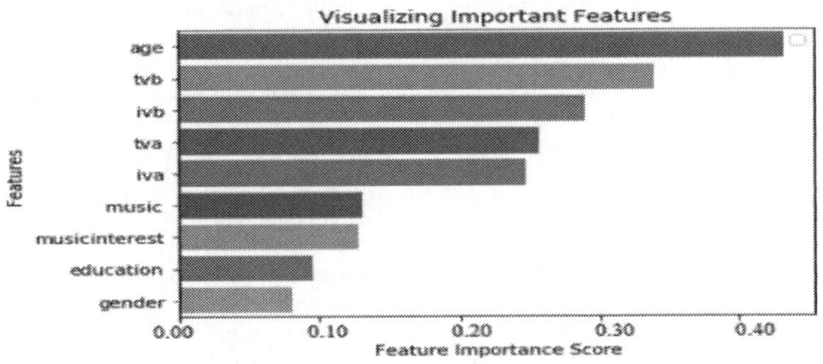

Figure 3. Visualizing Important Features

The data are collected from various people and loaded into a dataset. The dataset is pre-processed and used in classification algorithm. Pre-processing is the process of converting the raw data into feasible data. Here the data are converted from string data type to numeric data type. For example, gender has two choices male and female where it is converted into 0-male and 1-female in pre-processing data transformation. Now the dataset is fed into various machine learning algorithm for classification and prediction of music. In this process the dataset is randomly divided into two parts such as a training dataset and a testing dataset which is 70% and 30% respectively. Here the comparative analysis is done between the machine learning classifier algorithms such as decision tree, random forest, SVM, ANN, hybrid SVM-ANN based on their performance accuracy.

4. Machine Learning Techniques

Machine learning techniques are data analytic techniques that train computers to learn from experience. These algorithms learn information directly from the data. There are various machine learning techniques. Here five different algorithms were used for analysis and comparison.

4.1. Decision tree

The Decision tree classifier is supervised learning that continuously split according to a certain parameter. This classifier can be used for both regression and classification [6]. Nodes and leaves are the entities that are used to build a decision tree.

 https://doi.org/10.1051/itmconf/20213701014

The data is split through nodes and leaves are the outcomes. The implementation of decision tree occurs in such a way like the best attribute in the dataset is placed at root and splitting process is carried further like subset contains data. This process is repeated until leaf node occurs. Decision tree can be easily visualized and can be interpreted. The decision tree algorithm can be implemented using various algorithms such as ID3, CART, CHAID, MARS etc. In this study we build the model using CART algorithm which using gini index criterion for classification. LabelEncoder() function is used to convert the string data to the numeric values. Then the x and y variables are defined with the dataset whereas y assigned to the target attribute and x assigned to remaining attributes in the dataset. Now the data is spilt into training and test data with 70% and 30% respectively. The x and y variables are also assigned in accordance with test and training data as like x_train, x_test, y_test,y_train. The model is build with gini index criteria classifier. The decision tree classifier insist features like max_depth, max_features, max_leaf_nodes, min_impurity_split, min_sample_leaf, min_weight_fraction_leaf, presort, random_state, splitter etc. After the classification, prediction y_pred is applied with x_test dataset. Missclassified samples are identified and confusion matrix is built with y_test and y_pred. At the end decision tree is visualized.

4.2. Random Forest

The Random Forest is a supervised learning classification model that consists of many decision trees. Here bagging and feature randomness are used for building the tree and the prediction is more accurate than any individual tree. This algorithm creates decision trees on data samples and finally selects the best solution from the prediction on each tree using a voting strategy. The string attributes in dataset is converted into numeric values using labelencoder(). The dataset is split into train and test dataset percentage of 70% and 30% respectively. The classifier in this model is random forest classifier which has parameters n_estimators and the random state. In this model n_estimators is assigned to 10 and random_state to 30. The classifier is set fit to the x_train, y_train variables. With the x_test variable the prediction is processed and set to y_pred in random forest model. On comparing the test and prediction variable accuracy is determined. Confusion matrix is also visualized. New instances can also be fed and results can be obtained. Finally, the model predicts the music choice for music therapy.

4.3. Support Vector Machine

Support vector machine(SVM) is supervised learning which uses data points and creates a hyperplane that separates the tags. The standard SVM is used at binary classification but it can also be used for multiclass classification with one against one approach. The one versus one approach can be implemented with C-Support Vector Classification (SVC) algorithm which is based on libsvm. The dataset is divided into train and test data with 70% and 30% respectively[8]. The mathematical formulation provides a nonparametric clustering algorithm where it does not make any assumptions and works very fine to low dimensional data. This algorithm explores the parameter space starting from low value of gamma and high value of C with no outliers occurs. Then is it iteratively decreased on gamma to look for clusters. The data points are mapped from data space to ahigh dimensional features space using the kernel function [10]. The various parameters involved in the model are C, kernel(linear), gamma, coeff, degree, verbose, function shape etc. The SVM-SVC model predicts the choice of music for music therapy. New instances can also be fed and can obtain the prediction result. Confusion matrix is built and performance of the model is obtained.

4.4. Artificial Neural Network

Artificial Neural Network (ANN) is either supervised or unsupervised learning. This algorithm is used to model complex patterns and non-linear relationships. There is no restriction on input variables. ANN has its ability to learn hidden relation is data. The training process consists of forward and backward propagation. In ANN forward propagation is the inputs are taken with multiple weights and passed through a sigmoid function to obtain neuron output. Back propagation is for error calculation. The process is iterated for many iterations. The dataset is divided into test and train data in percentage of 30% and 70% respectively. x_train, x_test, y_train, y_test parameters are used to store the test and training data for classification and prediction. The Multi-layer Perceptron classifier optimizes the log-loss function using stochastic gradient descent. In the hidden layer the ith element gives the number of neurons in hidden layer. The rectified linear unit function is used as the activation function for the hidden layer and adam is used as solver for weight optimization and other parameters such as 'L2' penalty, alpha, learning rate are defined in the MLP classifier. If there is no any training loss for consecutive 10 iterations the iteration will be stopped. We can give any new instances and get the prediction of music as the result. The music choice is predicted and the confusion matrix is built which gives the performance measure of the model.

4.5. Hybrid SVM-ANN

Despite the disputes between SVM and ANN, their hybrid can prove to be the best classifier to solve problems. The hybrid machine learning could accumulate the advantages of standard algorithms and at the same time, they avoid their weaknesses to obtain the predictive target [11,12]. This hybrid technique is possible by dividing the process into two and apply it with respective algorithm. First labelencoder() is used to transform the string into numeric data. Total 9 attributes are used to classify the target attribute. Here by using the linear SVM the important features are extracted (i.e. Feature Extraction is carried out through SVM). The linearSVC uses one vs the rest strategy and the class supports both sparse and dense input to multiclass support. The parameters used in linearSVC are penalty, C value, dual. The penalty specified the norm where 'l2' is a standard one but 'l1' penalty is used in our model which leads to coefficient vectors that are sparse, the default loss function is squared_hinge, C parameter is the regularization parameter which must be positive, dual parameter is used for selecting the algorithm with correspondence to dual or primary optimization problem. In our model we prefer dual as False because our dataset has n_samples>n_features.

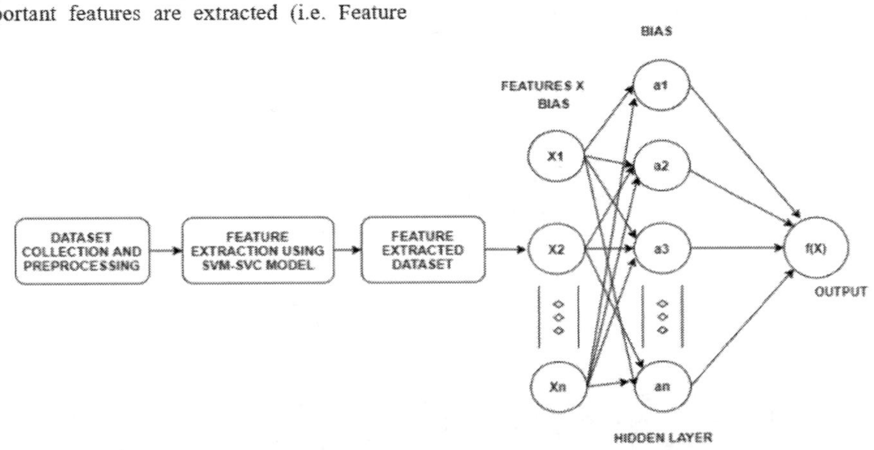

Figure 4. Diagrammatic representation of hybrid SVM-ANN

The features are extracted and prepared to fed into the final ANN model. The extracted feature consists of 6 attributes used for classification. The extracted feature attributes are age, self-selected music choice, VAS score before and after listening to both type of music. (i.e. self choice and therapist choice) and now the extracted features in the dataset is fed into the ANN model which uses MLP classifier to predict the music preference for music therapy. The dataset is divided into test and train data in percentage of 30% and 70% respectively. x_train, x_test, y_train, y_test parameters are used to store the test and training data for classification and prediction. The Multi-layer Perceptron classifier optimizes the log-loss function using stochastic gradient descent. In the hidden layer the ith element gives the number of neurons in hidden layer. The rectified linear unit function is used as the activation function for the hidden layer and adam is used as solver for weight optimization and other parameters such as 'L2' penalty, alpha, learning rate are defined in the MLP classifier. The music choice is predicted and the confusion matrix is built which gives the performance measure of the model. We can give any new instances and get the prediction of music as the result.

5. Result and discussion

There is a difference between the confusion matrix on binary classification and multi classification category in machine learning. Table 1 depicts the confusion matrix for multiclassification model. This illustrates the confusion matrix calculation based on three categories as such as in this study's target attribute. In our dataset there are three class labels namely, classical, hiphop and pop. There is no positive or negative classes here instead for this classification the true positive, true negative, false positive, false negative are determined for each of the classes. Actual and predicted determines the deviation in the target attributed what we expected and what we obtained. True shows that both expected and obtained result is same other false represents the deviation from expected.

Table 1. Confusion matrix of multicategory classification

		Actual		
		Classical(A)	Pop (B)	Hiphop(C)
Predicted	Classical(A)	True A	False A, Actually B	False A, Actually C
	Pop(B)	False B, Actually A	True B	False B, Actually C
	Hiphop(C)	False C, Actually A	False C, Actually B	True C

The table 2 illustrates the accuracy performance of various machine learning technique in music prediction. The performance metric is calculated with the help of measures like precision, recall, f1 score. From the study results, the hybrid approach is resulted with the higher accuracy than other models.

Table 2. Comparative performance metrics for various machine learning algorithms

Models	Accuracy
Decision tree	71
Random forest	76
SVM	82
ANN	70
Hybrid SVM-ANN	86

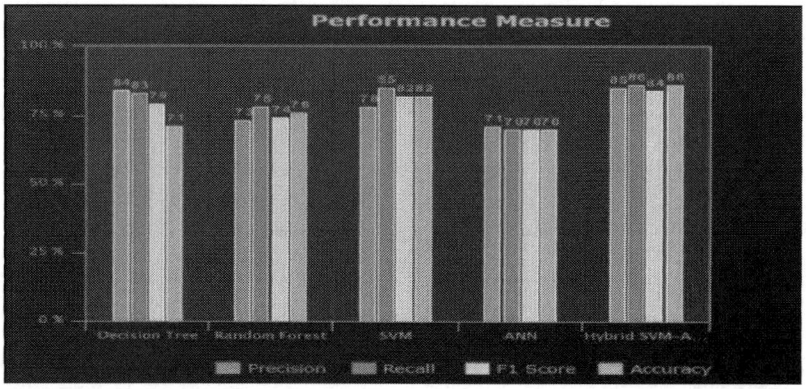

Figure 5. Diagrammatic representation of hybrid SVM-ANN

6. Conclusion and Future work

This paper proposed a classification and prediction on music choice for music therapy. The classification of music includes how personal factors and therapist aims determine the music choice. This paper also deals with comparative study on machine learning algorithms like decision tree, random forest, support vector machine, artificial neural network, hybrid SVM-ANN. The importance of music therapy and its beneficiary effects on both physical and mental health is taken on this research. Combining of two model will result in better performance than individual model. Music therapist could able to select or prefer music which concern their therapeutic aims and apply it on practicing mental betterment in people.

Further improvement in this research can be applied on playing the audio files that possess music for therapy after the prediction through automation.

The Effectiveness is music therapy can also employed for further research.

7. References

[1] Faus, S., A. Matas, and E. Elósegui. "Music and regaining calm when faced with academic stress." *Cogent Arts & Humanities* **6.1** (2019): 1634334.

[2] Radstaak, Mirjam, et al. "Music and psychophysiological recovery from stress." *Psychosomatic medicine* **76.7** (2014): 529-537.

[3] Hohmann, Louisa, et al. "Effects of music therapy and music-based interventions in the treatment of substance use disorders: A systematic review." *PloS one* **12.11** (2017): e0187363.

[4] Gallagher LM, Steele AL. *Music therapy with offenders in a substance abuse/mental illness treatment program.* Music Ther Perspect. **20** (2002): 117–122

[5] Raglio, Alfredo, et al. "Machine learning techniques to predict the effectiveness of music therapy: A randomized controlled trial." *Computer methods and programs in biomedicine* **185** (2020): 105160..

[6] Strobl, Carolin, James Malley, and Gerhard Tutz. "An introduction to recursive partitioning: rationale, application, and characteristics of classification and regression trees, bagging, and random forests." *Psychological methods* **14.4** (2009): 323..

[7] Jensen, Karl Kristoffer, Sølvi Ystad, and Richard Kronland-Martinet. *Computer Music Modeling and Retrieval. Sense of Sounds:* 4th International Symposium, CMMR 2007, Copenhagen, Denmark, August (2007),

[8] Musical Genre Classification Using Support Vector Machines and Audio Features. Article in TELKOMNIKA Indonesian Journal of Electrical Engineering Vol. **14** (2016).

[9] Fulzele, Prasenjeet, et al. "A hybrid model for music genre classification using LSTM and SVM." *2018 Eleventh International Conference on Contemporary Computing (IC3).* IEEE, (2018).

[10] Kar, A., Ahuja, C., & Mukherjee, A. Music Classification using DNN's (2015).

[11] International Journal of Engineering Development and Research - Hybrid SVM-ANN Classifier is used for Heart Disease Prediction System by S.Shylaja, R. Muralidharan, IJEDR, **7.3** (2019).

[12] Advances in Machine Learning Modeling Reviewing Hybrid and Ensemble Methods Sina Ardabili , Amir Mosavi and Annamaria R. Varkonyi-Koczy (2019) in preprints.

[13] Bradt, Joke, et al. "Music interventions for improving psychological and physical outcomes in cancer patients." *Cochrane Database of Systematic Reviews* **8** (2016).

[14] Uggla, L., et al. "Music therapy supported the health related quality of life for children undergoing haematopoietic stem cell transplants." Acta paediatrica **107**.(2018).

[15] Pereira, Ana Paula S., et al. "Music therapy and dance as gait rehabilitation in patients with parkinson disease: a review of evidence." Journal of geriatric psychiatry and neurology **32.1** (2019):

[16] Silverman, Michael J., and Jennifer Bibb. "Acute care mental health workers'assumptions and expectations of music therapy: A qualitative investigation." The Arts in Psychotherapy **59** (2018): 94-100.

[17] Aleixo, Mariângela Aparecida Rezende, Raquel Luiza Santos, and Marcia Cristina do Nascimento Dourado. "Efficacy of music therapy in the neuropsychiatric symptoms of dementia: systematic review." Jornal Brasileiro de Psiquiatria 66.1 (2017): 52-61.

A Survey on Healthcare Systems using Internet of Things

Chitra Ramesh[1*], and Jayalakshmi Sekar[2]

[1]Department of Computer Science, Immaculate College for Women, Cuddalore, India
[2]Department of Computer Applications, VISTAS, Chennai, India

Abstract: The Internet of Things allows things to become active users, facilitating interaction with things and the sharing of data between them. The most interesting issue in the science world, the public sector and industry in the IoT is unavoidable. It ensures a seamless relationship between doctors and patients that results in medical treatment with high quality results. This is accomplished by constant surveillance of patients through the use of sensors. The collected data is registered for potential uses and used for analytics. The analytical approach offers the opportunity for disease detection in healthcare results. This paper concerns the Internet of Things in healthcare and explores the different algorithms used in it. The system involved in analytics of healthcare and data sources involved in analytics are further clarified. Finally this paper demonstrates the Internet of Things and Big Data survey of healthcare systems with a reference table.

1 Introduction

The traditional internet makes communication with the number of small devices and individuals, through a comprehensive network of interrelated machine knowledge whereas IoT incorporates all sorts of connected things without the presence of human. The IoT acceptance and development in wireless communication Technologies allows the patient health issues to be streamed in real time to the caregivers [1]. In addition, mobile devices and various available sensors can measure the basic human physiological parameters such as Blood Pressure(BP), Heart Rate(HR) and Respiration Rate(RR) with a simple click. While it is still at an early stage of growth, corporations and businesses have increasingly accepted the power of IoT in their current processes, and have seen changes in both manufacturing and consumer experiences [2]. However the introduction of IoT Technologies into Healthcare has many challenges including data exchange between devices, data storage, security and privacy, data analysis, unified and ubiquitous access.

In order to connect, participate and cooperate on things, the Internet of Things (IoT) extends human individuality. Heterogeneous technologies of specialized protocols and algorithms have been steadily developed by IoT. The IoT plays a vital role in the global connectivity between millions of wired and wireless sensor systems, electronic devices such as TVs, refrigerators, etc., linked to the Internet [3]. In several sectors, such as agriculture [4, 5] , healthcare [6], entertainment [7], automotive [8], sports, home [9, 10], industrial appliances, business [11], etc., it then plays an enormous position. Fast precision, short expenditure, and dropping instance by better forecasting the futures are the explanations for the successful growth of the IoT. Using Wi-Max,Wi-Fi, IEEE 802.11, Bluetooth, Zigbee,etc., the IoT gadgets such as sensors and actuators is installed in physical devices such as home appliances, cars and smart manufacturing machines to track and transmit data using wireless networks.

Sensors that monitor physiological state such as electroencephalogram(EEC) [12], electrocardiogram(ECG), pressure rate, body temperature, pulse rate, blood pressure are implantable in the body or wearable. For individuals who had elders, kids alone in the house, monitoring ailing people is more comfortable. The user-connected sensor also tracks the user 24/7 and sends warnings to the family physicians, and emergency providers about the changing health status of the user. Significant attention from the global science community is required to meet the growing demands of better technology to address real-time problems in the production of broader data without losing security and privacy.

Big Data analytics [13, 14, 15] has emerged as a good technique for the extensive data process and boosts the decision-making of 6 V: length, variety, pace, veracity, authenticity, and volatility. Volume, which refers to the amount of information being used and evaluated to achieve the desired outcome, is the key attribute. Scanner, cellphone, sensor, digitiser, film, email, Social Network and the internet are all part of a torrent of data from diverse outlets. An information that has been collected, analyzed and interpreted. Text, pictures, multimedia and the mixture of each sort can be the form of data. For Big Data the quality of data is referred by the veracity which is good enough. To collect, interpret and support the Big Data poses a problem[16]. Cloud computing [17, 18] arises as a fundamental help to store and process resource information obtained from every place across the internet to solve the problems of Big Data. To access cloud-based services they are grouped as public cloud, private cloud and hybrid cloud. Third-parties can access, exchange and process the tools of the public cloud. The services accessible by the enterprise

*Corresponding Author: chitrashanmougam@gmail.com

are made possible by a private cloud. A hybrid cloud is often used to replace the internal data with public orchestration which incorporated both public and private clouds. Cloud applications or tools accessible to users on demand via the internet or Software as a Service(SaaS), Infrastructure as a Service(IaaS) and Platform as a Service(PaaS). The cost of cloud-based services or increased day-by-day as a volume of data to be processed, analysed, handled and stored in the cloud are overwhelmed. We are implementing fog computing [19, 20] the pre-processing of generated data to solve the issue, the fog computing is implemented at the edge of the cloud. Within millions of connected devices, Fog allows even data delivery, guarantees high scalability and safe data transaction. Big Data processing is a voluminous approach of data mining which involves both localization and cloud globalization of fog.

With the introduction of IoT and cloud technologies into the healthcare industry, healthcare providers are able to deliver healthcare facilities that are quicker, more effective and cheaper, resulting in better customer engagement. As a result, better hospital facilities, better customer engagement, and fewer paperwork for health workers are brought in.

The paper's outline is as follows: Section 2 focuses on IoT healthcare analytics. For Healthcare Analytics, section 3 explains multiple data sources. In section 4, various machine learning algorithms and their performance are discussed. Section 5 includes a literature review on healthcare systems and the comparative table on the quality and deficiency of the healthcare systems discussed is presented in Section 6. Finally conclusion is given in Section 7.

2 IOT IN HEALTHCARE ANALYTICS

Data processing is carried out by keeping electronic health records, updating clinicians with decision-making abilities with medical information obtained from imaging devices, and allowing people to take care of their health. The analysis aims to avoid patients suffering from chronic diseases.

Data review is conducted by holding electronic health records, medical information obtained from imaging equipment upgrades decision-making experts and allows patients to take care of their health [21]. The research helps keep people with debilitating conditions from suffering. Using analytics can enhance precision, earlier medical detection, allow customization and reduce costs by avoiding laboratory procedures, and also reduce the rate of infection. A blend of the medical sector and the IT industry is IoT Healthcare. The size of the IOT healthcare industry is projected to hit $12.4 billion by 2018, according to a report published by IDC.

Analytics can make use of pattern recognition and machine learning algorithms on wearable sensors. To obtain classification, the machine learning algorithm is used and helps to make decisions about the disease [22]. The machine learning algorithm is defined as the adoption of computational methods from trained data to detect and characterize consistencies and patterns.

3 HEALTHCARE ANALYTICS DATA SOURCES

3.1 Electronic Health Record

The digitised version of medical records for individuals is the Electronic Health Record. It offers medical services to people through the exchange of data with medical organizations [23]. It also makes it easier to edit documents for registered medical users. The data is processed and collected effectively, contributing to better decision support and quality control. They make the diagnosis of diseases easier.

3.2 Biomedical Image analysis

By providing anatomical structures of human beings, Medical Picture plays a critical function in health care [22]. The study of these photographs helps doctors and researchers to track illnesses and prepare care. This produces quantitative data from the images.

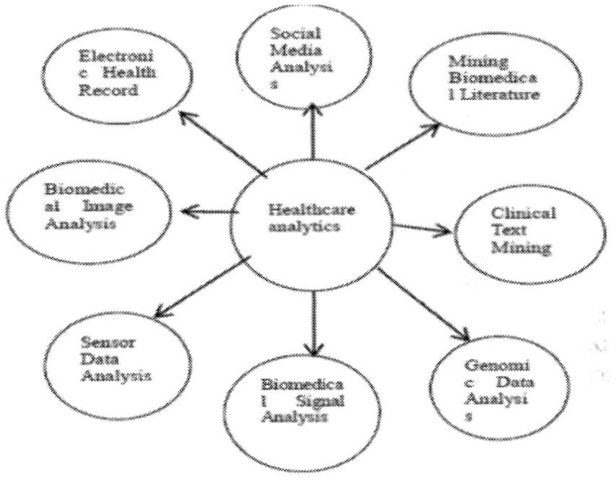

Fig.1. Healthcare analytics Data Sources

3.3 Sensor Data Analysis

Electroencephalogram sensors gather data from different areas of our body, such as electrocardiograms. This is used for real-time analytics that produce immense volumes of data leading to complexity overloading. These are regulated by the use of analytical instruments.

3.4 Biomedical Signal Analysis

Electroneurogram (ENG), electrogastrogram (EGG), phonocardiogram (PCG) etc. signs are used to assess medical disorders and help to determine an appropriate pathway of treatment. The assessment of these signals offers a visual examination of the human body. Invasively or non-invasively, these signals may be obtained. Any of the methods used for the study of these signals [24] are Singular Value Decomposition (SVD), wavelet transformation and Principal Component Analysis (PCA).

3.5 Genomic Data Analysis

Most diseases are hereditary in nature, but the association has not been completely developed between the genetic markers and the diseases. A extremely non-trivial job with a lot of technological difficulties is the transformation of genetic findings into personal medical care [25]. Analytical methods such as Normalization and Quality Control, Differential Expression Identification, Clustering and Labeling, Pathway and Gene Set Enrichment Analysis, etc., solve these problems.

3.6 Clinical Text Mining

Patient knowledge is in the form of an unstructured format of data that forms the foundation of healthcare. Because of the difficulty involved in translating unstructured data text to a standard format, it becomes impossible to analyze automatically. Natural Language Processing is the key technique for extracting information, including clinical details, from unformatted text.

3.7 Mining Biomedical Literature

Text mining techniques are used in biomedical applications for the storage, availability and accessibility of data resources. They provide the leading researchers in information discovery with the extraction, study, and summary of textual data [26]. Biomedical science profits from this by connecting textual evidence to biomedical pathways.

3.8 Social Media Analysis

The creation of multiple social media tools offers valuable details on different facets of healthcare regarding people's views. These are used to collect population health and public health reporting data. They concentrate on identifying comprehensive health patterns such as infectious disease outbreaks, monitoring adverse drug interaction data, and strengthening health-related activities.

4 MACHINE LEARNING ALGORITHMS

Different algorithms are used in healthcare analytics to classify diseases from collected data. An efficient way to classify diseases is given by a machine learning algorithm. When considering data mining, disorders such as heart disease, liver diseases, kidney disease prediction can be achieved successfully by using machine learning algorithms.

There are numerous machine learning algorithms that are successful in the identification of different diseases. Machine learning algorithms are self-learning algorithms through predictive processing, improve the precision of diagnosis. They are categorized as algorithms for supervised machine learning and unsupervised machine learning. Machine learning algorithms that are supervised make predictions based on the qualified sample collection. Whereas the unmonitored machine learning algorithm organizes data into clusters to define its structure and draw inferences without a named response from input data datasets [25].

4.1 Supervised Machine Learning Algorithms

Machine learning algorithms predict based on specified sample sets. It looks for data points within the value labels that are given. The supervised algorithm is composed of regression and classification. Classification, for example, predicts when the patient may experience a heart attack where an observation is consistently calculated as a regression.

4.2 Unsupervised Learning

There are no labels associated with data points. This machine learning algorithms organize the data into a series of clusters to clarify the architecture and to make composite data appear smooth and organized for analysis. It is used in bioinformatics, data mining for sequence and pattern extraction, medical imaging for image segmentation, computer vision for object recognition for sequence analysis and genetic clustering.

4.2.1 Decision Tree

The Decision Tree is used to arrange details in graphical form on the future results and end value of choices. For quick decision making, they are statistical models used. The description is carried out on the basis of an arrangement of characteristic values [27].

4.2.2 Classification of Naive Bayes

Naive Bayesian networks consist of directed acyclic graphs in which the non-observed node is represented by the parent and the nodes are represented by children. In comparison to parents, there is only one parent having multiple children and having independence between child nodes. The use of the robust Bayes classifier increases the precision of disease diagnosis by up to 78-84 percent for patients in intensive care.

4.2.3 Bayesian Theory of Networks

Bayesian networks often referred to as belief networks play a stronger function in the precision of the classification of diseases such as Hepatitis, Pima Indian,

https://doi.org/10.1051/itmconf/20213701015

Thyroid, etc. [28] These are graphic models that reflect unknown domain information.

4.2.4 Learning Based on Neural Network

Neural networks and multilayered neural networks are categorized into single layered neural networks. The full volume of information in the output unit is maintained by the single layered neural network. Hepatobiliary diseases can be identified using neural networks. Improvements in precision and speed are given by the single layer neural network.

4.2.5 Support Vector Machine

By constructing an N-dimensional hyper plane that is divided into two groups, SVM classifies details. A set of features that characterize a case called vector. Vectors close to the hyper plane are called vectors of support. The highest distance separation to the closest training data in any class is obtained by successful separation. The smaller the margin, the greater the generalization error. For the detection of heart diseases, SVM algorithms are more fitting. The comparison of different algorithms is seen in the table.

For the classification of the health data set and for the successful identification of diseases, machine learning algorithms are needed. Many classification algorithms exist. Taking into account the different parameters such as efficiency, speed, classification, tolerance rate, etc the performance of algorithms is analyzed. The comparison between different algorithms is seen in Table 1 below by considering these parameters. The description of health data sets by the Supervised Machine Learning Algorithm appears more detailed.

Table 1. Comparison of Learning Algorithms

Parameter	Decision Trees	Rule Based	Naïve Bayes	Neural Network	SVM
Efficiency	Medium	Medium	Low	High	Very High
Speed	High	Medium	Very High	Low	Medium
Classification	High	Medium	High	Very High	Very High
Event Attributes	High	Medium	Medium	Low	Very High
Tolerance	Medium	Medium	Low	Medium	High
Prevention	Medium	Medium	High	Low	Medium

In both linear and non-linear data sets, supervised machine learning algorithms perform well. In order to diagnose coronary disorders, renal failure, thyroid disease, diabetics, etc., SVM is more fitting. It needs more running time for the SVM, but the accuracy rate is good in disease detection. SVM decreases the error in classification and increases the output rate.

In general, in continuous and multidimensional functions, SVM and neural networks perform well. Prediction accuracy in SVM is improved when large datasets are used. On small data sets, the Naive Bayes classifier is less precise. The decision tree algorithm does not work well with data involving diagonal partitioning and it requires a complicated data representation due to replication problem. This needs extended running time. SVM performs well on knowledge comprising numerous features of input and output. It's really forgiving of meaningless characteristics. SVM performs best among the classifiers in general, offering better precision. The next SVM arrives, the better output of the neural networks. In performance terms, the table shows the distinction of different algorithms.

5 Literature Survey

A wireless device which was developed by Kovuru Chandu Chowdary [29] is used for calculating oxygen saturation, pulse rate, temperature and blood pressure. The fingerprint sensor is attached to the microcontroller to grant access only to the physician concerned. For visualisation to a web server, acquired data is received from a Raspberry Pi which functions as a sensor node of the provided device. Every 20 seconds, the calculated real time physiological parameters are modified. Zigbee module and GSM interfaces are also available in Raspberry Pi. Once the health data reaches threshold values, an emergency SMS will be sent to the mobile doctor.

E.T.Tan and Z.Abdul Halim [30] implement the AI system to forecast the possible risks such as kidney disease and diabetes as a result. The machine monitors three key fundamental signs, such as internal heat speed, pulse rate, and pulse, high-precision preparation and adjustment of sensor signs to meaningful yield, and showing, for example, the observation results on mobile apps. The technology structure of the medical care sector is broken into a front-end (introduction layer) and a back-end (information access layer). Throughout the front-end sheet, the device coordinates microcontroller-powered sensors to quantify the vital signals of a client and uses Arduino Nano and Intel Edison as discernible yields to travel across the known indicators. An interactive UI (GUI) or software framework is included in the front-end layer that allows the two patients and specialists to access and view the gathered information and data. In the back-end layer, the yield from the information acquisition system is placed on the cloud data set and the prior information expertise is evaluated. Synchronized sensors are connected to the Intel Edison point, and yield readings are sent to IBM Bluemix for the cloud information database and view. The accuracy of the generated model between a healthy human and patients with diabetes and kidney disease is 90.54 percent and 87.88 percent respectively.

An IR sensor has been developed for a smart heart rate control system by Puneet Bansal [31]. The heart rate signal values are built in the range of 1Hz to 3Hz by Arduino Uno processes and philtres. To send the values between Messsage Queuing Telemetry Transport (MQTT) protocol to node-red cloud, the Raspberry Pi is used as an IoT gateway. An SMS and email update is transmitted via the cloud to a physician or another registered individual reminding the person of the heart rate. The real-time heart rate level can be observed in Things Peak.

In order to make early provision for quickly approaching heart failure, AKM Jahangir et al [32] adopted a new tactile method by using a keen IoT that obtained input from the Body Area Sensor. The goal behind the current work was to create a genius coordinated IoT that implemented a lower power correspondence unit such that pulses could unnoticeably accumulate alongside internal heat levels using a mobile phone without blocking their usual everyday life. The sign preparation along with ML techniques was presented for sensor knowledge analysis for detecting and forecasting sudden coronary failure with greater accuracy.

A device has been developed by Vikas Vipapalapalli [33] for calculating the temperature, blood pressure and heart rate by using the Arduino Fio transmitter-receiver and XBee module. The instruments capture and exchange information seamlessly with each other and also archive the information, allowing data to be captured and analyzed. The graphical interface used by LabVIEW for the acquisition, retrieval and dissemination of physiological health data. As a URL is created by labVIEW, which can be accessed from anywhere, the provided device supports real-time monitoring.

Pack Yang et al [34] developed a compact gadget with a bio-detecting facial cover to track the tormenting power of a patient using electromyogram facial surface (sEMG). The wearable gadget serves as a remote sensor centre and is combined into an Internet of Things system for distant surveillance of pain. In order to allow the full recurrence range, up to eight sEMG channels could be examined inside the sensor centre at 1000 Hz and progressively transmitted via the entryway to the cloud worker. Expanding this both poor vitality use and wearing solace are considered by the portable gadget technique for long-term tracking. For continuous gushing of high-volume sEMG information, advanced sign preparation, deciphering, and representation to detail endless pain information remotely to guardians, a portable web application is developed. The cloud stage inside the device supervises interactive contact with the worker and the web application as an interface between the sensor centre and the applications. In summary, this study proposes an adaptable IoT framework for continuous biopotential control and a wearable response for programmed torment evaluation by external appearances.

The Healthcare System for the diagnosis and control of chikungunya virus was developed by Sood and Mahajan [35]. Wearable IoT monitors, such as fitness sensors, location sensors, opioid sensors, meteorological and environmental sensors, the data were collec**ted by sensors. For real-time processing and diagnosis of potentially compromised CHV users, the acquired data is transmitted to the fog and produces warnings to the cell phones of the patient. Every user's results and compiled medical records are processed in the cloud to measure each user's ORI to reflect their chance of transmitting and receiving the infection.

A. Alani [36] has developed a smart healthcare system based on IoT, which uses the Intel Galileo board to monitor blood pressure, pulse rate and temperature. It has an ethernet shield, a 32-bit CPU and Arduino Uno compatibility with its pins. The Arduino IDE is programmed on the Intel Galileo computer. Over the Xampp-based storage server, the board processes and uploads the sensor data. A doctor visualizes the data from a database server using a login ID and password.

Ashwini Gutta [37] has developed an IoT-based health management system for elderly people and chronic patient. A PC, an IoT server and Raspberry Pi are available for a device. The temperature, ECG and the pulse rate sensor data are moved to the Raspberry Pi and then loaded into the database server. The Message Queuing Telemetry Transport (MQTT) protocol is used for correspondence. The state of the patient is deemed serious once the number of email/ SMS notification reaches 3, then the doctor is informed by an email or SMS. A doctor can view clinical records remotely to assess the patient's current health status.

This literature review depicts detecting various chronic diseases by using different sensors, techniques like Machine Learning, Deep learning, Artificial Intelligence on different datasets. Each work has its own strength and weakness. Future work can be focused on the development of framework that detects Heart, Kidney and diabetics diseases using Raspberry Pi by applying Deep Learning algorithm. The framework would provide 2-way chat box so that victims can have direct communication with physician.

6 Comparison Table

Table 2 displays the comparative table for the procedures checked, sensors, equipment, network and data set used, along with the intensity and lack of the relevant work.

Table 2 – Comparison table on techniques, devices, senosrs, network and data set used in survey

Year	Techniques	Sensors used	Device	Network	Data set	Strength	Weakness	Ref.
2018	No algorithm	PR, Temp, BP, Finger Print	Raspberry Pi 3 Model B	Zigbee, Internet	Public data set	This scheme decreases time for action Patient in a low-cost hospital situation.	No optimization algorithm has been used by the machine and it needs to be evaluated on real time results.	[29]
2018	Machine Learning	Temp, BP, PRSensors	Arduino Nano and Intel Edison	Wi Fi	UCI ML Repository	The study provides a method for the normal prediction of possible risks, such as diabetes and kidney infection, for example.	The framework review initiated is a critical review that involves further clinical research and continuous knowledge in order to better authorise the ML concept prior to the coordination of the framework.	[30]
2018	Machine Learning	Finger Tip Sensor	Arduino Uno, Raspberry pi	Internet	Tested on 25 people	The real-time cardiac rate Values on Thingspeak are observed	The device must enhance the utilisation of power, scalability and cost effectiveness.	[31]
2019	Multisensory utilizing a systemwith smart IoT	PR, Temp Sensor	Arduino Uno™, Smart Phone, Wrist band	Bluetooth	Public Data sets	In comparison to ML calculations, the combination of sign handling viably understood the sudden cardiovascular malfunction with higher accuracy.	Since the system is not savvy, the estimate did not improve information security.	[32]
2016	Deep learning	BT, PR, BP Sensors	Arduino Fio	Internet	Self Data	Various critical patient parameters are tracked and distributed to doctors with limited time and low cost.	The ECG sensor can be added to the current device and several entities can also be checked.	[33]
2017	Big Data Processing and Deep Learning	sEMG sensor facial mask	Not mentione d	Wi Fi	Public Data Set	Low use of vitality and compatibility extends the potential for long-haul assessment to track facial cover and sensor centre.	On wide data sets, the device needs to be checked.	[34]
2017	Fuzzy- C means Algorithm	Health, Location, Environmental, Meteorological Sensors	Not mentioned	Internet	Symptom based data is generated	High bandwidth of the Performance, quick warning The generation	Consumption of electricity is High with the use of multiple cameras	[35]
2018	No Algorithm	BP, HR, Temp Sensors	Arduino Uno	Internet	Self Data	In order to save lives in emergency situations, this system is deployed at low cost.	No optimization algorithm has been used by the machine and it needs to be evaluated on real time results.	[36]
2018	Big Data Processing	Temp, HB, ECG, BP Sensors	Raspberry pi	Internet	Self Data	Concentrate on the health of older people and improved management of chronic diseases	For improved results, the optimization algorithm must be applied to	[37]

HR – Heart Rate, BP – Blood Pressure, PR – Pulse Rate, Temp – Temperature Rate, HB – Heart Beat, BO – Blood Oxygen, BR – Breathing, ECG - Electrocardiogram

7 Conclusion

Internet of Things analytics helps companies to capture and interpret sensor data in the environment, which in certain cases offers improved decision-making, lowers maintenance costs, etc., leading to technological growth. Analytics increases performance in the area of IoT healthcare by accurately detecting illnesses and supplying patients with improved medication. We illustrate the relevance of the Internet of Things in the healthcare sector in this article. Continuous wearable tracker tracking of patients helps them to self-care about their own wellbeing. Machine learning algorithms used in analytics help to efficiently classify diseases. Different machine learning algorithms for data classification are also supported. More importantly, we provided an elevated level of representation of different IoT-enabled medical care applications. Nevertheless, we recognize that the targets discovered for IoT in medical care are not efficiently feasible, and there are still various problems to be met and thus this area of research is getting more and more traction. We are also optimistic that from this synergistic approach will come the whole theory of the IoT and its full application of medical care and human prosperity.

References

1. Xu, Q., Ren, P., Song, H. & Du, Q. Security enhancement for IoT communications exposed to eavesdroppers with uncertain locations. *IEEE Access* **4**, 2840–2853 (2016).
2. Scuotto, V., Ferraris, A. & Bresciani, S. Internet of Things: Applications and challenges in smart cities: a case study of IBM smart city projects. *Bus. Process Manag. J.* **22**, 357–367 (2016).

3. Khan, M., Han, K. & Karthik, S. Designing Smart Control Systems Based on Internet of Things and Big Data Analytics. *Wirel. Pers. Commun.* **99**, 1683–1697 (2018).
4. Parvez, B., Haidri, R. A. & Verma, J. K. IoT in Agriculture. *2020 Int. Conf. Comput. Perform. Eval. ComPE 2020* 844–847 (2020) doi:10.1109/ComPE49325.2020.9200035.
5. Paulraj, G. J. L., Francis, S. A. J., Peter, J. D. & Jebadurai, I. J. Resource-aware virtual machine migration in IoT cloud. *Futur. Gener. Comput. Syst.* **85**, 173–183 (2018).
6. Wu, T., Wu, F., Redoute, J. M. & Yuce, M. R. An Autonomous Wireless Body Area Network Implementation Towards IoT Connected Healthcare Applications. *IEEE Access* **5**, 11413–11422 (2017).
7. Jalal, L., Popescu, V. & Murroni, M. IoT architecture for multisensorial media. *2017 Ieee Urucon, Urucon 2017* **2017-Decem**, 1–4 (2017).
8. Gowda, M. *et al.* Bringing IoT to sports analytics. *Proc. 14th USENIX Symp. Networked Syst. Des. Implementation, NSDI 2017* 499–513 (2017).
9. Kim, Y., Park, Y. & Choi, J. A study on the adoption of IoT smart home service: using Value-based Adoption Model. *Total Qual. Manag. Bus. Excell.* **28**, 1149–1165 (2017).
10. Jo, H. & Yoon, Y. I. Intelligent smart home energy efficiency model using artificial TensorFlow engine. *Human-centric Comput. Inf. Sci.* **8**, 1–18 (2018).
11. Jin, Y. & Ji, S. Mapping hotspots and emerging trends of business model innovation under networking in Internet of Things. *Eurasip J. Wirel. Commun. Netw.* **2018**, (2018).
12. Peng, H. *et al.* Secure and Energy-Efficient Data Transmission System Based on Chaotic Compressive Sensing in Body-To-Body Networks. *IEEE Trans. Biomed. Circuits Syst.* **11**, 558–573 (2017).
13. Wang, Y., Kung, L. A., Wang, W. Y. C. & Cegielski, C. G. An integrated big data analytics-enabled transformation model: Application to health care. *Inf. Manag.* **55**, 64–79 (2018).
14. Wang, Y., Kung, L. A. & Byrd, T. A. Big data analytics: Understanding its capabilities and potential benefits for healthcare organizations. *Technol. Forecast. Soc. Change* **126**, 3–13 (2018).
15. Yang, C., Yu, M., Hu, F., Jiang, Y. & Li, Y. Utilizing Cloud Computing to address big geospatial data challenges. *Comput. Environ. Urban Syst.* **61**, 120–128 (2017).
16. Yang, C., Huang, Q., Li, Z., Liu, K. & Hu, F. Big Data and cloud computing: innovation opportunities and challenges. *Int. J. Digit. Earth* **10**, 13–53 (2017).
17. Upadhyaya, J. & Ahuja, N. Higher Education : A Critical Survey and Innovative Model. *Int. Conf. I-SMAC (IoT Soc. Mobile, Anal. Cloud)* 137–140 (2017).
18. Stergiou, C., Psannis, K. E., Kim, B. G. & Gupta, B. Secure integration of IoT and Cloud Computing. *Futur. Gener. Comput. Syst.* **78**, 964–975 (2018).
19. Kraemer, F. A., Braten, A. E., Tamkittikhun, N. & Palma, D. Fog Computing in Healthcare-A Review and Discussion. *IEEE Access* **5**, 9206–9222 (2017).
20. Liu, X., Deng, R. H., Yang, Y., Tran, H. N. & Zhong, S. Hybrid privacy-preserving clinical decision support system in fog–cloud computing. *Futur. Gener. Comput. Syst.* **78**, 825–837 (2018).
21. Rahmani, A. M. *et al.* Smart e-Health Gateway: Bringing intelligence to Internet-of-Things based ubiquitous healthcare systems. *2015 12th Annu. IEEE Consum. Commun. Netw. Conf. CCNC 2015* 826–834 (2015) doi:10.1109/CCNC.2015.7158084.
22. Naqishbandi, T., Sheriff, I. C. & Qazi, S. Big Data, CEP and IoT : Redefining Holistic Healthcare Information Systems and Analytics. **4**, 613–618 (2015).
23. Riggins, F. J. & Wamba, S. F. Research directions on the adoption, usage, and impact of the internet of things through the use of big data analytics. *Proc. Annu. Hawaii Int. Conf. Syst. Sci.* **2015-March**, 1531–1540 (2015).
24. Tan, J. & Koo, S. G. M. A survey of technologies in internet of things. *Proc. - IEEE Int. Conf. Distrib. Comput. Sens. Syst. DCOSS 2014* 269–274 (2014) doi:10.1109/DCOSS.2014.45.
25. Madakam, S., Ramaswamy, R. & Tripathi, S. Internet of Things (IoT): A Literature Review. *J. Comput. Commun.* **03**, 164–173 (2015).
26. Zhang, G., Li, C., Zhang, Y., Xing, C. & Yang, J. SemanMedical: A kind of semantic medical monitoring system model based on the IoT sensors. *2012 IEEE 14th Int. Conf. e-Health Networking, Appl. Serv. Heal. 2012* 238–243 (2012) doi:10.1109/HealthCom.2012.6379414.
27. Yang, L., Ge, Y., Li, W., Rao, W. & Shen, W. A home mobile healthcare system for

wheelchair users. *Proc. 2014 IEEE 18th Int. Conf. Comput. Support. Coop. Work Des. CSCWD 2014* 609–614 (2014) doi:10.1109/CSCWD.2014.6846914.

28. Swan, M. Sensor mania! the internet of things, wearable computing, objective metrics, and the quantified self 2.0. *J. Sens. Actuator Networks* **1**, 217–253 (2012).
29. Chowdary, K. C., Lokesh Krishna, K., Prasad, K. L. & Thejesh, K. An efficient wireless health monitoring system. *Proc. Int. Conf. I-SMAC (IoT Soc. Mobile, Anal. Cloud), I-SMAC 2018* 373–377 (2019) doi:10.1109/I-SMAC.2018.8653716.
30. Tan, E. T. & Halim, Z. A. Health care Monitoring System and Analytics Based on Internet of Things Framework. *IETE J. Res.* **65**, 653–660 (2019).
31. Bansal, P., Malik, M. & Kundu, R. Smart heart rate monitoring system. *2018 IEEMA Eng. Infin. Conf. eTechNxT 2018* 1–4 (2018) doi:10.1109/ETECHNXT.2018.8385347.
32. Majumder, A. J. A., Elsaadany, Y. A., Young, R. & Ucci, D. R. An Energy Efficient Wearable Smart IoT System to Predict Cardiac Arrest. *Adv. Human-Computer Interact.* **2019**, (2019).
33. Vippalapalli, V. & Ananthula, S. Internet of things (IoT) based smart health care system. *Int. Conf. Signal Process. Commun. Power Embed. Syst. SCOPES 2016 - Proc.* 1229–1233 (2017) doi:10.1109/SCOPES.2016.7955637.
34. Yang, G. *et al.* IoT-Based Remote Pain Monitoring System: From Device to Cloud Platform. *IEEE J. Biomed. Heal. Informatics* **22**, 1711–1719 (2018).
35. Sood, S. K. & Mahajan, I. Wearable IoT sensor based healthcare system for identifying and controlling chikungunya virus. *Comput. Ind.* **91**, 33–44 (2017).
36. A. Alani, IoT Based Smart Healthcare System. Int. Journal of Engg Research and Development, Vol 14, Iss 1 (2018).
37. Ashwini Gutte, Ramkrishna Vadali, IoT based Health Monitoring System using Raspberry Pi, IEEE Int. Conf on Computing Communication Control and Automation (ICCUBEA) (2018).

Tcp Syn Flood Attack Detection and Prevention System using Adaptive Thresholding Method

Ramkumar B N[1,*] *and Subbulakshmi* T[1]

[1]School of Computer Science and Engineering, Vellore Institute of Technology, Chennai, Tamil Nadu, India

Abstract. Transmission Control Protocol Synchronized (SYN) flooding contributes to a major part of the Denial of service attacks (Dos) because of the easy to exploit nature of the TCP three way handshake mechanism. Attackers use this weakness to overflow the TCP queue of the server and make its re-sources consumed resulting it to be unavailable for the requests of legitimate users. So we are in need of a quick and precise defence mechanism to detect the TCP-SYN Flood attack. The main objective of the paper is to propose a detection and prevention mechanism of the TCP-SYN flood attack using adaptive thresholding. Adaptive threshold algorithm (ATA) is used to calculate dynamic threshold .Thus this algorithm helps to overcome the limitations of static thresholding like high false positive ratio and also alert users after violation of the threshold calculated by adaptive thresholding algorithm. The result of the suggested mechanism is very effective in the detection and prevention of the TCP SYN flood attack using adaptive thresholding algorithm.

1 Introduction

The recent advancements in technology and the widespread use of the internet have resulted in the need of the internet security at an alarming rate. Denial of service attacks (Dos) one of the dangerous attacks towards the computer network. Dos compromise the availability of the service which is a very important aspect in today's business world. The widespread use of TCP protocol and easy to exploit nature of the TCP three way handshake mechanism has resulted in Dos becoming more common among the cyber-attacks. This paper proposes an effective way of preventing the TCP SYN flood attack using the adaptive thresholding algorithm. This method detects the anomalous TCP requests by monitoring the rate of TCP SYN packets from the attacker to the system.

The TCP three way handshake mechanism is responsible for creating a TCP connection between a client and server. To create a TCP connection a client must send a synchronize flag packet (SYN) to the server as shown in Fig.1. After receiving the SYN packet sent by the client the server sends an acknowledgement flag for the synchronize packet (SYN-ACK) to the client. Upon receiving SYN-ACK flag from the server the client sends an acknowledgement flag to server. After these three stages a connection between both client and server is created and the transformation of data can begin now.

*Corresponding author: bnramkumar492@gmail.com

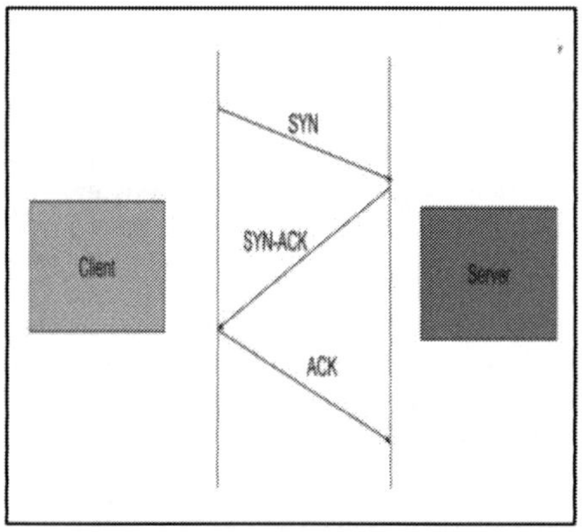

Fig.1 TCP three way handshake mechanism

To perform TCP SYN flood attack on servers, attackers exploit the half opened connection state of the server. This is the state where the server waits for the ACK flag from the client to create a connection. During this state the server would have already allocated memory resources to the client. For exploiting this behaviour attacker sends enormous amount of SYN flags to the server as shown in Fig.2 so that the system would allocate memory resources and wait for the ACK flag from the client which it would never receive. This results in opening of illegitimate half open connections and wastage of memory resources in this server until the session gets expired. During the attack if a legitimate user requests for a connection the server would not respond to the request as all the memory resources are allocated to the illegitimate request from the attacker.

To attack a server with TCP SYN flood attack the attacker sends a large amount of TCP SYN flag from different spoofed IP address to the server. The server takes these requests as legitimate and allocates memory and resources to these IP sources and sends an SYN-ACK flag to the client. The server would now wait in half open-ended state expecting the ACK flag from the client. The large amount of illegitimate SYN requests send by the attacker results in the overflow of the TCP backlog queue and create half opened connection until all system resources are depleted.

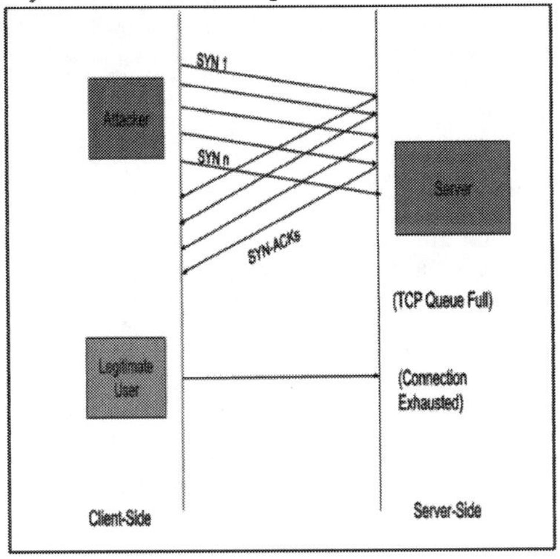

Fig.2 TCP SYN-Flood attack on a server by an attacker.

Due to the overflow of the TCP queue the request made by the legitimate user are not accepted by the server. The main motive of the TCP SYN flood attack is to affect the availability of the system which poses a major threat to the business aspect of the organization.

2 Related works

Several methods for detecting TCP SYN Flood attacks have been proposed. Some of the techniques proposed are the following:

A SYN flood detection system was proposed by Y.Ohista [1] considering the time variation of the incoming traffic. They modelled the arrival of the normal TCP SYN packets into normal distribution where the anomalous requests fail to follow the normal distribution. This method is quick to detect attacks but fails in the case of low time variation attacks because it also follows the normal distribution. H.wang [2] proposed a mechanism which detects the attack at routers instead of the victim's end. This detection mechanism uses a non-parametric CUSUM method to detect TCP SYN floods at a low computational cost. The mechanism considered the nature of the TCP FIN and RST flags to detect the Changepoint effectively. The proposed mechanism provides alerts during the detection of the attack and also reveals the flooding sources location. Schuba [3] proposed that the SYNKILL mechanism is capable for detecting the TCP SYN flood attacks. This mechanism classifies all the incoming packet's IP sources as good or bad. The addresses which are classified bad are sent a TCP-RST packets which would reveal whether the packet from the IP source is spoofed or not. Blazek [4] proposed a method which involves inspection of packets control bits during the observation period using the Cumulative Sum (CUSUM) mechanism for detecting the attack. This mechanism fails in case of a flash crowd where the number of requests over an observation period will be higher than the normal, resulting in a false positive result. Jin and Yeung [5] proposed a covariance analysis model to detect the SYN flooding attack .The mechanism stated that where the attack can be detected by inspecting the degree of correlation between the TCP SYN packets. They used the difference between the correlation of the normal traffic and the traffic during an attack to detect the on-going attack. Siris and Papagalou [6] explored the two statistical anomaly detection algorithms adaptive thresholding and Cumulative Sum algorithm for the detection of the TCP SYN flood attacks. They also provided the suggestion for the performance improvement of the above Changepoint detection algorithms. D. Kshirsagar [7] proposed a mechanism for detecting the TCP SYN flood attack by combining the thresholding and misuse detection approach. The results are measured in terms of CPU workload by comparing the CPU workload during and after detection of the attack.

K.Pai [8] proposed a system where the number of TCP SYN packets is taken as a metric to determine the flooding attack. The number of SYN packets in a particular interval of time from a source is greater than the threshold is considered as an attack. This system can detect the attack precisely but implementing the system for large networks is very difficult. The authors of [9] proposed an efficient methodology for preventing the TCP SYN flooding attacks using the iptables firewalls. They have also explained about various functionalities of the IPtables firewall and making firewall rules for preventing the attacks. The authors of [10] presented a detection mechanism where the SYN flood attack is detected by monitoring the anomalous TCP handshakes between the client and server. This method employs the Cumulative Sum (CUSUM) algorithm for detecting the change. Nakashima and Sueyoshi [11] proposed a method which uses the packet loss rate of the TCP packets as a metric to detect the difference between the normal and abnormal traffic flow during the attack. Lu et al. presented a new framework to effectively identify packets which are compromised. It uses a perimeter-security based Distributed Denial of Service (DDoS) prevention system to classify compromised packets at the router end [12]. The authors of [13] proposed real time architecture for DDos detection using cluster analysis. In this method, the authors extracted particular features from a DDoS architecture and selected variables based on the feature. Wei et al. [14] proposed a detection mechanism using the rank correlation (RCD) of the incoming traffic to detect the change between the normal and abnormal traffic

https://doi.org/10.1051/itmconf/20213701016

through the network. S.H.C.Haris et al. [15] proposed a method in which the network is monitored for anomalies in the payload of the incoming packets to detect TCP SYN flood attacks.

3 Environment Setup for the proposed architecture

Proposed TCP-SYN Flood attack detection and prevention system using adaptive thresholding method architecture is implemented in Linux operating systems.

3.1 Python

Python is a general purpose high level programming language which follows an object oriented programming approach. Python programming language is easy to learn and also provides the advantage of readability. Python is well known for its packages and modules which provides code reusability and program modularity. Python is coming pre-installed with operating systems like Linux and Mac. Since windows don't have Python pre-installed, it must be installed explicitly. In Windows, there's no universal library for installing Python, so it must be downloaded like all other GUI applications.

3.2 Scapy

Scapy is a python programming library supported by Python and its later versions. It is used for analysing the packets on the network. It has the ability to decode or forge packets, capture them, send them on the wire, and match requests and replies. It can also handle tasks like probing, unit tests, attacks, and network discovery, scanning, tracerouting [16]. This python library can be used to develop more advanced tools related to network security and ethical hacking. As scapy library is not included with Python libraries by default, scapy python libraries can be installed using pip for our proposed architecture.

3.3 IPtables

IPtables is user-space utility software that empowers a system administrator to configure the rules of IP packet filtering using the Linux kernel firewall, implemented as different Netfiltering modules. The filters are categorized in different tables, which consist of chains of rules for how to deal with network traffic packets [17].IPtables are pre-installed with newer versions of Linux operating systems. If not installed in the system, users can install it through apt-get from the repositories of their own flavours of Linux. IPtables can also be installed in Solaris operating system as a substitute for a firewall.

3.4 Hping3

Attack generation module uses hping3 tool for creating the TCP SYN Flood from the attack generation module to detection module.hping3 is a powerful to create Dos attack against systems [18]. It is one of the most common tools used for testing of the network security of an organization.hping3 tool should be executed for simulating the TCP SYN flood attack. Hping3 can be installed as a source tarball from the project website. Installation for Debian or Ubuntu operating system can be done either with apt-get or Synaptic Package Manager.

3.5 Smtplib

The architecture model is implemented in python so the proposed architecture requires the installation of python SMTP Libraries for sending mails to the administrator [19]. Python offers smtplib module, which creates an SMTP client session object which can be used to send emails. The smtplib module for sending mails can be installed through pip command. As it is supported with python it can be used in all types of operating systems including Windows and mac OS.

4 Proposed TCP-SYN Flood attack detection and prevention system.

Proposed TCP-SYN Flood attack detection and prevention system using adaptive thresholding method architecture is mainly based upon the anomalous TCP hand-shakes behaviour. The detection and prevention mechanism architecture is divided into five modules: the attack generation module, the detection module, prevention module, adaptive thresholding module and an alert module as shown in Fig. 3. The detection and prevention is done at the end of the victim's computer. The detection module in the architecture is responsible for detection and classification of the network traffic. The detection module collects all the necessary information for detection from the incoming network traffic and analyses the data and classifies whether traffic is normal or not using the detection rule. It analyses the information collected by the sniffing process and, according to the adaptive threshold value provided by the adaptive thresholding module, it makes the decision to classify a request into malicious or not.

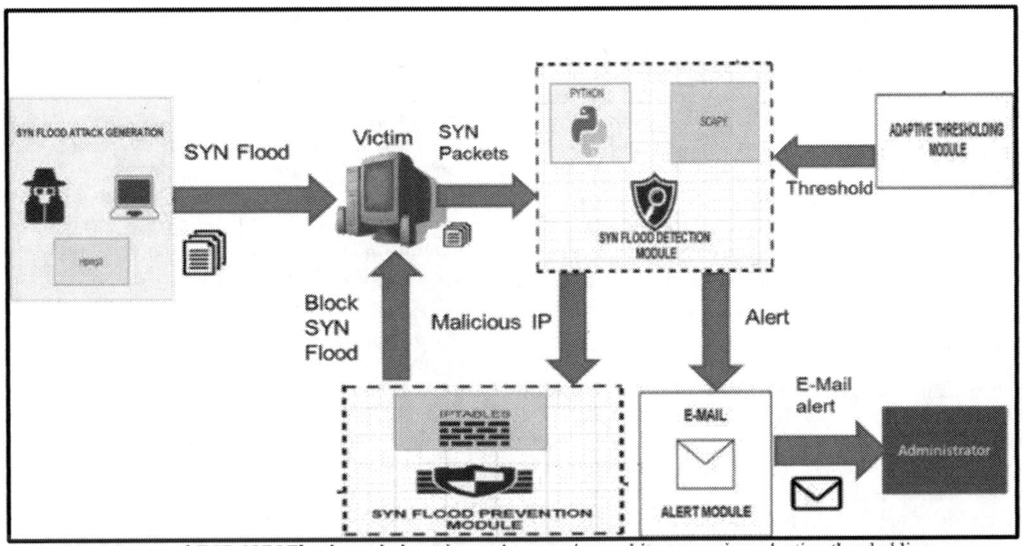

Fig.3 Proposed TCP-SYN Flood attack detection and prevention architecture using adaptive thresholding.

The prevention module is triggered when it gets the alert from the detection module. It uses the information from the alert messages to prevent the SYN flooding attack by using IPtables. The alert module uses Simple Mail Transfer protocol (SNMP) to alert the possibility of the on-going attack. Thus reducing the catastrophe of the flooding attack by deciding earlier whether to prevent the attack or to avoid false positive from the detection module.

4.1 Detection Module

As it is stated before, to detect Denial of Service (DOS) attacks, the module monitors the network traffic. So the proposed architecture needs information related to every network traffic from a IP source differentiate normal from abnormal conditions .The detection module needs two important metrics to detected whether a request is anomalous or not:

- The detection module requires an adaptive threshold for classifying the traffic.
- The number of TCP connections made by a source.

In this proposed detection method it uses the number of requests made by a single source within the specified time interval to classify whether a TCP SYN packet is malicious or not using the adaptive threshold generated by the algorithm. The detection module analyses the source IP address and number of requests made by the source. If the number of requests within a given interval of time exceeds than that of the threshold the detection mechanism classifies this as an anomalous request. As it is known, getting thousands of TCP-SYN requests within a minute from a single IP source is not usual.

4.1.1 Algorithm for detection of TCP-SYN flood attack.

The algorithm for detection of TCP SYN flood attack using the adaptive threshold value generated by adaptive thresholding algorithm and uses technologies like scapy and python to detect the ongoing attack is shown in Fig.4. It actively analyses the incoming TCP packets to the system to classify whether the packet is malicious or not.

1. Start the detection module.
2. Get the adaptive threshold value from the adaptive threshold algorithm.
3. Sniff the incoming packets using scapy.
4. Check for the packet protocol.
    1. If the packet has a TCP layer, proceed with the next step.
    2. Else ignore the packet.
5. Check for the number of packets sent by the source IP address within the time interval.
    1. If the number of requests exceeds the threshold, proceed with the next step.
    2. Else ignore the packet.
6. Trigger the alert module to send the details of the attack to the user.
7. Trigger the prevention module to stop the attack.
8. Stop if the program is closed.

Fig.4 TCP-SYN Flood attack detection algorithm

4.2 Adaptive thresholding module

The adaptive thresholding module provides the threshold value for the detection module based on adaptive thresholding algorithm. The threshold provided to the detection module should be very precise if not it would affect the performance of the system. The value of threshold is set adaptively based upon the seasonal, monthly or daily usage where an estimate of SYN packets mean is computed from recent measurements of network traffic. The threshold can also be calculate using the EWMA(Exponentially Weighted Mean Average) method .In Equation 1 ,yn is the count of SYN packets in the n-th interval of time, and $\bar{x}$n-1 is the mean rate of the traffic computed from measurements prior to n, then the anomalous traffic condition is

If $y_n \geq (\beta + 1)\, \bar{x}_{n-1}$ then the request is malicious (1)

Where $\beta > 0$ is the percentage of packets that can be allowed greater than the adaptive threshold value before classifying it as an anomalous traffic.

4.3 Prevention Module

The prevention module plays a vital role in the architecture as it prevents the attack from happening further. The prevention module is triggered by the detection module when it finds any anomalous TCP requests. The prevention module uses IPtables for blocking the requests from the malicious IP addresses. IPtables is user-space utility software that empowers a system administrator to configure the rules of IP packet filtering using the Linux kernel firewall, implemented as different Netfiltering modules. The filters are categorized in different tables, which consist of chains of rules for how to deal with network traffic packets [17].This detection module gets an IP address as an input from the detection module. This malicious address is blocked by the module from performing any request to the host. This is done by a OS system call by the prevention module to the operating system to add the drop action rule for the malicious IP address using IPtables.

4.3.1 Algorithm for prevention of TCP-SYN Flood attack.

The algorithm which is responsible for the prevention of TCP SYN flood attack is shown in Fig.5. It uses the detected malicious IP address from the detection module to prevent the attack. It uses technologies like python and IPtables to block the malicious requests.

1. Start the prevention module.
2. Check for triggers from the detection module.
   a) If malicious activity detected continue with next steps
   b) Else ignore the next steps.
3. Get the IP address sent by the detection module.
4. Block the IP address by adding it to the iptables rules by setting option DROP.
5. Check for a stop signal.
   a) If found, proceed with the next step.
   b) Else repeat from step-2.
6. Stop the prevention module.

Fig.5 Algorithm for alert module of the proposed architecture

4.4 Alert module

The alert module proposed in this architecture provides a real time solution to avoid catastrophic results from the actual flooding attack and also prevents from the business loss due to false positive results from the detection algorithm. It uses Simple Mail Transfer Protocol (SMTP) for sending the alert to the system administrator in real-time of the attack. So that administrators can act accordingly to prevent from effects of SYN flooding attack.

4.4.1 Algorithm for alerting the TCP-SYN flood attack

The algorithm for the alert module used in the proposed architecture is shown in Fig.6. It uses the detected malicious IP address from the detection module to alert the user about the attack. It uses common technologies like python, Simple Mail Transfer Protocol (SMTP) for sending email alerts to the administrators.

1. Start the alert mail.
2. Configure SMTP and relevant details like receiver's address.
3. Check for the trigger from the detection module.
   a) If triggered, execute following steps.
   b) Else ignore next steps.
4. Get the details of the attack from the detection module.
5. Wrap the details as a message with the protocol header of mail.
6. Send the mail to the system administrator about the attack.
7. Stop the alert module.

Fig.6 Algorithm for alert module of the proposed architecture

4.5 Attack generation module

The attack generation module in the proposed architecture is responsible for creating a real world attack scenario. This helps us in testing our detection, prevention and alert modules. The attack generation module sends enormous number of TCP SYN packets to the detection machine for testing the effectiveness of the proposed architectures algorithm. Attack generation module gets the IP address of the victim's machine to flood the machine with thousands of SYN packets per second. Attack generation module uses hping3 to generate TCP SYN flood to the detection machine for testing.

5 Implementation

To implement the proposed architecture, simulation of a real world scenario of TCP SYN flood attack is to be created. The attack generation module is launched to flood the detection machine. After launching the attack hping3 would have started sending numerous amounts of TCP SYN packets to the detection machine. The main program launches the execution of the detection, prevention, adaptive thresholding and alert module. The adaptive thresholding module returns the threshold value to the detection module. Now the detection module analyses the packet flow in the network and stores the information and counts the number of requests made by a single source. The detection module could detect the anomalous request being made by the hping3 tool by verifying with the thresholding amount and the number of requests made by a single source IP. Now this detection will trigger the prevention module. The prevention module works based on IPtables. It gets the malicious IP address as a input from the detection by module. Then this IP address is blocked by the prevention module by adding drop IPtables rule by invoking a system call to the operating system. This prevents the TCP-SYN flood attack from happening again from the malicious IP address. This prevention module triggers the alert module that sends the details of the attack to the system administrator via Email in real time. Then the administrator can act accordingly to the attack to avoid prevention of false positive results. In this way the TCP-SYN flood could be detected and prevented.

6 Results

The proposed architecture's experimental results and analysis of the TCP SYN flood attack detection and prevention system is stated in the below section. From the above modules of the architecture we can infer that.

6.1 Attack generation

Attack generation phase simply involves running the hping3 tool residing in the attacking module. This hping3 takes in the IP and port of the target machine along with the number of packets which the target has to be flooded with. As soon as the attack was started, the packets started getting produced and sent to the target machine.

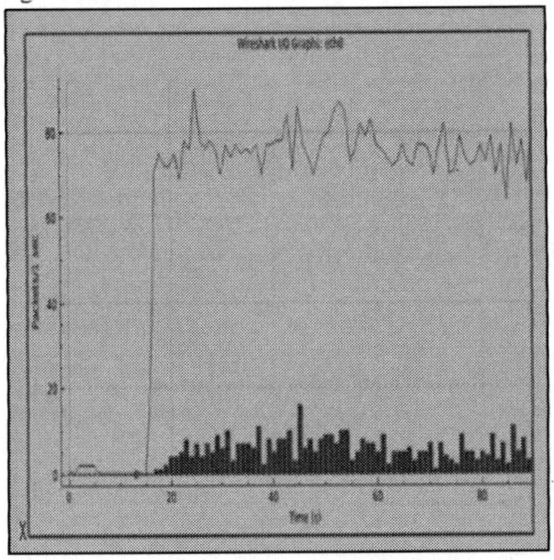

Fig.7 Attacker's network traffic graph during the SYN flood attack.

The I/O network graph of the attacker's machine during the TCP SYN flood attack is shown in Fig.7 which clearly states that the number of packets sent per second increases to an abnormally large value on the eth0 interface. Also from the rapid spike in the graph at around the 15 second mark as compared to the normal traffic which was captured before the attack was generated.

6.2 Detection and Prevention

The efficiency of this module is stated in terms of the CPU workload of the system before attack, in course of the attack and after detection of the attack. The CPU workload before attack, during attack and after TCP SYN flood detection and prevention by blocking the malicious IP address is shown in Fig 8 .The value of the of the CPU load under normal condition before the attack ranges from 7-12. The CPU load value observed during the attack ranges from 96-100 because of the malicious incoming requests. After detection of the attack the CPU load ranges from 7-13 which is similar to the CPU load of the system before the attack. Figure below clearly depicts that this system is capable of detecting and preventing TCP SYN Flood attack efficiently which also reduces the CPU's workload after detecting TCP SYN flood attack and restores the system back to the normal state prior to that of the attack.

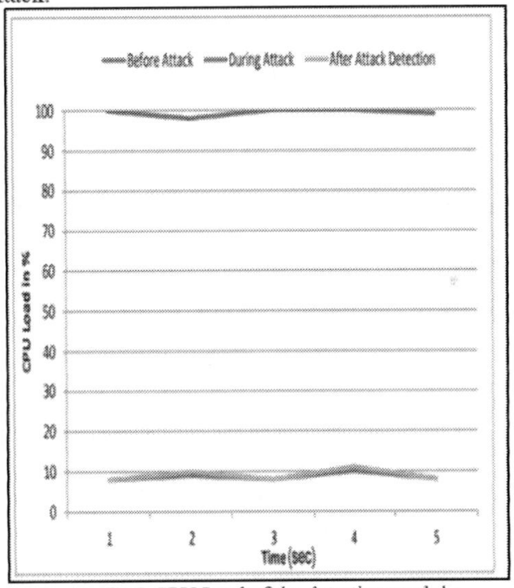

Fig.8 CPU Load of the detection module

The time taken by our proposed algorithm to detect and prevent TCP SYN flooding algorithm under variable SYN requests per seconds is shown in Table 1 when threshold is set to 300 SYN requests. So we can see the detection time reduces as the number of SYNs per second increases. This clearly states that our algorithm can effectively detect large SYN flood attacks.

Table.1 Time taken for detecting the SYN flood under different SYN rates

SYN s/ s	Detection Time
35	8.591
50	6.011
60	5.750
80	3.750
100	3.132

IPtables rules are updated after the detection of the SYN flood as shown in Table 2. The prevention module adds the malicious IP addresses to the IPtables which helps in preventing the attack happening further. The packets further after detection from the malicious source is dropped by the IPtables which prevents the TCP-SYN Flood attack.

Table.2 IPtables rules after prevention of the TCP SYN Flood attack.

Num	Target	Prot	Opt	Source	Destination
1	DROP	all	--	192.168.229.129 (attacker IP)	anywhere
2	DROP	all	--	192.168.229.234 (attacker IP)	anywhere

The results shows that our algorithm is very effective and efficient for detecting and preventing TCP-SYN Flood attacks at a higher rate also the performance of the algorithm can be increased to great extent by introducing and tuning various thresholding parameters according to the scenario.

7 Conclusion and Future Works

In this paper, the proposed architecture is capable of an effective and efficient detection and prevention of SYN flood attacks. It consists of four modules detection, adaptive thresholding, prevention and alert modules and the relevant metrics (IP source, number of requests made, time of previous request and SYN arrival rate). The detection module is efficient because it requires less computation and memory resources of the system .The adaptive thresholding module increases the efficiency of the detection module by providing a precise threshold. The prevention module works well in preventing the attack. Using this result, the proposed architecture described an attack detection and prevention method taking the IP address and number of requests made by a IP source as a metric to classify the incoming traffic .The results states that our proposed architecture can effectively detects and prevents SYN flooding attacks faster and precisely.

The future works are to optimize and add some additional parameters (e.g., variance, standard deviation and other statistical and network parameters) to the adaptive thresholding algorithm so the detection module can get more precise threshold values for detection hence improving the performance of the module. Also the proposed architecture can be extended for other denial of service attacks (HTTP Flooding, ICMP Flooding, UDP flooding) [20] using this adaptive threshold based detection algorithm.

8 References

1. Ohsita .Yuichi, Ata. Shingo, Murata, Masayuki, Detecting Distributed Denial-of-Service Attacks by Analyzing TCP SYN Packets Statistically, Ieice Transactions, Vol.4. (2006).
2. Wang, Haining, Zhang, Danlu, Detecting SYN flooding attacks. Proceedings of IEEE INFOCOM. Vol **3**, (2002).
3. Schuba. CL, Krsul. IV, Kuhn. MG, Spafford. EH, Sundaram. A, Zamboni D, Analysis of a denial of service attack on TCP. In Proceedings of the IEEE Symposium on Security and Privacy, (2017).
4. Blazek. RB, Kim. H, Rozovskii. B, Tartakovsky. A, A novel approach to detection of denial of service attacks via adaptive sequential and batch sequential change point detection methods. In Workshop on Information Assurance and Security, IEEE, June (2001).
5. Jin. S, Yeung. DS, A covariance analysis model for DDoS attack detection. In IEEE International Conference on Communications, Vol. **4**, 1882–1886, (2004).
6. Siris. VA, Papagalou. F, Application of anomaly detection algorithms for detecting SYN flooding attacks. In IEEE, GLOBECOM, December (2004)
7. D. Kshirsagar, S. Sawant, A. Rathod, S. Wathore, CPU Load Analysis & Minimization for TCP SYN Flood Detection ,Procedia Computer Science 85, (2006).
8. K. Pai, N. HR, A. Bhat, "Detection and Performance Evaluation of DoS/DDoS Attacks using SYN Flooding Attacks, International Journal of Computer Applications, (2014).
9. Mirzaie. Sara, Elyato. Alireza , Sarram. Mehdi Agha, Preventing of SYN Flood Attack with IPtables Firewall, ICCSN, (2010).
10. Bellaiche. Martine, Grégoire. Jean-Charles, SYN flooding attack detection by TCP handshake anomalies, Security and Communication Networks, Vol **5**, (2012).
11. Nakashima. T, Sueyoshi. T, Performance estimation of TCP under SYN flood attacks, First International Conference on Complex, Intelligent and Software Intensive Systems, (2007).
12. K. Lu, D. Wu, J. Fan, S. Todorovic, A. Nucci,Robust and efficient detection of DDoS attacks for large-scale Internet, Comput. Netw., vol. **51**,(2007).
13. K. Lee, J. Kim, K. H. Kwon, Y. Han, S. Kim, DDoS attack detection method using cluster analysis, Expert Syst. Appl, vol. **34**, (2008).

https://doi.org/10.1051/itmconf/20213701016

14. W. Wei, F. Chen, Y. Xia, G. Jin, A rank correlation based detection against distributed reflection DoS attacks, IEEE Commun. Lett, vol. **17**, Jan, (2013).
15. Haris. S, Ahmad, R. Badlishah, Abd. Ghani, Mohd, Waleed, Ghossoon, TCP SYN flood detection based on payload analysis, SCORED, (2010).
16. Rohith, R. Yadav, Rohith, Moharir, Minal, Shobha. G, SCAPY- A powerful interactive packet manIPulation program, (2018).
17. Othman, Mohamed, Kermanian, Mostafa, Detecting and preventing peer-to-peer connections by Linux IPtables, ITSIM, (2008).
18. Kaur. H, Behal. S, Kumar. K, Characterization and comparison of distributed denial of service attack tools (IEEE), (2015).
19. Banday, M. Tariq, Qadri, Jameel, Shah, A Practical Study of E-mail Communication through SMTP, (2010).
20. Zlomislic, Vinko, Fertalj, Krešimir, Sruk, Vlado, Denial of service attacks: An over-view. Iberian Conference on Information Systems and Technologies, CISTI, (2014).

A Deep Learning Model to predict the Industry Readiness of Engineering Students Community

Ashok Murugesan[1], Kumar Ramasamy[2], Umadevi Ashok[3] and Revathy Pandian[4]*

[1]Assistant Professor/CSE, Rajalakshmi Institute of Technology, Chennai.

[2] Professor/CSE, Dhirajlal Gandhi College of Technology, Salem.

[3] Assistant Professor/MBA, SRMValliammai Engineering College, Chennai.

[4] Professor/MBA, Velammal Engineering College, Chennai.

Abstract:Industry readiness of Engineering students community is a big challenge in the recent campus recruitments. 21st century skills are completely mapped with the technical and non – technical knowledge background of the engineering graduates. In this paper the work narrated the process of identifying the parameters for skill assessment of the candidates and derived a learner model using deep learning framework. Further the model can be used to predict the employability readiness of candidates.

1. Introduction

Campus placements are the vital key indicator for the students to choose their professional course of study. After enrolling in the course, the orientation of students towards job profiling depends on the job market. In this work, we are considering engineering students community. During their pre final and final year courses of study, their complete starvation sticks to placement oriented programs. Their preparation comprises of both technical as well as non technical domains. Such programs are focusing very rarely in problem solving domain. With respect to the statistics, 10% of students landing in high band packages, 70% of students landing in average band packages and 20% students are moving for higher studies or exiting from the campus without placements.

Our work is ready to concentrate with 20% of students exiting campus without placements. From their second year of course onwards, 21st century skill based parameters must be explained and train them using the outcome of the model derived. A parameterized framework can be designed using the convolutional neural network and analyze the outcome of the students in the regular intervals. This will enhance the placement opportunity for the identified community.

Ronal et al[1] provided the results of their tool by supporting self learning with course engagement. Kaveri et al[2] correlated the parameters for learner's engagements. Allison et al[3] illustrated personal learning portfolio of different types of candidates. Zahra et al[4] conducted an experimental investigation with 30 students and conveyed the behaviour of learners with their emotions using non survey based system. Lin et al[5] performed a comparative survey of different types of feedback based learner models. Rajendran et al[6] discussed the frustrations of learners when the academic experts providing feedback to them. In turn he also conveyed about the role of stress in different types of learning environment. Tran and Meachean[7] summarized the routines for enhancing the learner's experience using extended learning systems. They have incorporated test, analytics, administration and flipped learning in their framework. Muljo et al[8] produced a prototype for online learning pertaining to higher education. Technology acceptance model designed by them had yielded a fruitful measure in 21st century skills of the students. Rahila et al[9] depicted a model to predict the growth of learner's performance by assessing the attributes of the learners in the weekly assessments. Ananthi and Nazreen [10] illustrated the user's experience in learning management systems. Also they have identified the behavioral analytics of the interdisciplinary users of the system. Justian and Esther[11] listed the systematic survey of learning teachnologies in terms of learning environments, learning process and learning outcome. Bart et al[12] created a learning frameworks which evaluates the students community using design and evidence based research. Through this they are analysing the positive impact of learner's attitudes.

*Corresponding author: ashok.m@ritchennai.edu.in

2. Literature Survey

An observation[1] of course attainment was mapped with self learning in a regulated fashion. Sample size of the experiment was restricted to 263 learners. They had used four parameters to assess self learning. It had reflected a positive correlation with the attempt. They had a zero significance in learner behavior management. A hypothesis[2] based approach was elaborated with learner's engagement. Sample size of the experiment was restricted to 59 users. They had used thirteen parameters for learner analysis. They had extracted four parameters as mandate through feature extraction. Their empirical results demonstrated the outcome of the framed hypothesis.

A complete evaluation[3] of learner behavior while engaging a specific course was discussed. Sample size of the experiment was restricted to 750 users. They had used five parameters for learner's engagement analysis. They had derived the descriptive statistics of aforementioned five attributes. A fuzzy based approach[4]followed to analyze the emotions of learners. They had designed two questionnaires to retrieve the attributes of learners. Sample size of the experiment was restricted to 32 users. They had used six parameters to narrate their research flow.

Three different data models[5] were focused. They were linked with analytics based front end. Six different parameters were analysed to measure the learning strategies.70K online learners were tracked. Seven different parameters were identified for the learners and projected in the dashboard. An UI[6] based interactive tool was discussed. It had received comments from all the learners and respond with the learners had frustrated experience.600+ school students were involved in this exercise. It witnessed the outcome of the learner's attempt on the static goals.

Extended learning systems[7] was built with four different zones. Learner's behavior was assessed .Integrated curriculum design was focused. Adaptive course management had issued its provision. A model for flipped learning to incorporate the aforementioned three parameters. Sample size of the experiment was restricted to 60 learners. A prototype[8] for online based higher education was discussed. Five parameters were designed to analyze the learner in online mode. Sample size of the experiment was restricted to 200 users. They have tried to solve the specific challenges in online learning too.

Three different machine learning[9] algorithms were incorporated to predict the test performance of the learners. They had used six variables to trace the history for the learning. They had repeated the experiment for 16 weeks to gather the data for complete analysis. They had incorporated both online and offline activities. Online activities served as the catalyst for the research. Size of the experiment was restricted to 99 users. User experience[10] in the learning domain played a vital role for complete involvement and engagement in course learning. Adaptive learning based data were fed for the learner based on the learning trend. They have employed data mining techniques to analyze the user experience and recommend the upgrades for the subsequent learning domain.

A systematic literaturel[11] review about the learning strategies, evaluations and implementation beds were discussed. They had organized the literature through three different sections. Section 1 narrated the methods of evaluation. Section 2 illustrated the behavioural aspects of learner in the learning domain. Section 3 depicted the limitations and recommendations of the analytics in technology based learning enhancement. Learning[12] analytics were captured in terms of learning strategies, framework of evaluation and predictive modeling. Sample size of the experiment was restricted to 1076 users. They had suggested topics to improve the phase of scoring. They had elaborated the process of strategy shifting from types of learning groups.

3. Methodology

3.1. Identification of Parameters

The parameters responsible for industry readiness were Productivity Rate, Self Learning, Responsible Engagement and Skill for different types of learning community. To create the framework, the parameters were treated as Usage Quotients (UQ). Every Parameter was described as the ratio of two or more numerical attributes. It would be easy to perform computations or predict using the data sets.

Productivity(PR) = Time spent for Professional growth per day/14 hours

Self Learning Quotient(SLQ) = A+B where A & B reflects(1)

A= % of syllabus covered through self learning
B= % of content beyond the syllabus learning

Responsible Engaging Quotient (REQ) = C/D (2)

C=Total Number of Professional Activities Participated

D=Total Number of Professional Activities conducted per Academic Year

Skill Quotient(SQ) = (E+F)/G (3)

 https://doi.org/10.1051/itmconf/20213701017

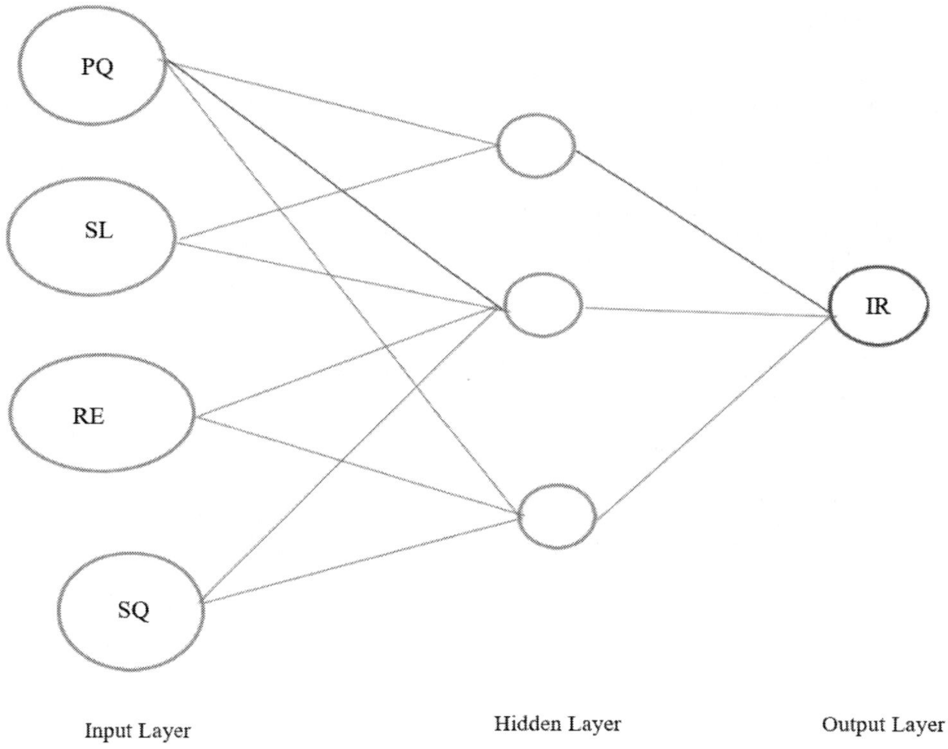

Fig 1 . ANN Model

E=Number of Problems narrated properly

F=Number of Problems methodology suggested

G=Total Number of Problems provided as a part of Placement Curriculum

3.2. Parametric Framework

Industry Readiness Quotient (IRQ) was defined as the variant for storing the influence of (1), (2), and (3)

$$IRQ = PQ+SLQ+REQ+SQ \quad (4)$$

The results of (4) can be expressed as continuous function and narrated as follows.

f (IRQ) = {IRQ>=2, Subject has the impact of Placed Category; Otherwise, Nonplaced Category.

3.3. Learner Modeling

900 students per year were involved in campus placements. The model is trained using 10years data. The datasets belong to the non autonomous private engineering college affiliated to a state university in Tamilnadu.

4. Experimental Investigation

The following table 1 represents the Actual statistics and table 2 represents the comparative study.

Year	Total Students	%of Placed Candidates	%of Non Placed Candidates
2011	895	74	13
2012	900	82	8
2013	890	79	20
2014	892	67	20
2015	888	84	12
2016	894	92	7
2017	891	89	10
2018	892	88	11
2019	897	93	6
2020	893	91	8

Table 1. Actual Statistics

Year	Total Students	Actual number Of students placed	Outcome of the parametric model	Outcome of the learner model
2011	895	74	68.08	66.6
2012	900	82	75.44	73.8
2013	890	79	72.68	71.1
2014	892	67	61.64	60.3
2015	888	84	77.28	75.6
2016	894	92	84.64	82.8
2017	891	89	81.88	80.1
2018	892	88	80.96	79.2
2019	897	93	85.56	83.7
2020	893	91	83.72	81.9

Table 2. Comparative Study

Actual statistics of students as mentioned in Table 1.1 are compared with parametric model and learner model. The accuracy of the learner model during training phase reached 92%.The parametric evaluation also reached 90% accuracy when compared with the actual statistics in Table 1.2.Students opted for higher studies are omitted from this investigation.

11% variation observed when we had compared the results of our model with Skill Assessment Software of XYZ Private Limited. By considering the legal issues, the name of the company was renamed.

Low level and high level indicators[1,5] of granularity could be anlaysed by our parametric model in terms of their engaging parameters. Different types of learner modeling [2,6] could be estimated through the proposed parametric and outcome based models. Self regulation parameters[3,7] could be correlated with the proposed parameters and feature extraction could be done. Emotional modeling[4,8] could be enhanced with the proposed parametric modeling.

5. Conclusion & Future Work

The learner model will be rebuilt with different architectures of neural network. The issues of over fitting and under fitting on the designed model need to be analysed with larger datasets. Through this system, the portion of non-placed students can be filtered and properly monitored during their initial course of study itself.

References

[1] R. Antonio Pérez-Álvarez, J. Maldonado-Mahauad, K. Sharma, D. Sapunar-Opazo and M. Pérez-Sanagustín, "Characterizing Learners' Engagement in MOOCs: AnObservational Case Study Using the NoteMyProgress Tool for Supporting Self-Regulation," in IEEE Transactions on Learning Technologies, vol. **13**, no. 4, pp. 676-688, 1 Oct.-Dec. 2020, doi: 10.1109/TLT.2020.3003220.

[2] Kaveri, S. Gunasekar, D. Gupta and M. Pratap, "Decoding Engagement in MOOCs: An Indian Learner Perspective," 2016 IEEE Eighth International Conference on Technology for Education (T4E), Mumbai, 2016, pp. 100-105, doi: 10.1109/T4E.2016.027.

[3] Allison Littlejohn, Nina Hood, Colin Milligan, Paige Mustain,Learning in MOOCs: Motivations and self-regulated learning in MOOCs,The Internet and HigherEducation,Volume**29**,2016,Pages 40-48,ISSN 1096-7516,https://doi.org/10.1016/j.iheduc.2015.12.003.

[4] Z. Karamimehr, M. M. Sepehri and S. Sibdari, "Automatic Method to Identify E-Learner Emotions Using Behavioral Cues," in IEEE Transactions on Learning Technologies, vol. **13**, no. 4, pp. 762-776, 1 Oct.-Dec. 2020, doi: 10.1109/TLT.2020.3020497.

[5] K. Liu, S. Tatinati and A. W. H. Khong, "Context-Based Data Model for Effective Real-Time Learning Analytics," in IEEE Transactions on Learning Technologies, vol. **13**, no. 4, pp. 790-803, 1 Oct.-Dec. 2020, doi: 10.1109/TLT.2020.3027441.

[6] R. Rajendran, S. Iyer and S. Murthy, "Personalized Affective Feedback to AddressStudents' Frustration in ITS," in IEEE Transactions on Learning Technologies, vol.**12**, no. 1, pp. 87-97, 1 Jan.-March 2019, doi: 10.1109/TLT.2018.2807447.

[7] T. P. Tran and D. Meacheam, "Enhancing Learners' Experience Through ExtendingLearning Systems," in IEEE Transactions on Learning Technologies, vol. **13**, no. 3, pp. 540-551, 1 July-Sept. 2020, doi: 10.1109/TLT.2020.2989333.

[8] H. H. Muljo, A. S. Perbangsa and B. Pardamean, "Online learning prototype for higher education," 2017 International Conference on Information Management andTechnology (ICIMTech), Yogyakarta, 2017, pp. 49-53, doi: 10.1109/ICIMTech.2017.8273510.

[9] R. Umer, T. Susnjak, A. Mathrani and S. Suriadi, "A learning analytics approach: Using online weekly student engagement data to make predictions on student performance," 2018 International Conference on Computing, Electronic and ElectricalEngineering (ICE Cube), Quetta, 2018, pp. 1-5, doi:

 https://doi.org/10.1051/itmconf/20213701017

10.1109/ICECUBE.2018.8610959.

[10] Sheshasaayee and M. N. Bee, "Evaluating user experience in moodle learning management systems," 2017 International Conference on Innovative Mechanisms for Industry Applications (ICIMIA), Bangalore, 2017, pp. 735-738, doi: 10.1109/ICIMIA.2017.7975562.

[11] J. Knobbout and E. Van Der Stappen, "Where is the Learning in Learning Analytics? A Systematic Literature Review on the Operationalization of Learning-Related Constructs in the Evaluation of Learning Analytics Interventions," in IEEETransactions on Learning Technologies, vol. **13**, no. 3, pp. 631-645, 1 July-Sept. 2020, doi: 10.1109/TLT.2020.2999970.

[12] Rienties B., Cross S., Zdrahal Z. (2017) Implementing a Learning Analytics Intervention and Evaluation Framework: What Works?. In: Kei Daniel B. (eds) Big Data and Learning Analytics in Higher Education. Springer, Cham. https://doi.org/10.1007/978-3-319-06520-5_10

Internet of Things in Agriculture to Revolutionize Traditional Agricultural Industry

Dr. Kirubanand VB[1,], Dr. Rohini V[2] and Laxmankumar. V[3]*

[1,2] Department of Computer Science, CHRIST (Deemed to be University) Bangalore-560029
[3]SimplyTurn Tech, Hyderabad-500085

Abstract. Agriculture is familiarly called "Farming". Agriculture is the basic art to cultivate food which is a necessary need to every living individual. Agriculture needs the practice of science for cultivating the soil factors and growing crops. In traditional farming, it includes more labor work and less yield quantity. This demerit can be overcome by the modern farming techniques which makes use of the advanced technology and focuses on maximizing the yield and maintaining the quality. Earlier the farmers used to figure out the type of the soil based on their suspicion and they would never think of the humidity, temperature, climatic condition and especially the level of water. IoT is trying to overcome all these factors by helping to assemble the information. This paper focuses on the soil moisture and soil type which lets the farmer know about the type of crops to be grown.

1 Introduction

Agriculture is the backbone of India. Food and food grains are essential for the livelihood. The demand for food is a challenging task for every being, as there is rapid climate change, extreme weather conditions and environmental impacts which impacts on farming practices. The ever-growing global population would touch around 9.6 billion by 2050[14,17]. So, to feed this immense population, the agriculture industry needs to embrace IoT. Smart IoT technology farming can help farmers minimise waste produced and increase productivity that can come from the amount of fertiliser used. So, Smart Farming is essentially a hi-tech food growing device [6, 7,9,11]. This application of modern ICT (Information and communication technologies) has been introduced into agriculture.

Nitrogen Phosphorus and Potassium (NPK) are important to the plants. Plants need these to grow healthy. The plant will fail without any one of the nutrients. The leaf growth in the plant is taken care of nitrogen. The root, flower and fruit development are taken care by Phosphorous. The overall plant growth is taken care by Potassium in a good way. The farmers should know the NPK value of the plant and fertilise them accordingly. High Nitrogen is required for Leafy vegetables. Vegetable and fruit pants should be rich in Phosphorous. Potassium is needed for all type of plants. Based on the purpose of the plant, the needed nutrients should be given to him to yield well. For flowering plants, it is advisable to give fertilisers rich in Phosphorous to give good blooms. Agriculturist normally recommend the fertilisers based on the NPK value of the soil.

Along with NPK, to the soil the soil should be tested for its acid and alkaline properties. The manure should be given to the soil as the microbes decompose the material, releasing CO_2 which forms carbonic acid. Humus decomposition releases organic acids also, which is good manure for the plants along with the cow dungs and hen waste. Natural organics like plant waste, animal waste and human waste are good for the soil, to become rich in nutrients.

Elemental sulfur can be given to the soil, which decomposes with Water and Oxygen and forms the nutrient for the soil in the presence of sulfur-oxidizing bacteria. Iron and aluminum, react with water to produce Hydrogen, which is a good manure for the soil, to improve the plant growth. By applying acids to the fertilisers given to the plants thru the soil, reduces the pH content of the soil, which helps in the growth of the plant. Ammonium is liquefied by the bacteria present in the soil and give nitrate and hydrogen ions to ensure the fertility of the soil.

Soil with carbonates or lime are very resistant to acidify, as their pH is very high. The soil needs a low pH range to grow the plants in a good condition. The advice of not using plastic in the soil is same as it does not degrade, in turn, which makes the soil to degrade and not feasible to grow plants in the soil. pH is an important factor for plant growth. The level of pH should be maintained according to the type of the plant to have high yield. The paper mainly concentrates on the pH of the soil to ensure the plant growth in a healthy way.

*Corresponding author: kirubanand.vb@christuniversity.in

Vegetable plants like carrot, beetroot, tomato, brinjal, Ladies finger, green leaves, cucumber, grows well in the field, where the pH of the soil is between 7 and 8. Coffee, tea, olive, orange, arrowwood, black berry are the alkaline tolerant plants, which grow in a hilly area, where the alkaline content is more in the soil and of course the pH of the soil is more.

The papers is divided into parts as follows: introduction, System analysis, System Design, Test plan, Experiment, Results and conclusion.

1.1 IOT Technology benefiting today's modern farming Industry

Precision farming is a process or a practice that makes the farming procedure more accurate and controlled for raising livestock and growing crops [17, 18]. It has been one of the most popular IoT applications in the agricultural sector in recent years and a large number of companies have started using this technique around the world [9-12]. Agricultural drones are a very good example of Agriculture for IoT applications. Today, agriculture has become one of the main industries in which drones can be integrated. There are two types of drones, namely- Ground-based and Aerial-based.

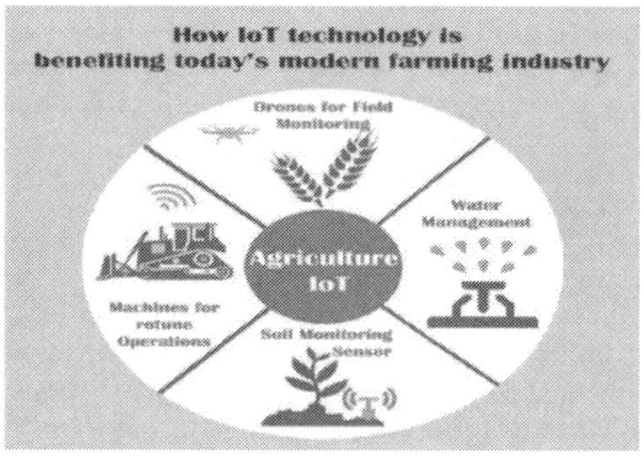

Fig. 1. Use of IOT in agriculture

Drones are used in agriculture for various reasons crop health assessment, irrigation planting, and soil analysis. The drone technology will give a cutting-edge makeover to the agriculture industry by utilizing systems and arranging the continuous information assortment [18]. The drone innovation will give a bleeding edge makeover to the horticulture business. Information is gathered by the drone like: helpful bits of knowledge on different factors, for example, plant tallying, plant wellbeing lists, plant tallness estimation, and furthermore. Fig. 2 depicts the applications of IOT in industry.

1.1.1 Livestock Monitoring

It helps the farmers to collect data regarding the location, wellbeing, and health of cattle's. It helps Farmers to gather information with respect to the area, wellbeing, and strength of their cattle [15,17]. The information encourages them in distinguishing the state of their cattle ie. Discovering the cattle are sick and preventing the spread of the illness to the whole cattle. The sensors empower farmers with more clarity to focus. [8,10-13]

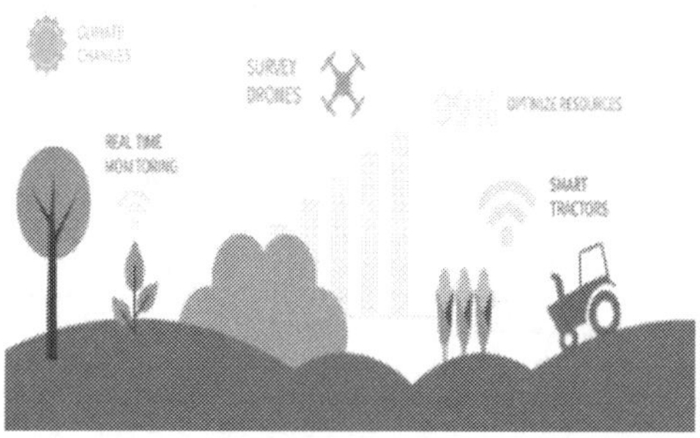

Fig. 2. IOT Applications in Industry

Figure 2 depicts the Applications of IOT in the various industries. It depicts the versatility of the IOT in the industrial purposes.

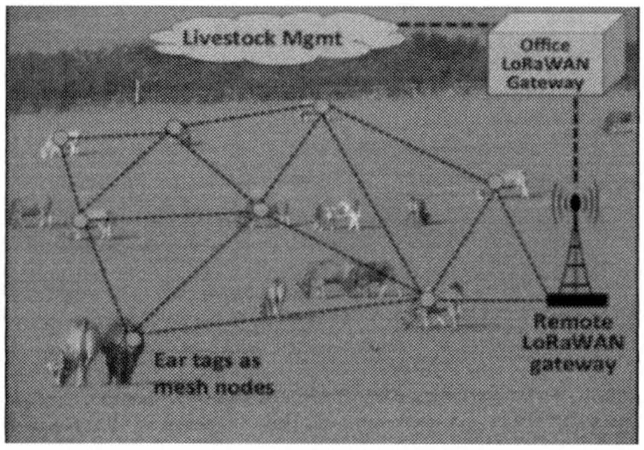

Fig. 3. LoRA & LPWA

LoRA WAN (Low power Wide Area (LPWA) networking protocol de-signed to wirelessly connect battery operated 'Things' to the internet [10, 15, 19]. **LoRa**, is the physical layer or the (wireless) modulization which creates the long range communication link.

Greenhouse farming, technique enhances the yield of crops, vegetables, fruits and much more in the soil. Greenhouses control environmental parameters in two ways: Manual intervention and proportional control mechanism. In these techniques, different sensors are used to measure the environment as indicated by the plant necessity and are utilized for control-ling the environment in a smart greenhouse. [8,18]

In this process, the Cloud Server helps in the handling of information and it applies control activity [10,16,17]. Therefore, farmers get benefited with minimal and almost no manual intervention. Challenges for Urban people growing their own vegetables are:

- Lack of space with adequate sunlight
- Lack of availability: While no one is there in particular location for a long period of time
- Lack of farming knowledge

1.1.2 Hydroponic types

Fig. 4. Nutrient Film Technique (NFT)

NFT, suitable for leafy Vegetables, is depicted in Figure 4.

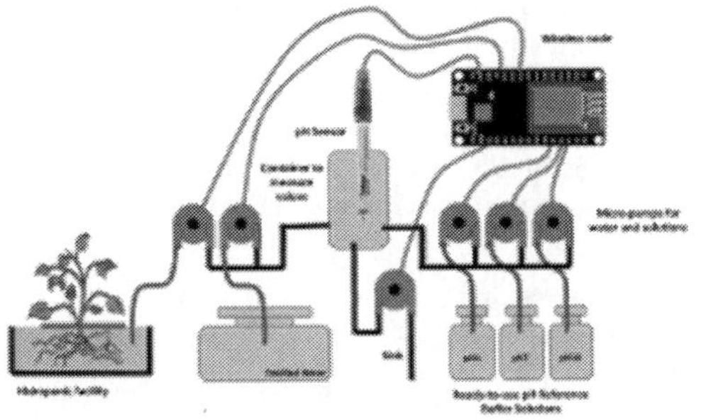

Fig 5. Hydroponic Precision Farming

In Hydroponic Precision Farming, the auto calibrated pH sensor is placed, to read the soil fertility and the LCD display to display the results. The Hydroponic Precision Farming system is depicted in Fig 5.

1.2 Soil Fertility using IOT

Adequate manure can assist plants creating great yields in good amount, to improve the quality and number of harvests, estimation of N (nitrogen), P (phosphorous) and K (potassium) substance of soil is important to decide how much compost is expected to add to the soil to build crop fruitfulness [1,2]. The pH sensor created can distinguish the degrees of NPK in the soil and afterward appropriately required composts can be added to the soil. These three component supplements advance the development of the plant in various manners - Nitrogen advances the development of leaves and vegetation, Phosphorus advances root and development and Potassium advances blossoming, fruiting and keeps guidelines of supplement and water in the plant cell [3]. Soil can be acid, neutral or antacid, as indicated by its pH value. [4]

pH strongly impacts the accessibility of supplements and the nearness of microorganisms and plants in the soil. For instance, growths incline toward acidic conditions though most microorganisms, particularly those providing supplements to the plants, have an inclination for tolerably acidic or slight antacid soils. Truth be told, in unequivocally acidic conditions, nitrogen fixing and the mineralization of vegetable lingering is diminished. Plants ingest the supplements broken up in the soil water and the supplement dissolvability relies to a great extent upon the pH esteem [5]. pH of the soil is measured by the hydrogen content in the soil. which check the acidity of the soil. Consequently, the Dirt can be corrosive, nonpartisan or soluble, as indicated by its pH esteem.[4] Most plants lean toward a pH run from 5.5 to 7.5; yet a few species groups incline toward progressively corrosive or soluble soils [12]. However, each plant re-quires a specific scope of pH, for ideal development. Each plant needs components in different amounts.

2 System Analysis

2.1 System Architecture

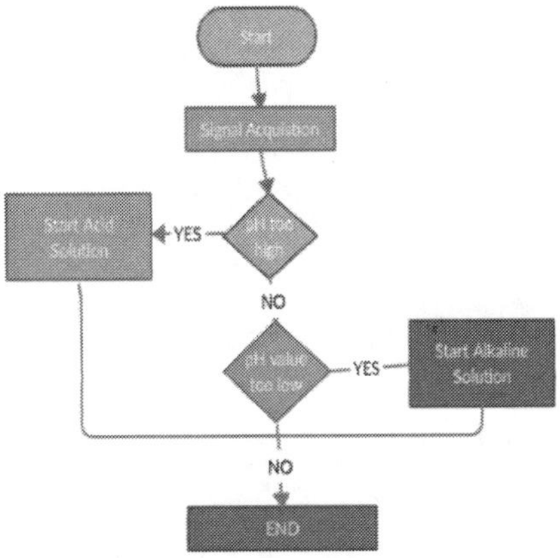

Fig. 6. Proposed System Architecture

The proposed architecture for the system is depicted in Fig. 6. The pH sensors given in the soil gets the information and process the signals. The **pH** of a solution measures the hydrogen ion concentration, which in turn is a measure of its acidity. Pure water dissociates slightly into equal concentrations of hydrogen and hydroxyl (OH^-) ions.

The best pH for plants is typically between 5.5 and 6.5, though some plants may need more acidic or more alkaline soils. The soil nutrients are strongly bonded with the pH of the soil, where pH controls the efficiency of the fertilizers available in it. Additionally, soil nutrients are tied strongly with pH of soil; in fact, pH control maximizes the efficiency of fertilizers by controlling nutrient available in it. pH of soil affects the bacteria, virus and harmful toxic elements, present in the soil. It is a notification for the Agriculturist to know about the health of the soil.

The pH sensor functions as a soil pH meter, which works for calculating the pH in the soil. The Arduino microcontroller acts as a controller, receiver and processor, which gets the input from the soil and receives the input and processes it according to the reading of the pH sensor.
The pH range in soil is given in table 1, which are acidic in nature. The level indicates the acidity of the soil and it can be balanced by adding the required nutrients.

Table 1. pH range in soil

Denomination	pH range
Ultra acidic	< 3.5
Extremely acidic	3.5–4.4
Very strongly acidic	4.5–5.0
Strongly acidic	5.1–5.5
Moderately acidic	5.6–6.0
Slightly acidic	6.1–6.5
Neutral	6.6–7.3
Slightly alkaline	7.4–7.8
Moderately alkaline	7.9–8.4
Strongly alkaline	8.5–9.0
Very strongly alkaline	> 9.0

.

As the lime contents are more in the alkaline rich soil, it is not suitable to grow plants and do vegetation in that type of soil. Alkaline rich soil can be treated by giving them rich manures, but it is treatable in some cases and not in other cases based on their environment. If the alkaline content is created based on the calcium carbonate rich, it is difficult to cure them. Rather than, the soil can be used for some other purpose like making lime stone etc.,

If the soil is rich in pH, based on the denomination given in the table 1, the Acid solution is started to give to the soil, to increase the fertility of the soil. Giving the only needed nutrients to the soil to ensure it will be good for the soil. If the soil is high in pH, add pulverized limestone to the soil. Limestone acts as neutralizer for the acidity of the soil. It consists of calcium and magnesium carbonate or calcium carbonate.

If the soil is poor in pH, alkaline solution is given to the soil for fertilizing. Chemicals like Sphagnum peat, elemental sulfur, aluminum sulfate, iron sulfate, acidifying nitrogen, and organic mulches will help the soil to be rich in nutrients. This enhances the fertility of the soil and gives good yield to the farmers.

An excellent way to enrich the pH in small areas is to give high nutrient coco peat for the plants, which are basically rich in NPK. Adding Sphagnum peat to the small garden area will be economical than to a bigger area of land, to increase the pH of the soil in it.

2.2 Requirement Specification

Some of the system requirements for the experiment are: Arduino Uno Board, pH Sensor Board, pH sensor electrode, Water level sensor (Probs), and Serial Monitor. Table 2 specifies the requirements used for the experiment.

Table 2. Requirement specification for the experiment

Component	Units Used	Specifications
Arduino Board	1	UNO
pH Sensor Board	1	Provides temperature Analog Output (LM35) and Digital Output (DS18B20)
pH sensor electrode	1	Glass which senses the hydrogen ion concentration

3 System Design

Defining the elements plays a vital role. Designing consists of numerous elements they are: architecture design, specifying the modules, components and most importantly the way of approach

3.1 Modular Specification

Module 1 – Collecting different samples of the soil and extracting water out of it, so that the pH of the water can be tested in different types of soil water collected.
Module 2 – This module determines the pH value of the soil by pH sensor electrode.
Module 3 – This module deter-mines the NPK value of the soil by using the pH values determined

3.2 Component Description

3.2.1 Arduino Uno Board

The Arduino UNO is a microcontroller board and is is a free open source available for experiments in small scale. . Sets of digital and analogue input / output (I / O) pins are given on the frame. It is based on the ATmega328P (datasheet). It has 14 digital in-put/output pins, 6 analog inputs, a 16 MHz ceramic resonator, a USB connection, a power jack, an ICSP header and a reset button. The board has 14 computerised pins, 6 analogue pins, and is programmable using a form B USB cable with the Arduino IDE. It acknowledges the voltage which ranges from 7 and 20 volts [17, 20].

3.2.2 pH Sensor Board

The analog pH sensor, designed for Arduino controllers can be utilized as a socket and plays a role for measuring pH estimation with no extra circuit. It consists of a LED which works as the Power Indicator, a BNC connector and PH2.0 sensor to interface. To utilize it, we simply associate the pH sensor with the BNC connector, and attach the PH2.0 interface into the simple information port of any Arduino controller.[20]

3.2.3 pH Sensor Electrode

The pH electrodes are made of a special compositional glass that detects the concentration of hydrogen ions. This glass is usually made of alkali metal ions in the research solution that undergo an ion exchange reaction with the hydrogen ions to produce a potential difference. The bulb is filled with an acid solution (e.g. 0.1 molL-1 HCl). [4,20]. Fig 7 shows a pH sensor Electrode.

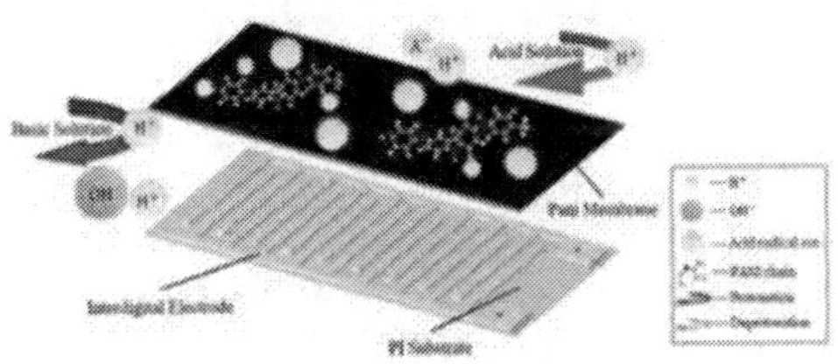

Fig. 7 pH sensor Electrode

3.2.3 Serial Monitor

The serial monitor is a bridge between Arduino board and the computer. It is used for sending and receiving the text messages , correcting the errors and to pass the commends from the system to the board using the keyboard. [4, 20]

4 Test Plan

To make sure that the pH sensor electrode is working fine and the pH values of the soil are getting determined.

•To determine the correct pH values of the soil as the determination of the NPK levels in the soil depends on the levels of the pH value of the soil.
•To ensure that pH value of the soil lies between the ranges of 0-14 as the standard scale of pH is from 0-14.

4.1 Testing Environment

A testing domain is a setup of soft-ware and hardware for the testing groups to execute experiments. As such, it facilitates execution of experiments with optimised facilities, ap-plications and network. Test Environment comprises components that help test execution with programming, equipment and system design.

4.1.1 Factors for designing Test Environment are as follows

- Determine if the test condition needs documenting so as to take back-ups.
- Check the network configuration.
- Identify the necessary server operating system, databases and other components.
- Identify the number of li-censes required by the test group.
- The area from where soil belongs to.
- The pH level of the soil from which NPK levels can be determined

The nutrients that are required by crops in the largest amounts are N, P and K, which are important for their growth.

4.2 Test Plan

The test plan is that the project should be checked for all types of soils. The pH level determined from the soil water can be used to deter-mine NPK level/values of the soil. By determining the NPK level of the soil, we can actually understand the fertility of the soil and accordingly crops can be grown on the field.

4.3 Test Cases and Results

The tests when performed on different types of soil depicted different results. Basically, we cannot predict the pH level of soil as soil differs from region to region. But most commonly, it is found as-

 https://doi.org/10.1051/itmconf/20213701018

SOIL ACIDITY	NITROGEN	PHOSPHATE	POTASH	FERTILIZER WASTED
Extremely Acid-4.5pH	30%	23%	33%	71.34%
Very Strong Acid-5.0pH	53%	34%	52%	53.67%
StronglyAcid-5.5pH	77%	46%	77%	32.69%
Medium Acid-6.0pH	80%	52%	100%	19.67%
Ntural-7.0pH	100%	100%	100%	00.0%

Fig. 8. NPK balance sheet

5 Experiment

5.1 Sketch

Sketch is a file consisting of the Arduino IDE. Pre-1.0 sketches were saved by Arduino Software (IDE) with the .pde extension. A minimum programme in Arduino C / C++ consists of two functions only:

setup():

The setup() function is used when the Sketch starts or resets. This features is enhanced to set or reset the features names. It is utilized to introduce the factors, information and change the modes of the pins and utilise different libraries used in the sketch.

loop():

By the completion of the setup() work, the loop() Study is regularly carried out in the central curriculum. It monitors the reading until the work is completed or reset.
The Arduino sheets contain a light-producing diode (LED) and a current restricting resistor associated between pin 13 and ground, which is useful for the component testing some features of the program correctly.. This program is generally stacked into another Arduino board by the maker

5.2 Algorithm

The algorithm for the proposed architecture is depicted below:

- The sensors placed in the field senses the features. (signal Acquisition)
- System checks the pH value.
- If it is high start the Acid solution for the field/plant
- If it is low, start the Alkaline Solution.
- Stop

5.3 Procedure

The procedure to measure the pH in the soil was done as follows:

i)

i) Take a jar of 1 liter. Put the soil for ¾ of the jar. Add distilled water for the remaining space in the jar
ii) Cap jar and shake the soil vigorously a few times.
iii) Let mixture stand 10 minutes to dissolve the salts in the soil.
iv) Calibrate the pH sensor with a pH 7 and à pH 10 buffer solution.
v) Remove the cap and place the pH sensor into the wet soil slurry.
vi) Measure pH and record measurement.

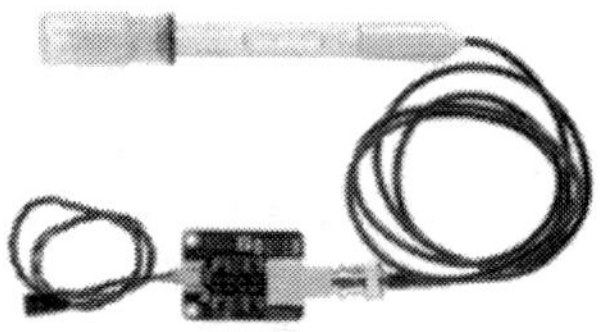

Fig.9 pH sensor connected with LCD

The figure 8 depicts the set up used to connect the ph sensor with the LCD. The sensor senses the pH value in the soil and is displayed in the LCD for the easy reference of the use. This helps the user to know the fertility of the soil.

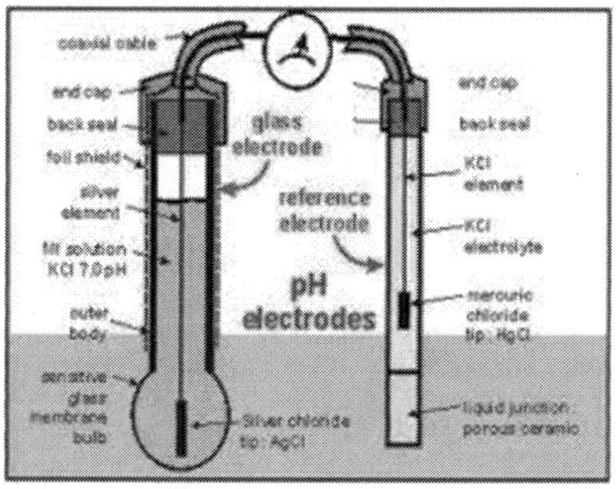

Fig. 10 working of the pH sensor

5.4 Output

If the same soil sample tested twice gives a difference of 0.5, then it is a good technique to be used and provides a high confidence in results.The output displays the values of the pH which is tested over the soil water. Firstly the output values shown on the serial monitor and which is accompanied with the average value. It is being displayed on the LCD.

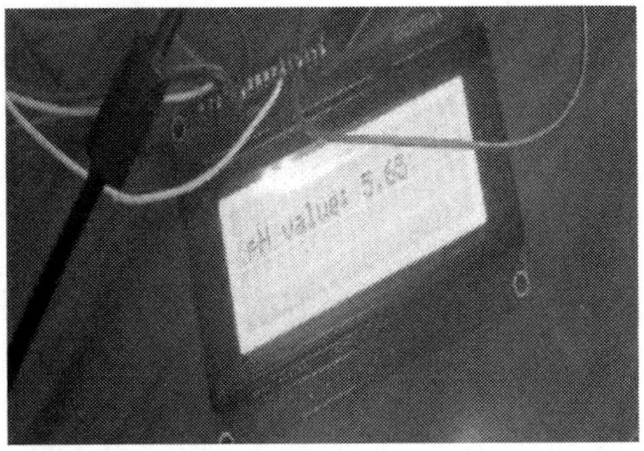

Fig. 11. Output in LCD Display

Figure 10 depicts the value in the pH sensor. Based on the value in the LCD display, the owner can act accordingly.

6 Conclusion and Future Scope

By checking the NPK in the soil of a land can give an accurate result on the soil. The functioning of the soil can be determined. It is advisable to use the vegetation based on the pH value of the soil to yield a good result and crop. The vegetation chart based on the acid value is available with the local people of that area or the agricultural research office in that area. Typically, these major nutrients are the first that the soil requires. Plants require huge quantities to expand and to live. Furthermore, these elements can naturally leach from the soil with the weather particularly during rainy or hot seasons.

The development of new methods of improving crop yield and handling are made possible with the help of IoT technology. In addition to targeting traditional, large-scale farming operations, IoT-based smart farming technologies could be new levers for uplifting other and or popular agricultural patterns as organic farming, family farming, particular livestock and/or crops and enhancing highly transparent agriculture. It helps the soil to be rich in nutrients and the farmers also to be benefited.

References

1. N. Singh and A. Shaligram, "D. NPK Measurement in Soil and Automatic Soil Fertilizer Dispensing Ro-bot", International Journal of Engineering Research & Technology, vol. 3, no. 7, pp. 635-637, 2014.
2. Y. Kulkarni, K. K. Warhade and S. Bahekar, "Primary nutrients determination in the soil using UV spectroscopy", International Journal of Emerging Engineering Research and Technology, vol. 2, pp. 198-204, 2014.
3. D. Q. Huo et al., "Recent Advances on Optical Detection Methods and Techniques for Cell-based Micro-fluidic Systems", Chinese Journal of Analytical Chemistry, vol. 38, no. 9, pp. 1357-1365, 2010.
4. D. V. Ramane, S. S. Patil and A. Shaligram, "Detection Of Npk Nutrients Of Soil Using Fiber Optic Sensor", International Journal of Re-search in Advent Technology (E-ISSN: 2321-9637) Special Issue National Conference "ACGT, pp. 13-14, 2015.
5. Ashwini, Humaira Hameed, Kavya Karkera, Neha Sherigar; Mr. Avinash N J "Soil Nutrient Tester and Air Purification System", 2019 International Journal of Engineering Research & Technology (IJERT) ISSN: 2278-0181.
6. Ayaz, Muhammad & Uddin, Ammad & Sharif, Zubair & Mansour, Ali & Aggoune, elHadi. (2019). Internet-of-Things (IoT)-Based Smart Agriculture: Toward Making the Fields Talk. IEEE Access. PP. 1-1. 10.1109/ACCESS.2019.2932609.
7. Verdouw, Cor & Wolfert, Sjaak & Tekinerdogan, Bedir. (2016). Inter-net of Things in agriculture. CAB Reviews. 11. 1-12. 10.1079/PAVSNNR201611035.
8. Wolfert, S., Ge, L., Verdouw, C., & Bogaardt, M.-J. (2017). Big Data in Smart Farming – A review. Agricultural Systems, 153, 69–80. https://doi.org/10.1016/j.agsy.2017.01.023
9. Review of Internet of Things towards sustainable development in agriculture. (2020). Journal of Critical Reviews, 7(03). https://doi.org/10.31838/jcr.07.03.28
10. Stočes, Michal & Vaněk, Jiří & Masner, Jan & Pavlík, J.. (2016). Internet of Things (IoT) in Agriculture - Selected Aspects. Agris on-line Papers in Economics and Informatics. VIII. 83-88. 10.7160/aol.2016.080108.
11. S. Jaiganesh, K. Gunaseelan and V. Ellappan, "IOT agriculture to improve food and farming technology," 2017 Conference on Emerging Devices and Smart Systems (ICEDSS), Tiruchengode, 2017, pp. 260-266, doi: 10.1109/ICEDSS.2017.8073690.
12. Doshi, J., Patel, T., & Bharti, S. kumar. (2019). Smart Farming using IoT, a solution for optimally monitoring farming conditions. Procedia Computer Science, 160, 746–751. https://doi.org/10.1016/j.procs.2019.11.016
13. Farooq, M. S., Riaz, S., Abid, A., Umer, T., & Zikria, Y. B. (2020). Role of IoT Technology in Agriculture: A Systematic Literature Review. Electronics, 9(2), 319. https://doi.org/10.3390/electronics9020319
14. G. Sushanth and S. Sujatha, "IOT Based Smart Agriculture System," 2018 International Conference on Wireless Communications, Signal Processing and Networking (WiSP-NET), Chennai, 2018, pp. 1-4, doi: 10.1109/WiSPNET.2018.8538702.
15. S. R. Prathibha, A. Hongal and M. P. Jyothi, "IOT Based Monitoring System in Smart Agriculture," 2017 International Conference on Recent Advances in Electronics and Communication Technology

(ICRAECT), Bangalore, 2017, pp. 81-84, doi: 10.1109/ICRAECT.2017.52.

16. R. N. Rao and B. Sridhar, "IoT based smart crop-field monitoring and automation irrigation system," 2018 2nd International Conference on Inventive Systems and Control (ICISC), Coimbatore, 2018, pp. 478-483, doi: 10.1109/ICISC.2018.8399118.
17. R. Dagar, S. Som and S. K. Khatri, "Smart Farming – IoT in Agriculture," 2018 International Conference on Inventive Research in Computing Applications (ICIRCA), Coimbatore, 2018, pp. 1052-1056, doi: 10.1109/ICIRCA.2018.8597264.
18. Article: https://www.biz4intellia.com/blog/5-applications-of-iot-in-agriculture/
19. IoT and understanding article: https://www.iotforall.com/what-is-iot-simple-explanation/
20. IoT equipments understanding: https://www.arm.com/glossary/iot-devices

https://doi.org/10.1051/itmconf/20213701019

Virtual AI Assistant for Person with Partial Vision Impairment

Rohith Raghavan[1*], *Vishodhan* Krishnan[1], *Hitesh* Nishad[1] and *Bushra* Shaikh[1]
[1]Department of Information Technology, SIES Graduate School of Technology, Navi Mumbai, India

Abstract: Smartphones help us with almost every activity and task nowadays. The features and hardware of the phone can be leveraged to make apps for online payment, content consumption, and creation, accessibility, etc. These devices can also be used to help and assist visually challenged and guide them in their daily activities. As the visually challenged sometimes face difficulty in sensing the objects or humans in the surroundings, they require guidance or help in recognizing objects, human faces, reading text, and other activities. Hence, this Android application has been proposed to help and assist people with partial vision impairment. The application will make use of technologies like face detection, object and text recognition, barcode scanner, and a basic voice-based chatbot which can be used to execute basic commands implemented through Deep Learning, Artificial Intelligence, and Machine Learning. The application will be able to detect the number of faces, recognize the object in the camera frame of the application, read out the text from newspapers, documents, etc, and open the link detected from the barcode, all given as output to the user in the form of voice.

1 Introduction

A Normal person without any disabilities have no issues with daily work in their life. But, on the other hand, it is difficult for a partially blind person to carry out daily tasks. Actions like reading texts, identifying objects cannot be performed by them due to their disability. Making Braille versions of every text is an expensive and tedious task. Also, recognizing objects from a distance is not possible for a visually challenged person. Although there are several applications to help and assist the visually challenged, they offer only some features, making the person install a handful of applications for that. So, to overcome the current issues faced by a visually challenged person, we have developed this application that offers convenience and assistance to the visually challenged person. The application offers text, object recognition, and face detection to identify text, objects, and humans. It also offers a chatbot so that the visually challenged person can interact with the bot for basic information and activities

.

*Corresponding author:rohith.raghavan17@siesgst.ac.in

2 Literature Review

We studied and went through the following research papers listed below to get more knowledge and ideas about the implementation of our project.

Tosun et al [1] discussed the process and the algorithms involved for real-time object detection.They also compared the various algorithms like YOLOv2, SSD, and faster R- CNN in terms of accuracy.The paper explained the ML algorithms in brief. YOLOv2 provided better accuracy and ran on even low fps with a GPU processor.

Tembhurne et al. [2] studied the implementation of a voice assistant for visually challenged. The paper discussed the various modules which can be implemented in the voice assistant like calls, messages, TTS, OCR, etc. The paper also talks about using Maps API for navigation.

Dahiya et al. [3] elaborates on the R-CNN algorithm in detail and also compares the accuracy and computational time of R- CNN and faster R-CNN combined with resnet-50. The paper also discusses the data preprocessing steps required for feeding the data into the machine learning model. The framework proposed in the paper claims an accuracy of 92%.

Ahmed et al. [4] discussed using RNN (recurrent neural network) and CNN (convolutional neural network) for obstacle avoidance and way-finding. Their work using CNN proved helpful to implement object detection using CNN-based algorithms.

Gianani et al [5] described real-time object detection implemented using OpenCV and also determining the position of the object using Euclidean distance. The paper also guides the user to the objects through voice output. The paper explains object detection using the SSD framework and MobileNet architecture which has an accuracy of 99.61%. This system is designed to work in an indoor environment.

Kukade et al. [6] focused on Speech-to-Text, Text-to-Speech, Optical Character Recognition, and voice assistance and the proposed system to implement the same. The paper also discussed the ways of it

Shishir et al. [7] explained object recognition using Tensorflow ML API along with the implementation of it. They included informative flowcharts for understanding the process behind it. They also explained the working of OCR and object recognition. This implementation provided accuracy of over 80%.

Karthik et al. [8] provided an overview of the OCR algorithm and the hindrances faced while the text is being extracted. They also share the idea of using Raspberry Pi instead of a mobile phone to capture images. The paper also talks about the future scope of using a GPS location tracker for guidance.

Singh et al. [9] proposed an Android application which offers text recognition, speech recognition, image recognition and a chatbot for the user to interact with the application. The paper proposed using Google Cloud APIs (various APIs which can be used to automate tasks) and Google Dialogflow (a natural language understanding platform on which chatbot can be implemented) to implement various modules instead of training deep learning models to perform various activities.

Sharma et al. [10] focuses on implementing a system offering face recognition, text-to-speech and object recognition on a web browser which can be opened on a mobile device. The paper also talks about adding a feature to add unknown faces to the database at the tap of a button for future reference. The proposed system also has a fairly simple and user-friendly UI designed specifically for the visually impaired.

Jakhete et al. [11] discussed about using Single Shot Detector (SSD) Algorithm to implement Object Detection in an Android Application. The paper lists other object recognition algorithms and mentioned the steps to implement SSD algorithm on an Android application.

3 Existing System

In this section, we are discussing the features of certain applications available on the Play Store

Supersense[12]– it is an application that assists the visually challenged and the features provided by it are Object recognition, Face recognition, and text recognition Sullivan+[13]–This application also serves the same purpose this provides Object Recognition to describe images, Face recognition and text recognition.

Envision AI [14]– This application also serves the same purpose and provides the features that are Face recognition, object recognition

LetSeeApp[15]– This application is also for the same purpose and provides the features of text recognition to read visiting cards as well as credit and debit cards

The above-mentioned applications provide more or less similar features (the links to these applications are provided in the references section)

4 Proposed System

An Android-based application based on technology and innovation promises to academically empower visually challenged by freeing them of their dependence on visuals by providing the information through an app.

This application aims to provide better functionality in an app that makes a partially blind user use it for navigation, identification, recognition, and also gaining information of the outer world. Some of them are listed below:

•The app will contain a chatbot such that we will be asking questions about time, weather, or any other kind to obtain information or asking to perform certain actions the user desires.
•It will detect the objects in real-time and provide the necessary information to the user.
•The app will also contain a barcode scanner which will help the user to get information about certain products.
•The app can also help the user detect human faces so that the user can understand human presence in the surrounding and also the number of people in the room.

https://doi.org/10.1051/itmconf/20213701019

•This application will have a text reader which will be used to read the text out loud to the user.

Using the app, the person can get help and guidance in day-to-day tasks and activities.

APP NAME	FR`	OR`	TR`	CHATBOT
SUPERSENSE	YES	YES	NO	NO
SULLIVAN +	NO	YES	YES	NO
LETSEEAPP	NO	YES	YES	NO
ENVISON AI	YES	YES	NO	NO
OUR APPLICATION	YES	YES	YES	YES

Table 1 Comparison of features provided by each application

NOTE- FR`- FACE RECOGNITION; OR`- OBJECT RECOGNITION; TR`-TEXT RECOGNITION

NOTE- As the algorithms used in the other applications are unknown to us .We have contacted their developers but haven't got any responses yet. So, we have done the comparison based on features provided

5 Methods

Face Detection:

Face detection is a computer technology that is used to detect human faces in images, videos, or in real-time video. Face detection is a broad technology that just marks or labels the human face identified by the application. The key difference between face detection and recognition is that face detection just identifies the face whereas face recognition will also label the person's name, gender, age, or other attributes. Face detection can be implemented in various fields - security, biometrics, entertainment, law enforcement, etc.

Basic face detection can be achieved through OpenCV whereas real-time face detection or face detection in different conditions can be achieved using machine learning or deep learning. The face detection algorithms start searching for human eyes in the frame as it is the easiest to detect.

It then searches for other factors like eyebrows, nose, ears, and iris. When the algorithm finds the factors in the image in the frame, it then applies additional tests and then confirms the detection of the face by labelling the face with a rectangular box.

Real-time face detection involves motion; hence traditional algorithms cannot be applied. So, advanced machine learning and deep learning algorithms are used to create models which can detect faces in real-time in various scenarios.

Object Recognition:

Object recognition is the technique to recognize and label an object detected in an image, video, or real-time. Object recognition is achieved using machine learning and deep learning. Object recognition algorithms take the frame from the camera as the input and then apply a bounding box of a specific size to the image and check for the object in the image. If the object is found in the image, the algorithm will recognize the object. There are two steps to object recognition –image classification and object localization. Image classification predicts the class of the object in an image. Whereas object localization identifies one or more objects in the image and drawing the bounding boxes. The object detection algorithm will combine both of the tasks and will classify the objects in the image.

Text Recognition:

Text recognition is the technique to detect and identify the text which is in printed, handwritten or digital format. Text recognition technology converts the text in different forms to digital form. It is also called OCR (Optical Character Recognition). Several APIs exist for various platforms which can be used to implement OCR.

For recognizing typed or printed text on objects or books, the user has to open the application on his smartphone and then select the required option. The application will identify the text and convert it to digital form. The text will then be read out to the user.

Chatbot:

Chatbots are AI-based computer programs that can simulate a human conversation. They are also called digital assistants as the chatbots can be used to do actions and commands given by the user. A chatbot can process the human conversation, reply to commands and queries or can solve user FAQs as well.

The key modules behind a chatbot are artificial intelligence, natural language processing, user-defined rules, and machine learning which are required to process the commands or messages sent by the user and deliver the required feedback.

Chatbots are of two types- task-oriented and data-driven. Task-oriented chatbots are designed for a single purpose and only generate automated responses. Their interaction is specific and restricted to only FAQs or basic questions.

The answers to the queries are already defined in task-oriented chatbots. Hence, they can only handle and process basic queries and are the most commonly used in websites and apps for uer queries.

https://doi.org/10.1051/itmconf/20213701019

Data-driven chatbots or virtual assistants are more interactive, sophisticated, and advanced than data-driven ones.
These chatbots use NLP, NLU, and ML to learn from the user's queries and responses. These chatbots analyze and use past user interaction data and behavior to provide responses or feedback to the user's queries. Hence, data-driven chatbots get better, efficient, and precise over time. Amazon Alexa, Google Assistant, and Apple's Siri are examples of data-driven chatbots.

Implementation

For implementing object recognition and face detection, we have chosen to use Tensorflow Lite (Google's open-source deep learning framework designed for on-device processing) framework in our proposed system.

TensorFlow Lite was chosen as other frameworks like Keras (an open-source library that provides a Python interface for artificial neural networks) and PyTorch (an open-source ML library designed for NLP and Computer Vision) do not offer Lite versions for low-end devices like smartphones.

TF (Tensorflow Lite) also offers various pre-trained models with commonly used algorithms and datasets for out-of-the-box usage in projects and applications.Several algorithms like You Only Look Once (YOLO) algorithm [16], Single Shot Detector (SSD) [17] and Region-based Convolutional Neural Network (R-CNN) [18] among others can be used to implement realtime object recognition.

We chose the SSD algorithm for our project as it offers a fair trade-off between speed and accuracy over other algorithms which offered either of these parameters.

The following table shows the speed and accuracy comparisons.

Table 2. Speed and Accuracy comparison among object detection algorithms.

Method	mAP	FPS	Size	Boxes	Input
Fast R-CNN	73.2	7	1	6000	1000X 600
Fast YOLO	52.7	155	1	98	448X 448
YOLO-VGG16	66.4	21	1	98	448X 448
SSD300	74.3	46	1	8732	300X 300
SSD512	76.8	19	1	24564	512X 512
SSD300	74.3	59	8	8732	300X 300
SSD512	76.8	22	8	24564	512X 512

The SSD algorithm also performs better when it comes to detecting objects of different shapes and sizes. This is evident from the comparison graph which shows the difference.

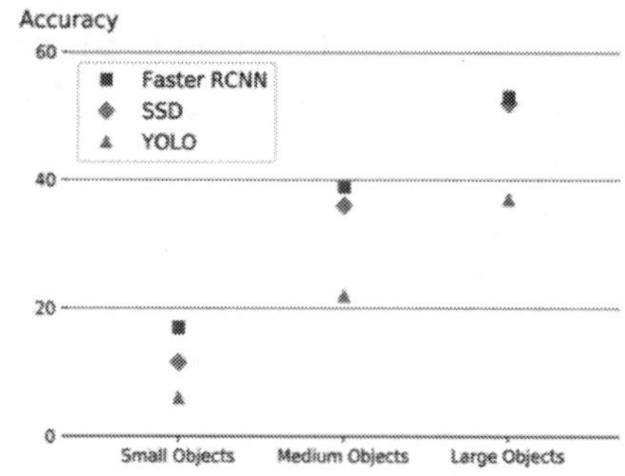

Fig 1 Algorithm's performance over objects of different sizes.

We have used the Tensorflow Object Detection API model which uses SSD mobilenet v1. This model is trained over the MS-COCO[19] dataset. The COCO dataset is a massive object detection dataset which has 330,000 images with over 200,000 labelled images consisting of 80 various object categories.

Real-time face detection can be implemented using algorithms like Multi-Task Cascaded Convolutional Neural Network (MT- CNN)[20], Google FaceNet[21] algorithm, and using the OpenCV Haar Cascade[22] and OpenCV Dlib[23] toolkits.

The FaceNet algorithm performed better among the others as it had a maximum accuracy of 99.63%. So, we chose to implement the FaceNet algorithm designed for low-power devices, the MobileFaceNet[24] model for face detection. The MobileFaceNet model offered better speed over others as is evident from the below graph.

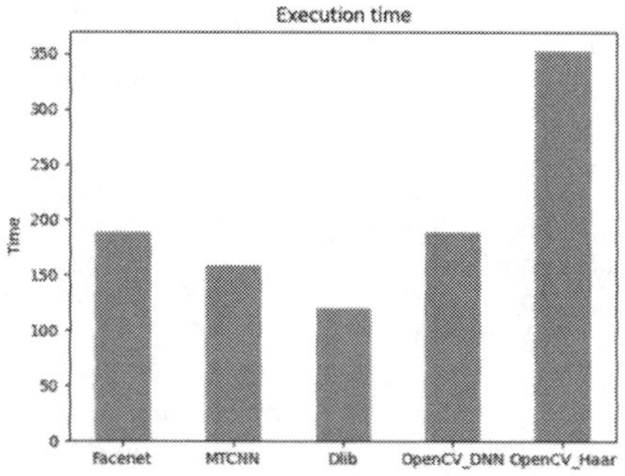

Fig 2 Comparison between face detection algorithms w.r.t time

https://doi.org/10.1051/itmconf/20213701019

The reason behind the fast performance is that the global average pooling layer has been replaced by a depth-wise convolutional layer which improves performance on face detection and recognition.

For the proposed system, we have used the MobileFaceNet model trained over the Labelled Faces in the Wild (LFW)[25] dataset. The LFW dataset contains over 13,000 human faces captured in various angles and orientations.

Real-time text recognition has been implemented using Text recognition API from Google's ML Kit (Google's machine learning for Android devices in the form of a mobile SDK) which provides various libraries for implementing computer vision-related recognitions. Google's ML Kit website offers prebuilt APIs and packages which can be imported into our application to implement text recognition.

For implementing the chatbot, we have used AIML chatbot which uses Python packages like Pyttsx3 (An offline Python Text to Speech conversion library (TTS)), nltk (Natural Language Toolkit, a package of libraries and programs written in Python for processing natural language), chatterbot to provide feedback to the user as per the queries asked. Implementing the chatbot requires natural language processing and artificial intelligence for it to give replies and perform actions. The chatbot will read the command from the user, detect the keywords in the command, and then will perform the action as programmed by the developer.

The barcode scanner has been implemented using Google ML Kit's Barcode API. The API can directly be used in the application by importing the app dependencies and package.

6 Results

Object Recognition

The object recognition module has been implemented successfully

- Accuracy of 90 %
- Average run time of 1.3 seconds.

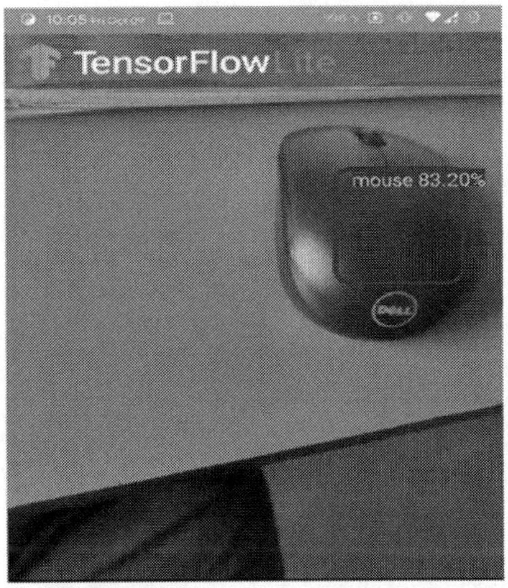

Fig 3 Detecting objects

Text Recognition

Text Recognition module has been implemented successfully

- Accuracy of 90 %
- Average run time of 1.4 seconds

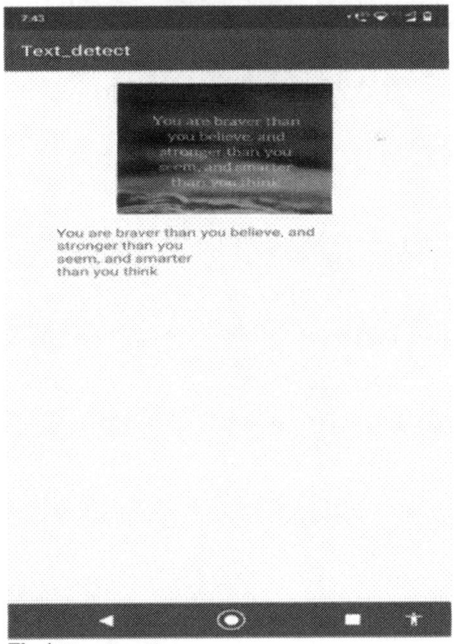

Fig 4 Text Detection

https://doi.org/10.1051/itmconf/20213701019

Face Recognition

Face Recognition module has been implemented successfully

- Accuracy of 85%
- Average run time of 1.2 seconds

Fig 5 Detecting human face

Chatbot

Voice-based chatbot has been implemented successfully

- Offline feature's like calling, asking for date and time are working properly
- Online features like asking for weather, temperature, information on certain products are working properly if provided suitable internet connectivity

Fig 6 User command to chatbot via speech

Fig 7 Chatbot performs the required action

Barcode Scanner

Barcode Scanner has been integrated successfully

- The information on product is provided correctly with a sufficient internet connectivity

Fig 8 Barcode detection

Performance of our application compared to others

A. OBJECT RECOGNITION

1. Average time taken to perform one task(tested 20 times)

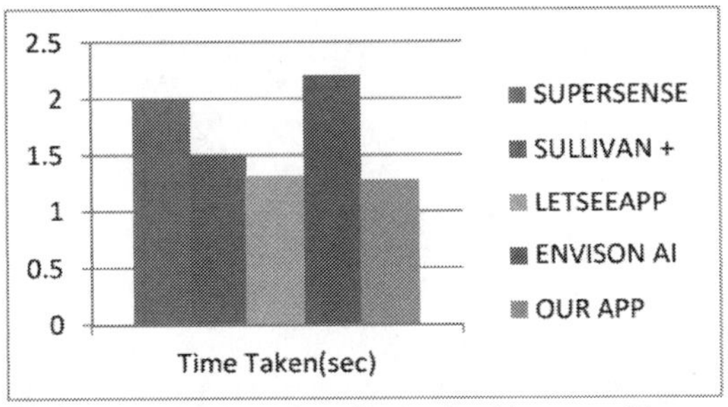

Fig 9 –Time taken by each application for object recognition

2. Accuracy- providing correct output (tested 20 times)

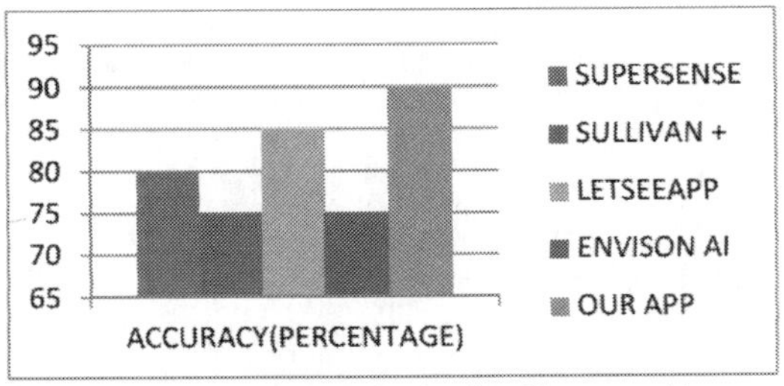

Fig 10 –Accuracy of each application for object recognition

B. FACE RECOGNITION

1. Average time taken to perform one task(tested 20 times)

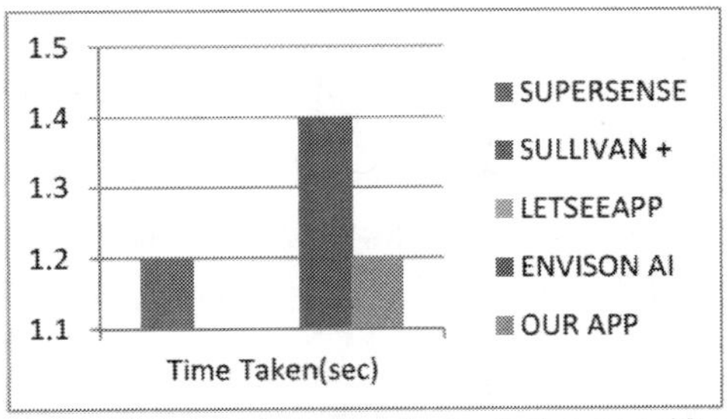

Fig 11 –Time taken by each application for face recognition

2. Accuracy- providing correct output (tested 20 times)

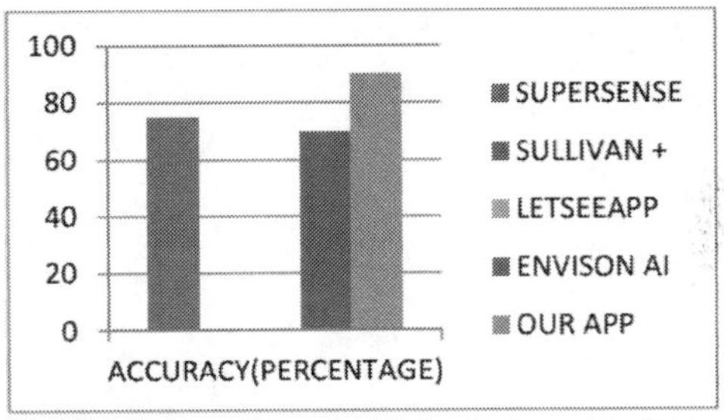

Fig 12 –Accuracy of each application for face recognition

C. TEXT RECOGNITION

1. Average time taken to perform one task(tested 20 times)

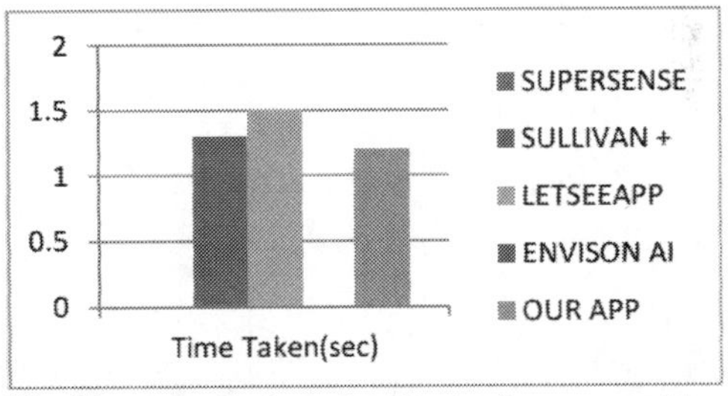

Fig 13 –Time taken by each application for text recognition

2. Accuracy- providing correct output (tested 20 times)

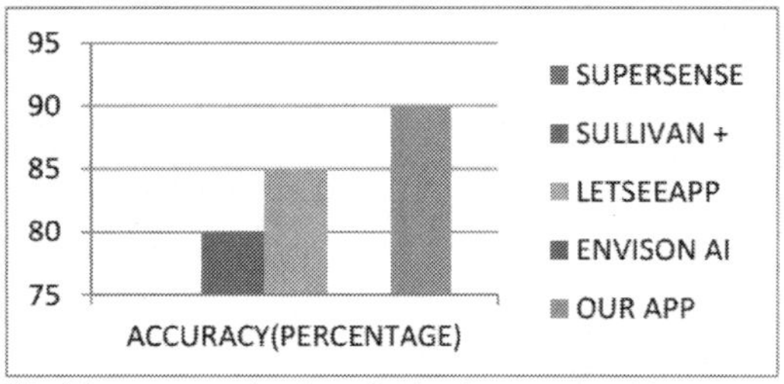

Fig 14- Accuracy of each application for text recognition

Note – the testing was done indoor under tube light (unnatural but ample brightness) and may vary according to the surrounding environmental condition

7 Discussion

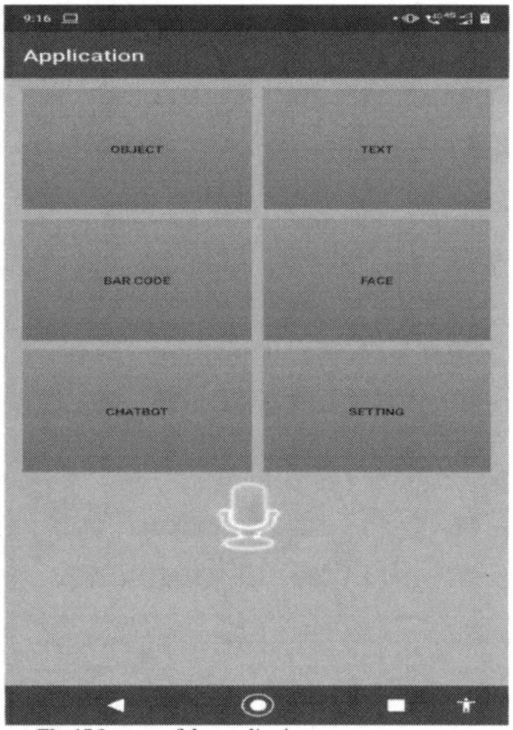

Fig 15 Layout of the application

The system (referred to as the android app hereafter) consists of 5 modules-- real-time face detection, real-time object and text recognition, barcode scanner, and chatbot. Each of these modules can be easily accessed from the android app with the click of a button. The UI of the application has been designed to be user-friendly for partially blind.

The working of the android app and its modules are explained below.

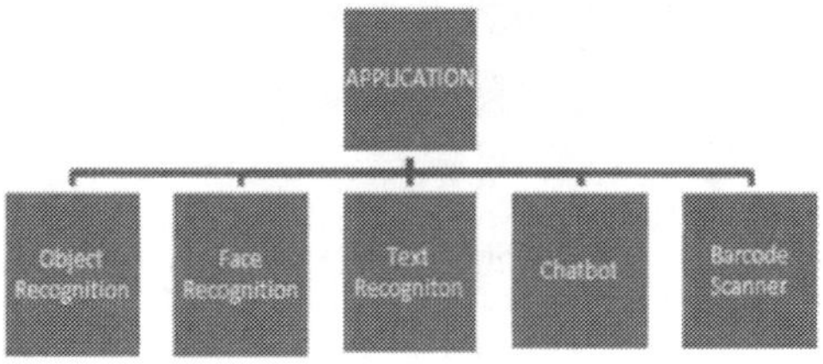

Fig 16 Application working flowchart

Modules used in ChatBot component of the Application.

The chatbot only requires the smartphone's microphone and Internet access. It offers some useful functionalities achieved through techniques and libraries mentioned below:
Pysttsx3 – It is a Python text-to-speech convertor that even works offline. We have implemented this module in our project to provide offline text-to-speech conversion.

This module provides us many features like

- TTS conversion without Internet
- Option to choose different voices
- Change speed or pitch of speech
- Easy-to-use and feature-rich API

Speech Recognition – A technique that is used to identify the queries of the user and convey it to the application. Which in turn will start the process it was requested to perform. This works in such a way that a keyword is associated with a particular action and when the keyword is spoken by the user the action will take place. Google Speech-to-Text has been used for speech recognition. We have implemented this for Natural Language Processing in our project

Natural Language Processing (NLP) – It is broadly defined as the automatic manipulation of natural language, like speech and text, by software. Natural language refers to the way humans normally communicate with each other.
This module is used in our project so that user can communicate with their device as they communicate with fellow human beings.

Datetime – This module is used to provide the date and time to the chatbot. This module works offline as its working on the data received from the device on which it runs. We have implemented this in our Project so that the user can ask the device for the time and date whenever needed the data from this module is given to another module (Pyttsx3)

Web Browser- The user can browse through the web using only voice commands given to the chatbot. We have configured this module in such a way that it can be used to gain information, play music (via an API to access YouTube), provide us with weather report(via an API to access The Weather Channel), and get news updates(via an API to access Times of India) and to get information on various topics we have also linked it with Wikipedia Module.

Working of Text Recognition:

Text Recognition API: Google ML Kit is a set of APIs and tools which can be used to deploy and automate certain applications like text recognition, barcode scanning, pose detection, etc. We have used the text recognition API in our project. The API will first use OCR to detect the text shown on the camera frame. It will then split the text into lines and the lines will be split into words.
These words will then be sent to the API for recognition and the recognized words will be spoken to the user using Google Text-to-Speech (TTS). Google TTS is available by default on all Android devices. Text recognition only requires the smartphone's camera access.

Working of Barcode Scanner:

Barcode API: Google ML Kit also offers a barcode API which can be used to scan barcodes and QR codes. The API will detect for any QR code/barcode displayed on the camera preview frame.

After detection, the QR Code/barcode will be read by the API to detect the embedded information or URL. The app will automatically open the URL link or will read out the information from the barcode using Google TTS. The barcode API only requires the smartphone's camera access.

Working of Face Recognition

For implementing face detection, we have used the MobileFaceNet model, which is an extremely efficient CNN model. The model is just 4.0MB in size and is designed for smartphones and embedded systems. The face detection process starts with detecting the human's faces in the real-time camera preview frame. The image is then warped using the detected landmarks like eyes, nose, jaws, eyebrows, etc and the face is captured. This image of the face is then processed and resized to be fed as input to the Deep Learning model.

The application is designed to capture preview frames at a resolution of 800*600px. The preview frame, if horizontal in orientation is rotated vertically and is cropped to 400*300px which removes the background and retains only the human body.

This image is then rescaled to 112*112px to be used as input for the MobileFaceNet model. On feeding the image, the model looks for the face in the image by matching the face features. It then creates a bounding box when it detects the face and is highlighted. The number of faces detected in the frame by the app is then outputted to the user orally using TTS functionality. This can be useful for the user to know the number of people in the room or a certain place. Face detection requires only the smartphone's camera access.

Working of Object Recognition

The android app has the object recognition feature where the user can point at an object and the app will recognize the object in frame and will output the object name to the user using Text-to-Speech. Object recognition has been implemented using SSD neural network. When the user points to an object, the frame is cropped to 600* 800 and is inputted into the model till the whole frame is covered. Based on the confidence level set by the user, the model creates multiple boxes with different aspect ratios throughout the image and tries to detect the object.

The accuracy of detection of the object depends on the confidence level. Once the object is detected, it then creates a box over the detected object with a label. The name of the object is then read to the user using TTS. The object recognition requires only the smartphone's camera access.

8 Conclusion

The proposed android application is designed to help and guide the partially blind in their daily tasks when needed. The application has 5 main components, namely- text recognition, object recognition, face detection, chatbot, and barcode scanner. The text and object recognition, barcode scanner, face detection, and chatbot are working as proposed and intended. Several changes in the text-to-speech module and the output are yet to be implemented which will be added in the coming months. This application is intended to work in indoor and outdoor conditions provided there is a good lighting condition.

Acknowledgment

This work is supported by the Department of Information Technology, SIES Graduate School of Technology. This Project is also supported by the Head of the Department Dr. K. Lakshmisudha and Project Guide Prof. Bushra Shaikh(co-author).

References

1. S. Tosun and E. Karaarslan, "Real-Time Object Detection Application for Visually Impaired People: Third Eye," 2018 International Conference on Artificial Intelligence and Data Processing (IDAP), Malatya, Turkey
2. N. M. Tembhurne, S. V. Vaidya, A. Shiekh, S. Dravyakar, "Voice Assistant for Visually Impaired People", International Research Journal of Engineering and Technology (IRJET).
3. D. Dahiya, H. Gupta and M. K. Dutta, "A Deep Learning based Real-Time Assistive Framework for Visually Impaired," 2020 International Conference on Contemporary Computing and Applications (IC3A), Lucknow, India, 2020
4. F. Ahmed, M. S. Mahmud, and M. Yeasin, "RNN and CNN for Way-Finding and Obstacle Avoidance for Visually Impaired," 2019 2nd International Conference on Data Intelligence and Security (ICDIS), South Padre Island, TX, USA, 2019
5. S. Gianani, A. Mehta, T. Motwani, and R. Shende, "JUVO - An Aid for the Visually Impaired", 2018 International Conference on Smart City and Emerging Technology (ICSCET), Mumbai
6. R. Kukade, R. Fengse, K. Rodge, S. Ransing, V. Lomte, "Virtual Personal Assistant for the Blind",2018 International Journal of Computer Science and Technology (IJCST), Bali, Indonesia.
7. M. A. Khan Shishir, S. Rashid Fahim, F. M. Habib, and T. Farah, "Eye Assistant: Using a mobile application to help the visually impaired," 2019 1st International Conference on Advances in Science, Engineering and Robotics Technology (ICASERT), Dhaka, Bangladesh
8. A.Karthik, V.K.Raja and S.Prabakaran, "Voice Assistance for Visually Impaired People," 2018 International Conference on Communication, Computing and Internet of Things , Chennai, India
9. G. Singh, K. Takhtani, O. Kandale, N. Dadhwal," A Smart Personal AI Assistant for Visually Impaired People", Vol 7, Issue 6, pg.1450-54, International Research Journal of Engineering and Technology (IRJET)
10. V. Sharma, V. M. Singh, S. Thanneeru," Virtual Assistant for Visually Impaired", (April 19, 2020). Available at SSRN:https://ssrn.com/abstract=3580035
11. S. A. Jakhete, P. Bagmar, A. Dorle, A. Rajurkar and P. Pimplikar, "Object Recognition App for Visually Impaired," 2019 IEEE Pune Section International Conference (PuneCon), Pune, India, 2019
12. Supersense:_ https://play.google.com/store/apps/details?id=com.mediate. supersense&hl=en_IN&gl=US
13. Sullivan+: https://play.google.com/store/apps/details?id=tuat.kr.suliva n&hl=en_IN&gl=US
14. EnvisionAI:_ https://play.google.com/store/apps/details?id=com.letsenvis ion.envisionai&hl=en_IN&gl=US
15. LetSeeApp:_ https://play.google.com/store/apps/details?id=com.letseeapp.letseeapp&hl=en_IN&gl=US
16. "You Only Look Once: Unified, Real-Time Object Detection": [1506.02640] You Only Look Once: Unified, Real-Time Object Detection (arxiv.org)
17. SSD: Single Shot MultiBox Detector": [1512.02325] SSD: Single Shot MultiBox Detector (arxiv.org)
18. "Faster R-CNN: Towards Real-Time Object Detection with Region Proposal Networks":[1506.01497] Faster R- CNN: Towards Real-Time Object Detection with Region Proposal Networks (arxiv.org)
19. "COCO - Common Objects in Context":COCO - Common Objects in Context (cocodataset.org)
20. " Joint Face Detection and Alignment using Multi-task Cascaded Convolutional Networks" [1604.02878] Joint Face Detection and Alignment using Multi-task Cascaded Convolutional Networks (arxiv.org)
21. "FaceNet: A Unified Embedding for Face Recognition and Clustering": [1503.03832] FaceNet: A Unified Embedding for Face Recognition and Clustering (arxiv.org)
22. GitHub - opencv/opencv: Open Source Computer Vision Library
23. GitHub - ageitgey/face_recognition: The world's simplest facial recognition API for Python and the command line
24. " MobileFaceNets: Efficient CNNs for Accurate Real-Time Face Verification on Mobile Devices": [1804.07573] MobileFaceNets: Efficient CNNs for Accurate Real-Time Face Verification on Mobile Devices (arxiv.org)
25. LFW Face Dataset: http://vis-www.cs.umass.edu/lfw

https://doi.org/10.1051/itmconf/20213701020

CODOC: COVID – 19 PATIENT TRACKER

Umamaheswari E[1*], *Gazal* Madan[1], and *Chinmayee* Das[1], *Shashank* Sridhar[1], and *Ajay* DM[1]

[1]School of Computer Science and Engineering, Vellore Institute of Technology, Chennai (VIT), India

Abstract. The COVID-19 corona virus has affected 81.9 million people with 1.79 million deaths in the entire world and has transformed into a global pandemic. The disease has effected 10.2 million people in India with about 148K deaths till December 2020. The COVID-19 virus is a highly infectious disease, which mainly spreads through the droplets generated by the cough, sneeze or exhale of an infected person. The disease spreads rapidly in crowded and poorly ventilated places. The virus quickly contaminates on floors and surfaces, people touching their eyes, nose and mouth after their contact with these contaminated surfaces get infected with this viral disease. As there is no cure and a vaccine currently for COVID-19, tracing, testing, tracking and treating are the most efficient tools to combat it. In order to track patients the existing methods implement fall detectors and routers which are effective in tracking patients in smaller areas. In order to overcome this drawback, a patient tracking technique COVID-Doctor (CODOC) has been proposed in this work which implements Global Positioning System (GPS) and multiple sensors to track patients in larger areas.

1 Introduction

The personal health monitoring of every individual is considered very important because of the rising health problems in today's world. COVID-19 scenario has made a tremendous increase in patients which is resulting in a vicious cycle. COVID-19, a rather infectious disease, made its first appearance in December,'19 in Wuhan [1-3], China. The Coronavirus (the cause of the disease) made its way around the world within a few months and was declared a global pandemic by the World Health Organisation (WHO) [4]. The symptoms were mostly low to moderate in majority of the population which including dry cough, fever, and loss of taste or smell, etc. However, in the elderly and people with cardiovascular disease, diabetes, chronic respiratory diseases, etc., the symptoms were rather serious, even resulting in the death of the individual in most cases. The lack of proper isolation and tracking of the numbers has resulted in a higher growth rate of patients. With the predicted second wave of COVID-19 as well as the development of new variants of Corona Virus, social distancing has become a new normal. In these circumstances, regular physical medical visits for a check-up might not be the most efficient option during this time [5-7]. In this situation, there is an urgent need for social distancing, staying from crowd and lockdown is essential to stop its spread. However, if online facilities and health tracking were made easier, it could result in an improvement in the patients' health conditions. Globally, there have been 79.8 million confirmed cases of COVID-19, including 1.75 million deaths as of December 2020 [8-10]. The following system is motivated by the world-wide pandemic, to handle the current situation as well as, tackle any major health concern in the future.

This paper proposes a system that continuously monitors the patient through a mobile application (app) that is connected to a wristband that continuously monitors the temperature, heartbeat rate, and blood pressure and is continuously updated in the application without any delay. Whenever any discrepancy is found in the patient's health or if the patient has come in contact with a COVID positive person, an alarm is triggered in the patients and the patients care takers mobile phone. Then the patient is automatically assigned to a doctor and the treatment is carried over the application itself. This system is described as the following:-

Continuous monitoring of the patient's health through an app which is connected to a wrist-band that continuously monitors the temperature, heartbeat rate, and Blood pressure and is updated in the app right away. The smart band can calculate temperature anytime without sitting or standing in an appropriate posture. It monitors the body temperature using a temperature sensor, the heart-rate monitor is based on Photo Plethysmography (PPG), which basically uses green LED lights paired with light-sensitive photodiodes to illuminate the skin and measures changes in light absorption and there also exists a cuff-less method called Pulse Transit Time (PTT). PTT is the time it takes for one pulse to travel from the heart to a peripheral point on the body (in this case, the wrist). It requires a PPG sensor to detect the pulse on the wrist and an ECG sensor to know detect the heart-beat. Any discrepancy is found in the patient's health, is immediately notified to the

* Corresponding author: umamaheswari.e@vit.ac.in

patient and the family members, an appointment with the doctor is immediately scheduled and continuous health updates are sent to the doctor from then till the patient cures. Though there are already many android patient tracking applications, most of them either measure the different health conditions of patients or only track the location of patients, if there is any contact with a COVID positive person. On contrary, CODOC contains the features of health monitoring using sensors as well as proper information delivery to the hospital database. The presence of a system for continuous storage of health data of an individual makes individual, resulting in the prediction of major health conditions. This system also features the emergency option, through which alerts can be sent concerning serious ill-health conditions to the nearest health facility for the faster health service delivery. This also ensures proper effective treatment of the patient while assuring higher rates of survival. The presence of a thermal sensor, a pulse rate monitoring system provide efficient information regarding the patients' health.

2 System Description

The CODOC system, an online application is an integrated approach of linking the GPS enabled Android app with wrist band towards continuous surveillance of suspected/confirmed COVID-19 patients, wherein timely medical intervention would assist in reduction of morbidity and mortality. As a step towards tele-medicine the proposed system utilizes the latest available technology for integration of android app through enabled GPS as sharing mode of information for alert generation.

The CODOC system, an online application is an integrated approach of linking the GPS enabled Android app with wrist band towards continuous surveillance of suspected/confirmed COVID-19 patients, wherein timely medical intervention would assist in reduction of morbidity and mortality. As a step towards tele-medicine the proposed system utilizes the latest available technology for integration of android app through enabled GPS as sharing mode of information for alert generation. The prerequisite for successful system configuration is an android based device with GPS, its registration on the said app and a sensor sensitive wrist band for completion of loop, the details of the proposed system are as follows:

The CODOC system is designed to introduce the health care infrastructure and support the covid patients. The following technologies have been used in the system. It is an effective integrated system consisting of an android app linked with wristband which monitors covid and non-covid patients. The system is embedded with a feature that can prevent thousands of people from coming in contact with Corona virus by providing timely treatment.

2.1 Monitoring Device

The monitoring device is designed as a wrist band with sensors namely light sensitive photodiodes, PPG, ECG and bio-impedance which detect the temperature, heartbeat and blood pressure of the patients registered in the app. In addition to regular monitoring of vitals of the patients, the device is programmed to generate alert in case of abnormalities read in any of the measurable parameters and an alert is triggered which flows not only to the registered android device but also to the emergency contact devices registered and also the associated concerned medical doctor for necessary intervention. The device used for monitoring the covered patient is in the form of a wristband which has the ability to detect the temperature, heartbeat and blood pressure of the patient which is regularly updated in the Android app whenever any abnormality is found. The Android app automatically

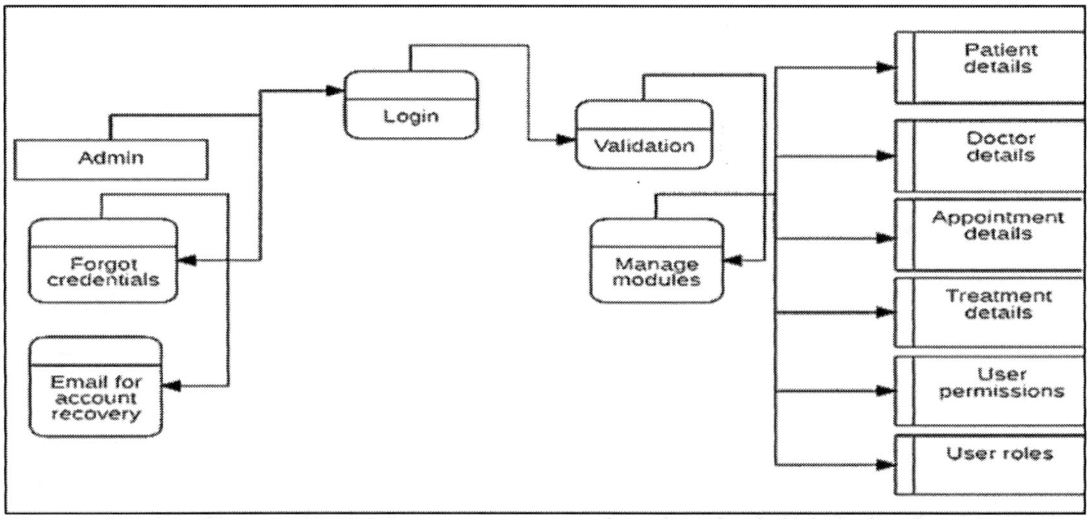

Fig. 1. Architecture diagram of CODOC

The prerequisite for successful system configuration is an android based device with GPS, its registration on the said app and a sensor sensitive wrist band for completion of loop, the details of the proposed system are as follows:

connects to the list of doctors by sending an alert and immediately a doctor is assigned over the patient.

These sensors helps in tracking the patient's health through monitoring all the essential requirement of determining any risk to health. In case of any discrepancy these sensors shows variations due to which the problem is detected.

2.2 Alarms and Locations

Automatic alarms in emergency situations are triggered with the beep in the Android app, through which the family members and emergency contacts as well as the doctor gets intimated simultaneously and the location information of the patient is sent to all. This feature enables the system to know the location of the patient within the safety parameters and update the same in the Android app. With this regular update spreading of Corona virus can be reduced. And the patient can be saved within the time limit and as soon as these updates are done in the doctor's app window as well as the patient's family members' app window the appointment will be automatically set with the doctor and over the call for further advice from doctor. The alert system is in fact a closed loop for alert generation and feedback facilitation from the expert for further management. It is envisaged not only for timely intervention for morbidity reduction but also as a contribution towards public health emergency control.

2.3 Doctor and family member's devices

As soon as an alert message from the patient's device is transmitted to the doctors and his or her family members' device. The option to accept or reject the appointment is flashed on the family members device along with an alarm and as accepted or rejected by the family member is transmitted the same to the doctor. When the appointment is accepted the appointment ID, date for the appointment, timings for the appointment, doctor's name, his qualifications, patient's details such as name, gender, age And his previous disease history is set with the date and time alarm on patients as well as doctors CODOC app.

This will facilitate the training doctor to have a complete details of patient to initiate the management and plan for any further intervention, if required. The patient is under constant surveillance. If the patient or his caretakers opt for online consultation with doctor online prescription after the audio or video call is generated. This consultation is updated in the patient's previous history record.

3 System Implementation

To ensure continuous surveillance of subject registration in the CODOC app is essential. Following the completion of registration procedure which requires demographic details of the subject along with essential information regarding any existing co morbidity and treatment details a unique ID is generated by the system which is co-shared with registered subject and the associated medical professional at same point of time.

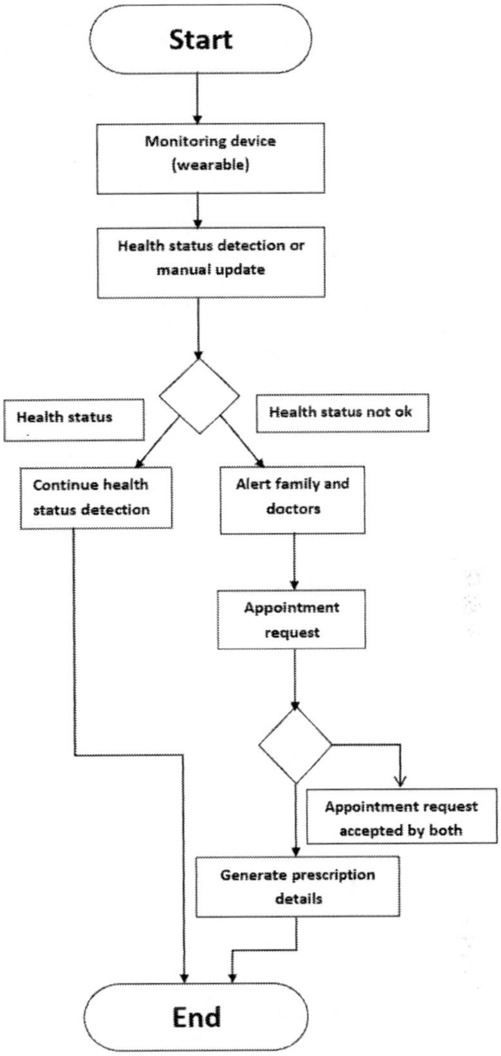

Fig. 2. Working process of CODOC

The prerequisite for the subject to be under continuous surveillance is regular wearing of wrist band with active connection to the registered mobile through GPS. The inbuilt alert system gets activated once there any discrepancy is cited for any of the values under surveillance as detailed above. The actionable alert is referred to the registered subject mobile, emergency contact device and designated medical professional.

The architectural diagram depicted in figure 1, gives an overall architectural view of CODOC and the application allows the admin to have all the controls, to edit the details, manage the modules, and grant permissions to the users. The Login module would be for both the patients and doctors or the health staff. Appointment details will have both upcoming appointments and the history of the past appointments with their respective details. Treatment details will have the medical history of the patient and the current treatments. In case a user forgets the login credentials, he/she will also have the option for recovery of their account.

The app further facilitates the appointment with the treatment doctor for a video conference or telephonic call depending upon the option agreed by the registered subject along with generation of prescription for effective timely intervention. Not only for emergency, but even for routine follow up visit the app can be of immense utility and a move towards digital health by effective telemedicine, a lesson learnt from the on-going COVID-19 pandemic. The screen shots of CODOC and the working process of the proposed system is represented in figure 2.

Once the patient registers himself on the app and wears the wristband. He is under constant surveillance. As soon as any discrepancy is found in patient's health an alert is given to the family and patient and appointment is made with doctor with patient's details. The patient is monitored by the doctor through online prescription and check- ups. These regular check-ups and prescription generation are regularly updated in patient's history.

4 Results

In order to design and develop a user friendly and user preferred application, a survey was carried out in order to understand the user preferences. The survey also helped in finding out people's opinion on CODOC mobile application being incorporated with a wrist band. This survey helped us to decide if people were interested in such an innovation. The survey was carried out among 80 people aged 15-70. The questions in the survey were- the device preferred by the people such as Locket, Mobile phone or Wrist watch/band, in order to make their family informed immediately of their health conditions with an affirmative and a negative option and if they want an automated doctor appointment at that very point of time where a positive and negative option was given. The results of this survey as presented in figure 3, were that 66.3% people chose a wrist watch, 27.5% favouring mobile phone and rest 6.2% wanted with the locket. When it came to informing the victim's health conditions to the family, as presented in figure 4, 95% had optimistic opinion. As presented in figure 5, about 86.3% preferred generating an automatic doctor's appointment on a contrary 13.7% didn't want an automatic doctor appointment to be generated.

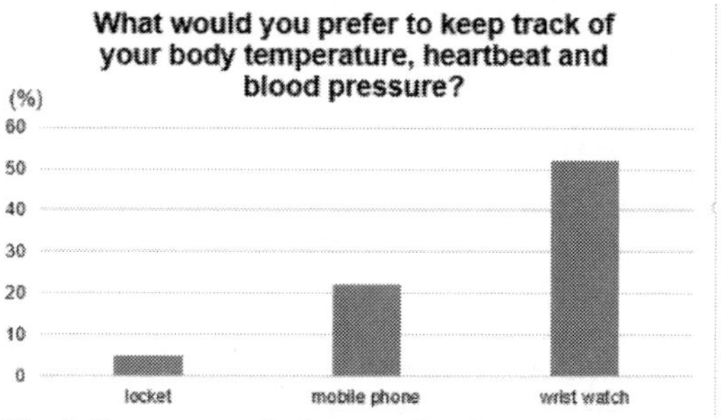

Fig. 3. Survey results for question 1

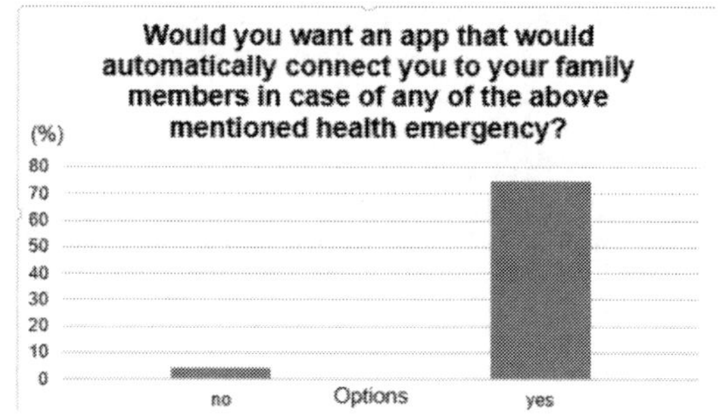

Fig. 4. Survey results for question 2

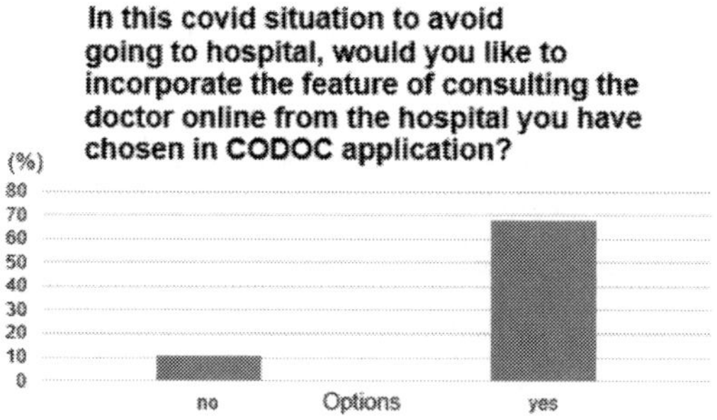

Fig. 5. Survey results for question 3

The CODOC application connected according to the survey results was incorporated with a wristband. The CODOC in a wrist band was tested in a home environment with where an elderly patient wore the wrist band continuously and the family members along with nearest hospital were connected through CODOC application. The results of the experiment were fruitful and the application performed as expected. In an instance of some disparity was found with regard to patient's health, the doctor and family members were notified concerning the same and the victim was treated over the telephonic conversation immediately as per the option of telemedicine. If health condition was so serious an automatic appointment was generated. The implementation of the CODOC application has been represented in the below figures.

https://doi.org/10.1051/itmconf/20213701020

Figure 6. CODOC Patient wrist band with sensor

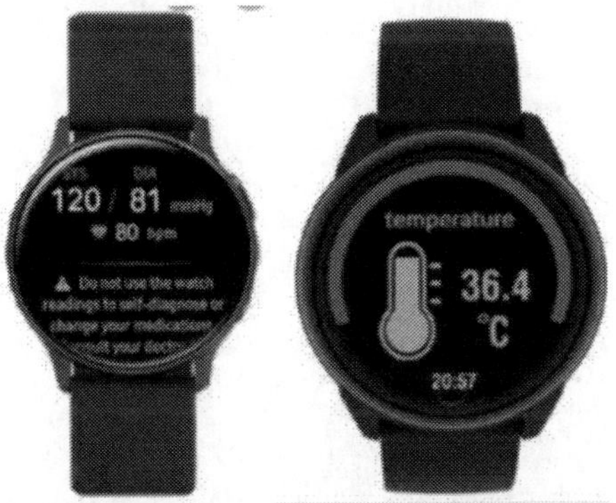

Fig. 7. Patient health monitoring results

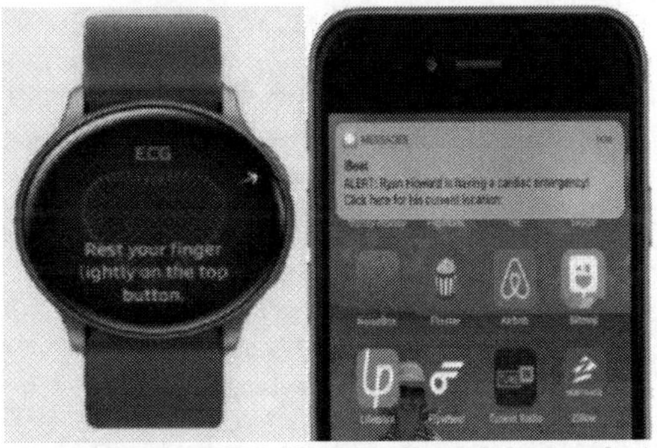

Figure 8. CODOC Patient family and doctor alert

The figure 6 describes the temperature sensor attached to a wristband in order to monitor the body temperature of the patient. They allow monitoring of important health parameters which are helpful in detecting temperature sickness of the person using it in any posture. The figure 7, describes the user interface on the wearable device, which shows the result of the continuous health monitoring. Left watch: Shows the result of heart rate with systole and diastole reading on the wearable device along with a notification of not using these readings for self-diagnosis. Right watch: Shows the body temperature of the use in degree Celsius. The figure 8, describes the alert system included in the CODOC application. This helps in detection of any major discrepancies in health monitoring as mentioned above regarding the heart rate, temperature or blood pressure. The alert system automatically sends a notification to the doctor and family members to accept, emergency medical service for the patient's treatment.

In order to calculate the efficiency of CODOC, experiments were conducted to analyse the time taken for alerts to reach family member devices and the time taken for the alerts to reach the doctor's device has been noted. Twenty sets of experiments have been conducted for each device i.e., 3 family devices and one doctor device and the results state that the alerts in order to reach the family member devices on an average took about 158 seconds, the experimental results have been presented in figure 9, figure 10 and figure 11. The experiments results show that it took about 116 seconds for the alert to reach the doctor's device as presented in figure 12.

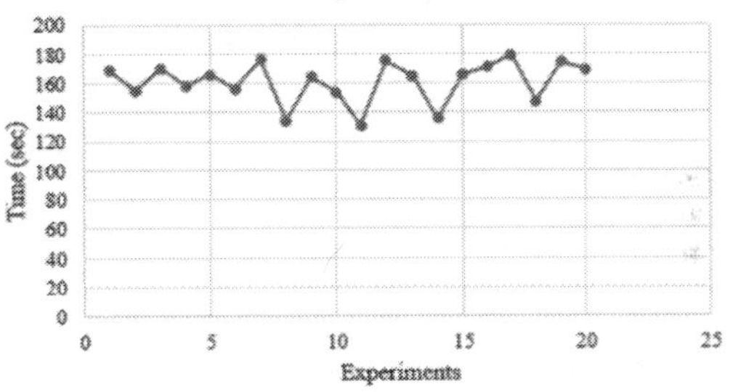

Figure 9. Alert Delivery time for Family Device 1

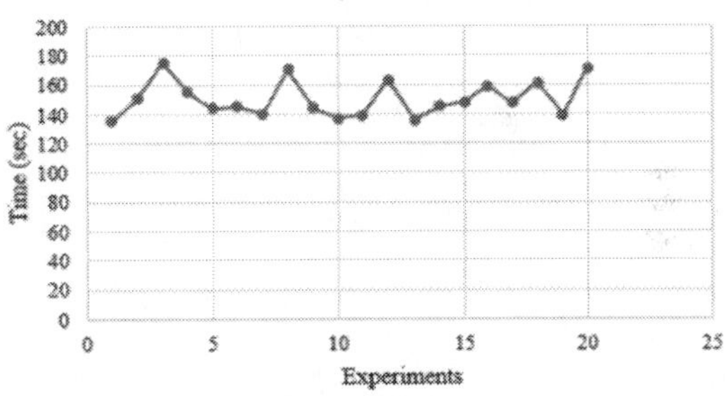

Figure 10. Alert Delivery time for Family Device 2

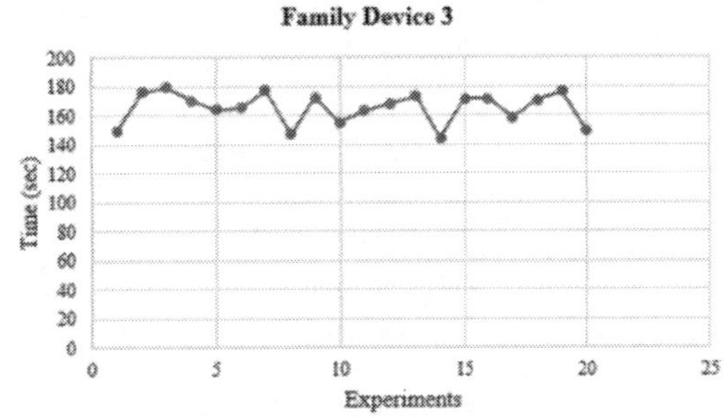

Figure 11. Alert Delivery time for Family Device 3

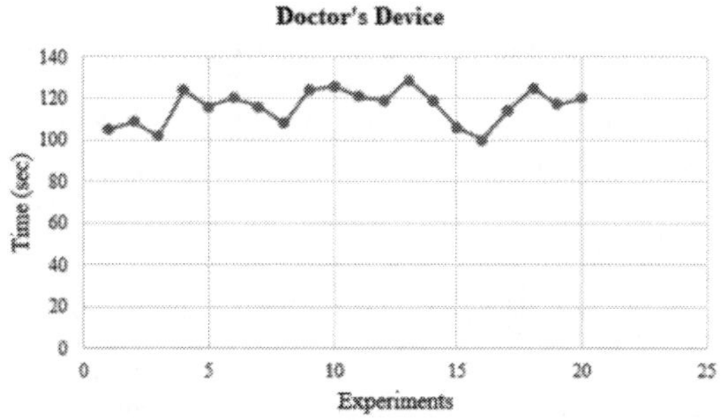

Figure 12. Alert Delivery time for Doctor's Device

The reason for the variation in the delivery time of the alert with respect to family member device and doctor device was the difference in the internet connections of the devices. The doctor's device was connected to a Wi-Fi connection in the hospital with 250Mbps speed and while the family devices were utilising the mobile internet speeds ranging from 10-15Mbps and the current location of a family member also played a major in the delivery time of the alert. More efficient results are to be expected in the extension of the project which would overcome the scenario of time delay due to variations in internet signals.

5 Conclusion and Future work

The world population is facing a global pandemic caused by the COVID-19 and it has endangered the human like never before. The disease has effected 10.2 million people in India with about 148K deaths till December 2020. The COVID-19 virus is a highly infectious disease, which mainly spreads when a person comes in close contact with an affected person. The disease spreads rapidly in crowded and poorly ventilated places which has forced the global population to live isolated, follow social distancing guidelines. The health concerns developed in the current situation and the alarming rate of increase in the number of COVID-19 patients which brings out the necessity of such telemedicine and online tracking, treatment of patients with other medical emergencies. The CODOC project is user-centric and was developed focusing on the problems faced related to medical visits and maintenance during the current situation. The proposed CODOC system ensures an increase in treatment efficiency and reduces the spread of the virus, by enabling digital services, maintaining the requirement of social-distancing and proper health monitoring of the patients. Patients and individuals suffering from the diseases can use this system for medical efficiency and personal well-being. Also, the CODOC feature of notifying in case of emergency using alerts to the nearest available medical services allows acquirement of medical services at the earliest. CODOC solves issues related to hospital management by providing them a full-fledged system. In the extended version of the CODOC application is aimed at reducing the latency delay in the alert mechanism and also addition of user customisations are to be implemented in the CODOC application.

References

1. Swain, Kaliprasanna, Prasad, M, Sahoo, Jagannath & Sahoo, S, Palai, Gopinath. (2016). *Patient Monitoring System using CAN and Android,* Journal of Research in Science, Technology, Engineering and Management (JoRSTEM). 2. 101-105 (2016)
2. T. Reza, S. B. A. Shoilee, S. M. Akhand, M. M. Khan, *Development of android based pulse monitoring system*, Second International Conference on Electrical, Computer and Communication Technologies (ICECCT), Coimbatore, 2017, pp. 1-7 (2017)
3. P. Hii, W. Chung, *A Comprehensive Ubiquitous Healthcare Solution on an Android Mobile Device*, Sensors (14248220), Vol. 11, Issue 7, pp.6799-6815, (2011)
4. P. Szakacs-Simon, S. A. Moraru, L. Perniu, *Android application developed to extend health monitoring device range and real-time patient tracking,* IEEE 9th International Conference on Computational Cybernetics (ICCC), Tihany (2013)
5. Kamel, Mohammed B., George, Loay, *Remote Patient Tracking and Monitoring System*, IAENG International Journal of Computer Science and Mobile Computing. 2. 88-94 (2013)
6. Gamboa, Hugo, Silva, Filipe, Plácido da Silva, Hugo, *Patient tracking system*. 1-2. 10.4108/ICST.PERVASIVEHEALTH2010.8916. . (2010).
7. Horgan D, Hackett J, Westphalen C, B, Kalra D, Richer E, Romao M, Andreu A, L, Lal J, A, Bernini C, Tumiene B, Boccia S, Montserrat, *A: Digitalisation and COVID-19:* The Perfect Storm. Biomed Hub (2020)
8. Organization WH. Coronavirus disease 2019 (COVID-19) Situation Report - 70. 30 March (2020).
9. Sang-Joong Jung, RistroMyllyla, Wan-Young Chung, *Wireless Machine-to-Machine Healthcare Solution Using Android Mobile Devices in Global Networks*, IEEE Sensors, vol. 13, no.5, pp.1419-1424, May (2013).
10. Zhonghua Liu Xing Bing Xue Za Zhi, *Epidemiology Working Group for NCIP Epidemic Response, Chinese Centre for Disease Control and Prevention. [The epidemiological characteristics of an outbreak of 2019 novel coronavirus diseases (COVID-19) in China]*, 10; 41(2):145-151. Chinese. DOI: 10.3760/cma.j.issn.0254-6450.2020.02.003. PMID: 32064853, Feb (2020)

 https://doi.org/10.1051/itmconf/20213701021

Memory Utilization and Machine Learning Techniques for Compiler Optimization

Shreyas Madhav A V[1], *Siddarth* Singaravel [1], and *Karmel* A[1,*]

[1]School of Computer Science and Engineering, Vellore Institute of Technology, Chennai, 600127, Tamilnadu, India.

Abstract. Compiler optimization techniques allow developers to achieve peak performance with low-cost hardware and are of prime importance in the field of efficient computing strategies. The realm of compiler suites that possess and apply efficient optimization methods provide a wide array of beneficial attributes that help programs execute efficiently with low execution time and minimal memory utilization. Different compilers provide a certain degree of optimization possibilities and applying the appropriate optimization strategies to complex programs can have a significant impact on the overall performance of the system. This paper discusses methods of compiler optimization and covers significant advances in compiler optimization techniques that have been established over the years. This article aims to provide an overall survey of the cache optimization methods, multi memory allocation features and explore the scope of machine learning in compiler optimization to attain a sustainable computing experience for the developer and user.

1 Introduction

With the advent of modern computing technology, the scale and complexity of applications have increased vastly. This rapid expansion, in the scale in which applications are built, has given rise to a dire need for making maximum use of the computing resources present. The effectiveness and impact of any optimization technique is evaluated based upon the environment of compilation, architecture and application in which it is made use of. The selection and employment of the relevant optimization strategy plays a vital role in achieving maximum performance and requires the developer to have a good understanding of the system background. Proper optimizations lead to a drastic reduction in the cost of a product and help enhance it, without relying on higher speed hardware. This paper explores various optimization techniques based on memory allocation, cache-based optimization, ML based optimization in detail, to establish an overview of the current state of the technology. In order to work on optimizing compilers for better performance, it is essential for a programmer to understand the basic functioning and nature of a general compiler.

A Compiler translates high level language (source code) written by humans into low level language (machine code) by passing through various compilation phases such as Lexical Analysis, Syntax Analysis, Semantic Analysis, Intermediate Code Generation, Optimization and Code Generation. While the properties of different compilers provide different variations in their structure and architecture, these steps provide a more generalized working of a compiler. The lexical analyser (scanner) converts source code, which is a stream of characters, into a stream of tokens (object that describes a lexeme). This involves removing comments, checking indentation and removing whitespaces. Now the code is split into lexemes and thereby a sequence of tokens is created. This is the reason why lexical analysis is also known as tokenization. Scanners need to be concerned with issues such as case sensitivity and whether comments are nested, or this might lead to lexical errors.

Now the sequence of tokens generated by the scanner phase is used to create a structure called abstract syntax tree (AST) which is a tree that depicts the logical structure of a program. This Syntax Analysis is also called Parsing and the compiler checks for syntax error as well as the grammatical structure. Each node of the AST is stored as an object and has no cycles. This AST will be used by the next compiler phase – semantic analysis. In this phase, the compiler uses AST to detect any semantic (meaningful) errors, such as performing operations on an undeclared variable or using the same name for multiple variables. The pieces/nodes of the syntax tree are tied to form a semantic graph which clearly describes the values of its attributes. This analysis consists of three steps which are type checking, flow control and label checking. Type checking involves checking type match in functions and method calls, flow control checking discusses flow control structure and label checking validates usage of labels. The legality rules differ, according to the chosen language.

*Corresponding author: ~~karmel.a@vit.ac.in~~

https://doi.org/10.1051/itmconf/20213701021

After the completion of the above mentioned processes, the compiler will generate one or more intermediate forms which is essentially a flow graph made up of tuples. This can be considered as some intermediate representation (IR) or as a program for an abstract machine. This IR must be easy to generate and converse into target code. Intermediate code is machine independent so there is very little need to consider differences in compiler properties and optimization techniques are more viable to be applied to this code rather than the machine translated code. This phase discusses efficiency of code following the three important rules. Firstly, the final code shouldn't change the meaning of the original program. Secondly, speed up time should be more and shouldn't impact the overall time. Thirdly, it should focus on consuming fewer resources and less memory.

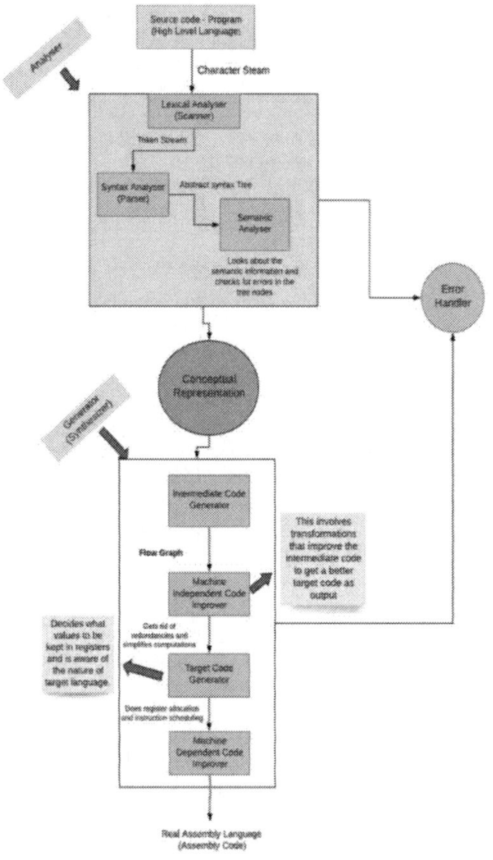

Fig. 1. Compilation Process

Optimization is classified into machine-dependent and machine independent techniques. In Machine Independent Optimization, the compiler accepts the IC (Intermediate code) and transforms it without any specific characteristic necessities required of the CPU registers and memory locations. The latter works on the generated target code according to the machine architecture involving CPU registers and have memory references. Optimization plays a crucial part over loops to save memory and CPU cycles. Loop optimization is done by moving the invariant code out of the loop and using cheaper expressions that reduces more CPU cycles without compromising the output.

This last compiler phase converts the optimized intermediate code to the machine code for that target machine. The final code should be efficient in such a way that memory and resource usage has been taken care of. After the assembly has been produced and we have the .asm file. This is then passed through an assembler and would generate the equivalent in binary. This binary code is written into a new file as an object (.o) file. Object files are machine code that are not executable and they need to be linked together to make it executable. Linker will make it executable as these are utility programs that compile the object code your backend generates.

2 Compiler Optimization

The main objective of compiler optimization methods and analysis is to get rid of all forms of possible redundant or inefficient blocks present in a computer program. It provides the system with the ability to manage the limited resources available and to efficiently maximize the output performance. The characteristics of a significant number of optimization techniques are inherent to a specific compiler and do not exhibit a general standard of properties. Hence, it is imperative to attain a complete understanding of the architectural dependencies of an application before deploying an optimization strategy to the code. The accuracy in achieving the intended purpose of the program and maintaining proper relations in the code are of primary importance. Several optimization techniques have been proposed in the past for different compilers and structures. A general categorization of the methods can be done on the basis of the levels of operation. This results in the categories - machine dependency, architecture based dependency and architecture independency. This paper investigates upon current mainstream strategies namely-Multi access memory based allocation, cache memory based optimization, machine learning techniques used for compiler optimization.

3 Multiple Memory Access Allocation

Multiple memory allocation is a relatively new method of optimization that works based upon the storing and retrieving of every instruction, onto multiple registers. A major issue faced in the field of embedded system design is the massive quantity and size of the programs. This optimization technique is a step towards resolving this issue. One of the main benefits that is provided by this method, for microprocessors, is the reduction in size of the executable code. Several data mapping and allocation strategies have been incorporated in this domain and varied approaches have been proposed in the past, for this technique.

Neil Johnson et.al [1] provided identification techniques in their proposed work for converting available groups of single instructions (load and store) and encapsulating them into multiple access instructions (load and store). This conversion of several instructions into a single multiple memory access instructions could be done efficiently by their proposed algorithm. It also guided the allocation and storage of resources in a manner that is precise, which previously is attributed only to opportunistic optimization. The implementation of their Solve MMA algorithm with the Value State Dependence based Graph framework is quite similar to the Simple offset algorithm for Assignment. Their model was analysed using an ARM thumb processor as the primary example and achieved significant code size reduction. Moving forward in this domain, the analysis done on the compatibility for combination of several instructions can be carried out in a more obsessive manner to identify and exploit more combinational opportunities. General purpose computations on graphic processing units can be optimized with the usage of novel optimizing compilers[2] that mitigate challenges of effectively utilizing the memory hierarchy present in the GPU and management of the parallelism. A functionally correct naive GPU kernel function is accepted by the compiler without considering the degree of performance optimization present. The memory access patterns are derived by applying code analysis methodologies to generate the appropriate kernel invocation parameters and kernel optimization techniques. Partition camping is eliminated by address offset insertion or remapping thread blocks. Memory bandwidth is improved using memory coalescing and vectorization.

An innovative way for the exploitation of re-usage of available data was proposed, with the main objective of reducing the number of operations carried out on the memory storage banks. The proposed approach [3] focusses primarily on the Coarse Grained Reconfigurable Architectures (CGRAs), which make use of a local multi bank memory. Figure 2 illustrates the typical architecture of CGRAs.

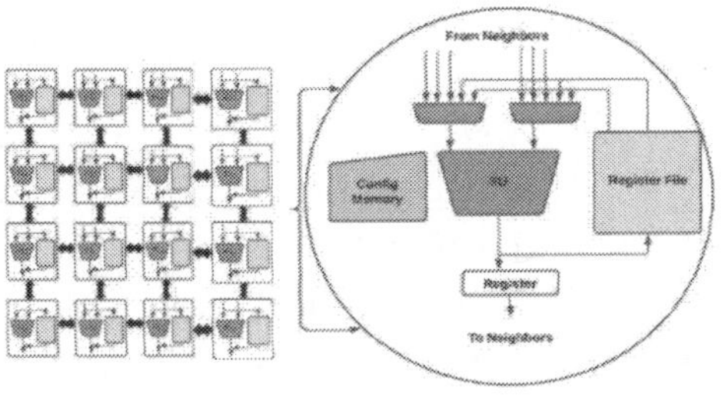

Fig. 2. CGRA Architecture

CGRAs are highly regarded for power efficiency and help reach peak performance metrics with optimal power. The entire architecture of the system is structured in a simple manner and the mapping of application holds the most complexity. The optimization technique used in this aims to solve the data mapping challenges faced in CGRAs where the same memory bank cannot be accessed simultaneously by two or more PEs. Their experimental results concluded their compilation approach that emphasizes on local memory awareness and generates mappings that provide a drastic increase in the performance when compared to typical scheduler does not focus on the memory allocation.

4 Cache Optimization Techniques

Making use of registers and cache memory to their maximum potential, in order to increase the efficiency of our computation is the core principle behind the process of cache optimization. The commutative and associative properties of the computations can be utilized to form a sequence of operations that retain the integrity of the results, while making best use of the memory resources available. The total number of instructions required can be reduced significantly by storing the frequently used data, close by. The use of an appropriate scheduling technique is also crucial in providing speedups to the performance. Activities such as reducing the hit time, increasing cache bandwidth, reducing the miss penalty, reducing the miss rate and reducing the miss penalty through parallelism.

A simple set of optimization techniques focussing on reuse of cache lines in order to find the favourable loop configurations was originally proposed by Steve Carr et al[4]. This ultimately optimizes the compiler by improving the data locality. The integration of loop fusion, reversal, permutation and distribution to provide an enhancement in data locality was one of the prime factors for the success of the optimization approach. This approach provides opportunities for improvements in machine independent programming, primarily in FORTRAN program optimizations. Cache locality optimization techniques focussed towards recursive programs have been developed in the past. Dynamic splicing and distinct function invocations by employing data effect annotations has helped recognize concurrency opportunities and reduction of reuse distance for faster data reuse access. Using non kernel lightweight threads, function invocations are executed in an interleaving manner and dependencies of a program are detected. Multicore execution is made possible by employing a nested fork join model.

With the expansion and rise of multithreaded architecture multithreading based solutions, adaptive optimizing compilations have come into existence that have extended optimization principles to multithreaded architectures. A data cache aware approach was proposed, that lowers the object conflicts misses by establishing a suitable data layout[5]. This approach serves as an expansion of the previously established cache based optimizations to multi-threaded architectures. CCDP application in single core architectures has been proven effective to achieve maximum performance. However the beneficial features are not exhibited when this method is employed in multi-architecture systems. The above mentioned approach has been a significant step forward in solving this problem by providing extensions to the existing CCDP. A general application scheduling approach as well as specific known application approaches have been demonstrated, employing the core principles of cache conscious data placement (CCDP). CCDP application in single core architectures has been proven effective to achieve maximum performance. However the beneficial features are not exhibited when this method is employed in multi-architecture systems. The above mentioned approach has been a significant step forward in solving this problem by providing extensions to the existing CCDP.

Another novel approach that aims to provide a combinational solution of both high level and low level optimization techniques to attain ideal results in multicore processors and parallel platforms[6]. Innovative approaches for parallel programs such as carrying out fundamental analytical strategies such as Side Effect as well as May-Happen-in-Parallel, combined with a Scalar Replacement method for eliminating the amount of load form the high level optimizations part of the proposed approach. Another major improvement in this paper is the introduction of a new Isolation consistency model that outperforms established models in terms of providing replacement transformation chances. Retaining the properties exhibited by Linear Scan, in terms of the time of compilation and memory efficiency forms the central idea of the novel approach that deals with allocation of registers. This is included in the low level optimization strategies part of the model. The experimental results of the research work considered the allocation phase on the Bipartite Liveness Graph as an optimization problem and the main motive of the assignment phase is the prevention of the total count of spill instructions. Data Locality based scheduling for an OpenCL compiler was implemented by intense analysis of the memory locations accessed in order to set up a proper scheduling strategy [7]. This procedure leads to the optimization in the order of memory access and its patterns. Previously established work did not take the relation between memory access patterns and work scheduling into account. The changes made to the system in this paper has a significant influence on the speedup and has been proven against standard benchmarks (Parboil and Rodinia).

5 Machine Learning Based Optimization Techniques

The growth of interest in the field of machine learning has increased exponentially, transforming the small computing domain to one of the world's most captivating fields in technology. The reason behind this rise in interest is due to the massive flexibility of the domain, with the ability to be employed in different applications, as a means of improving efficiency and accuracy of the original system. Machine learning is of prime interest in the field of compiler optimization as it has the capability to provide assistance to the programmer on the necessary code optimization techniques that need to be employed for maximum performance, in terms of both amount of power consumed and execution time. A machine learning model can help the programmer understand the best technique for optimization and also evaluate the existing techniques employed. The perspective of considering the finding of code improvement opportunities as an optimization problem has solidified the position of machine learning models as the provider of the optima. Compiler developers focus on minimizing execution time, reducing code footprint and the usage of minimal power as criteria to judge the effectiveness of changes made to the compiler heuristics and techniques applied for optimization. Machine learning algorithms can be configured and modified to make the appropriate decisions for the developers. This can be categorised into a two-step process. The first step constitutes the learning of the model by utilizing data provided for training. The second step is deployment of the model to the real time environment for predicting the optimization results for unknown programs. Compiler options are determined by the cost functions and compiler heuristics are extremely dependent on this. The quality metrics employed can differ based upon the optimizations goals. These metrics include execution time, energy consumption, code size etc. The implementation of a cost function is essential for the compiler to choose the best optimization option from a wide collection of techniques without running the program separately by incorporating each technique into compilation. Based on the efficiency score obtained from the model, we can decide upon whether we want to rectify the problem or not.

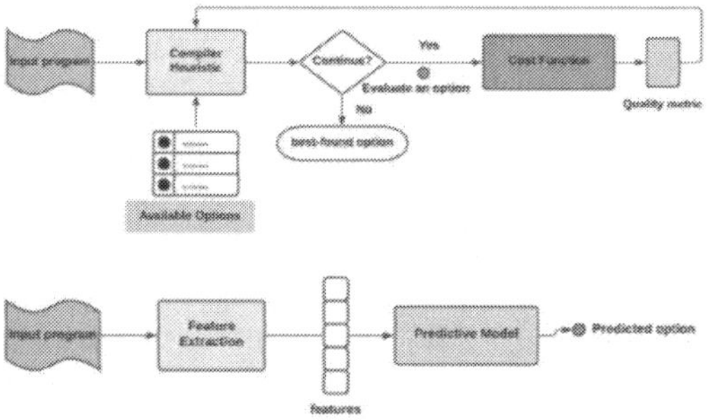

Fig. 3. Prediction Process

Decision trees have been the go to models for several optimization problems in the past [8]. Research conducted along these lines were geared towards identifying loop unroll factors[9], strategies for loop optimization and parallelization.[10], determining the viable and efficient algorithm implementations[11] and making decisions on GPU acceleration's profitability. The ability to interpret and visualize the process of decision trees makes it advantageous in multiple applications. Support Vector Machine[12] based classifications have been carried out in advanced cases for predicting optimizing opportunities[13].

Ensemble models involving SVMs[14], Decision Trees and K means Clustering were also tested out for this purpose and proved efficient in identifying optimization opportunities[15]. The machine learning model eases the burden of the programmer by automating the ideal optimization finding process. There is scope for research in determining whether the ML model can be extended to other compiler scenarios like analysis. With the rise of multicore architectures which provide parallel solutions to computationally intensive tasks, reinforcement learning based[16] and decision tree based[17] options for scheduling parallel sections of the programs are now gaining interest. The models have also been implemented for predicting the optimal number of threads[18] required for seamless execution of the program.

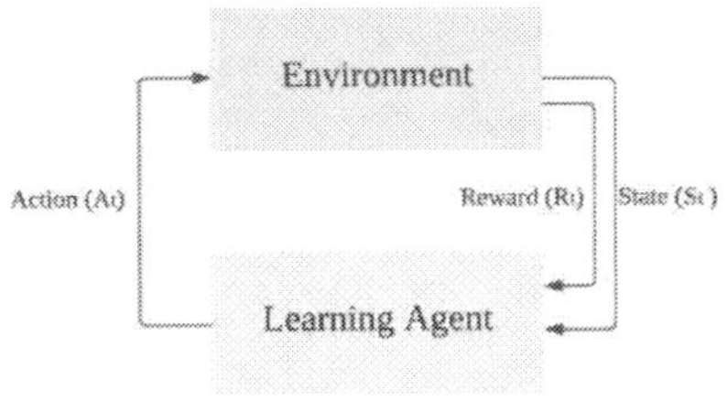

Fig. 4. Reinforcement Learning Model

The environment of execution plays a huge role in determining the effectiveness of an RL model. As the possible actions available to the model increases, the time to reach the convergence point to a proper solution also increases. The uncertainties in the environment in a dynamic computing system can be increased due to unpredictable disturbances. Today, RL algorithms have been combined with deep learning for learning of the value function in order to produce the most accurate results[19].

While static code feature extraction has no runtime overhead, it has its own set of vulnerabilities. The presence of data about code segments that are rarely executed can mislead the machine learning algorithm. Loop bound information can only be attained by the compiler during execution as it depends on the nature of user input. Resource content is not experienced in static code features and depends upon the available hardware. Hence dynamic compilers have been on the rise. The concept of ML has also been extended to the dynamic compiler by formulating a neuro-evolution based ANN network that predicts the order of benefits obtained when a code segment is being optimized [20]. Markov decision based phase ordering and current state code characterization are two of the main attributes of this approach, which is tested against standards set of Java benchmarks and shows significant increase in efficiency.

Extensive and exhaustive searching of compiler optimization techniques due to the large number of options present is infeasible on a large scale. This approach towards figuring out compiler optimizations is categorized as autotuning[21] or iterative compilation[22]. Iterative optimization techniques for compilers portray high efficiency and increased performance but are limited by the large cost of evaluations. A scalable approach, that is independent of compiler architecture, was implemented with the goal of increasing the speed of iterative optimization[23]. This novel approach reduces the number of evaluations using predictive modelling in order to locate the parts that are estimated to provide the maximum increase in performance. The model trains by modelling features and shape properties obtained by iterative evaluation, done on a defined set of programs. This model is implemented to figure out the focus points for optimization on a new set of programs. Such models have been proven to increase average speed up significantly. Attaining proper features is vital and time consuming, the generation of features automatically from the intermediate representation of the compiler [24]. Approaches towards solving this require huge grammar to be provided and work based upon greedy programming[25]. A powerful technique that has been explored limitedly is the implementation of Gaussian Processes for optimization [26].Considered one of the most renowned techniques before the advent of deep neural networks, this method is a robust way of approaching optimization problems when the training data is minimal[27]. The confidence interval provided can be used by the programmer to analyse the pros and cons of a specific implementation scenario.

6 Conclusion

Starting from the basics of a compiler to modern day compiler optimization techniques, The paper aims to provide an overall survey of the imperative optimization techniques that are of use today. The paper provides a brief overview of the properties of the memory based and machine learning techniques that have been employed for the purpose of compiler optimization. Multi memory allocation and cache based optimization techniques have been elaborated on in the category of memory based optimization techniques. Machine learning techniques for identification of potential opportunities of optimization have been provided with details of each approach surveyed. The extension of these techniques to different compilers and architectures have also been explored. Further research can look into the diverse possibilities of using machine learning algorithms to improve data locality and reuse to provide indirect optimizations to the compilers.

References

1. Johnson, Neil & Mycroft, Alan. (2004). Using Multiple Memory Access Instructions for Reducing Code Size. 2985. 265-280. 10.1007/978-3-540-24723-4_18.
2. Yi Yang, Ping Xiang, Jingfei Kong, and Huiyang Zhou. 2010. A GPGPU compiler for memory optimization and parallelism management. SIGPLAN Not. 45, 6 (June 2010), 86–97. DOI:https://doi.org/10.1145/1809028.1806606
3. Yongjoo Kim, Jongeun Lee, Aviral Shrivastava, and Yunheung Paek. 2011. Memory access optimization in compilation for coarse-grained reconfigurable architectures. <i>ACM Trans. Des. Autom. Electron. Syst.</i> 16, 4, Article 42 (October 2011), 27 pages. DOI:https://doi.org/10.1145/2003695.2003702
4. Steve Carr, Kathryn S. McKinley, and Chau-Wen Tseng. 1994. Compiler optimizations for improving

data locality. SIGPLAN Not. 29, 11 (November 1994), pp. 252-262.

5. Sarkar S., Tullsen D.M. (2008) Compiler Techniques for Reducing Data Cache Miss Rate on a Multithreaded Architecture. In: Stenström P., Dubois M., Katevenis M., Gupta R., Ungerer T. (eds) High Performance Embedded Architectures and Compilers. HiPEAC 2008. Lecture Notes in Computer Science, vol 4917. Springer, Berlin, Heidelberg. https://doi.org/10.1007/978-3-540-77560-7_24
6. Barik, Rajkishore. "Efficient optimization of memory accesses in parallel programs." (2010) Diss., Rice University. https://hdl.handle.net/1911/62060.
7. Hee-Seok Kim, Izzat El Hajj, John Stratton, Steven Lumetta and Wenmei Hwu. "Locality Centric Thread Scheduling for Bulk-synchronous Programming Models on CPU Architectures," in Proceedings of the 2015 International Symposium on Code Generation and Optimization (CGO '15), February 2015.
8. A. Monsifrot, F. Bodin, and R. Quiniou, "A machine learning approach to automatic production of compiler heuristics," in Proc. Int. Conf. Artif. Intell. Methodol. Syst. Appl., 2002, pp. 41–50.
9. H. Leather, E. Bonilla, and M. O'Boyle, "Automatic feature generation for machine learning based optimizing compilation," in Proc. 7th Annu. IEEE/ACM Int. Symp. Code Generat. Optim. (CGO), Mar. 2009, pp. 81–91
10. H. Yu and L. Rauchwerger, "Adaptive reduction parallelization techniques," in Proc. 14th Int. Conf. Supercomput. (ICS), 2000, pp. 66–77.
11. Y. Ding, J. Ansel, K. Veeramachaneni, X. Shen, U.-M. O'Reilly, and S. Amarasinghe, "Autotuning algorithmic choice for input sensitivity," in Proc. 36th ACM SIGPLAN Conf. Program. Lang. Design Implement. (PLDI), 2015, pp. 379–390.
12. Z. Wang and M. F. O'Boyle, "Mapping parallelism to multi-cores: A machine learning based approach," in Proc. 14th ACM SIGPLAN Symp. Principles Pract. Parallel Program. (PPoPP), 2009, pp. 75–84.
13. P. Zhang, J. Fang, T. Tang, C. Yang, and Z. Wang, "Auto-tuning streamed applications on Intel Xeon Phi," in Proc. 32nd IEEE Int. Parallel Distrib. Process. Symp. (IPDPS), 2018.
14. Z. Wang, G. Tournavitis, B. Franke, and M. F. P. O'Boyle, "Integrating profile-driven parallelism detection and machine-learningbased mapping," ACM Trans. Archit. Code Optim., vol. 11, no. 1, p. 2, 2014.
15. M. K. Emani and M. O'Boyle, "Celebrating diversity: A mixture of experts approach for runtime mapping in dynamic environments," in Proceedings of the 36th ACM SIGPLAN Conference on Programming Language Design and Implementation, ser. PLDI '15, 2015, pp. 499– 508.
16. M. K. Emani and M. O'Boyle, "Change detection based parallelism mapping: Exploiting offline models and online adaptation," in Languages and Compilers for Parallel Computing: 27th International Workshop (LCPC 2014), 2014, pp. 208–223.
17. Y. Zhang, M. Voss, and E. S. Rogers, "Runtime empirical selection of loop schedulers on hyperthreaded smps," in 19th IEEE International Parallel and Distributed Processing Symposium, ser. IPDPS '05, 2005.
18. Z. Wang and M. F. O'Boyle, "Mapping parallelism to multi-cores: A machine learning based approach," in Proceedings of the 14th ACM SIGPLAN Symposium on Principles and Practice of Parallel Programming, ser. PPoPP '09, 2009, pp. 75–84
19. Y. Li, "Deep reinforcement learning: An overview," CoRR, 2017.
20. Sameer Kulkarni and John Cavazos. 2012. Mitigating the compiler optimization phase-ordering problem using machine learning. In Proceedings of the ACM international conference on Object oriented programming systems languages and applications (OOPSLA '12). Association for Computing Machinery, New York, NY, USA, 147–162.

    DOI:https://doi.org/10.1145/2384616.2384628
21. K. Datta et al., "Stencil computation optimization and auto-tuning on state-ofthe-art multicore architectures," in Proc. ACM/IEEE Conf. Supercomput., Nov. 2008, pp. 1–12.
22. P. M. Knijnenburg, T. Kisuki, and M. F. O'Boyle, "Combined selection of tile sizes and unroll factors using iterative compilation," J. Supercomput., vol. 24, no. 1, pp. 43–67, 2003.
23. F. Agakov *et al.*, "Using machine learning to focus iterative optimization," *International Symposium on Code Generation and Optimization (CGO'06)*, New York, NY, USA, 2006, pp. 11 pp.-305, doi: 10.1109/CGO.2006.37.
24. M. Namolaru, A. Cohen, G. Fursin, A. Zaks, and A. Freund, "Practical aggregation of semantical program properties for machine learning based optimization," in Proc. Proc. Int. Conf. Compil. Archit. Synth. Embedded Syst. (CASES), 2010, pp. 197–206.
25. H. Leather, E. Bonilla, and M. O'Boyle, "Automatic feature generation for machine learning based optimizing compilation," in Proc. 7th Annu. IEEE/ACM Int. Symp. Code Generat. Optim. (CGO), Mar. 2009, pp. 81–91.
26. C. K. Williams and C. E. Rasmussen, "Gaussian processes for regression," in Advances in neural information processing systems, 1996, pp. 514–520.
27. C. E. Rasmussen and C. K. Williams, Gaussian processes for machine learning. MIT press Cambridge, 2006, vol. 1.

https://doi.org/10.1051/itmconf/20213701022

Speaker Recognition using Random Forest

Khadar Nawas K[1*], *Manish Kumar Barik*[1], *A Nayeemulla Khan*[1]

[1]*School of Computer Sciences and Engineering, Vellore Institute of Technology, Chennai, India*

Abstract. Speaker identification has become a mainstream technology in the field of machine learning that involves determining the identity of a speaker from his/her speech sample. A person's speech note contains many features that can be used to discriminate his/her identity. A model that can identify a speaker has wide applications such as biometric authentication, security, forensics and human-machine interaction. This paper implements a speaker identification system based on Random Forest as a classifier to identify the various speakers using MFCC and RPS as feature extraction techniques. The output obtained from the Random Forest classifier shows promising result. It is observed that the accuracy level is significantly higher in MFCC as compared to the RPS technique on the data taken from the well-known TIMIT corpus dataset.

1 Introduction

A lot of extensive research has been done in the past on the field of speaker recognition in order to achieve a common goal– efficiently identifying a speaker using his/her speech note[1][4]. Speaker Identification systems work on the fact that no two persons in the world have exactly the same voice. Each individual's voice is different as it is attributed to different biological features such as- pitch of the sound, vocal tract's length, sound frequency, the movement of the tongue against the palate, the shaping of the lips, the arrangement of teeth, etc[9].

A state-of-the-art speaker recognition system could really have a significant impact in the field of security. One important application of this would be to identify criminals through their telephonic conversations by listening to some vulnerable keywords[13]. This can eventually save a lot of police work.

Speaker recognition can further be categorized into speaker identification and speaker verification. This paper focuses on the implementation of speaker identification using Random Forest algorithm as a classifier. There are basically three phases in a speaker recognition system: Pre-processing the audio signals, feature extraction and the matching/classification phase. Pre-processing refers to the phase where the inputted audio signal is cleansed, i.e. the signal is made free of the environmental noise in order to efficiently recognize the significant voice signal. The phase of feature extraction involves generating the feature vectors for the given audio signal which constitutes the key characteristics of any individual's voice.

* Corresponding author: khadarnawas.k@vit.ac.in

After the corresponding feature vectors have been obtained for each speaker, these are then modelled to get the specific template for that speaker. Classification algorithms determine distinguishing characteristics among the speakers based on the features obtained. Lastly the testing stage matches the speaker's speech note to the existing classes. The match with the highest score is then assigned to the specified speaker.

The rest of the paper comprises of the below sections. Section II describes the related works and background study for this paper. Section III discusses the dataset used for the model experiment. Section IV illustrates the pre-processing phase of the voice signal, Section V discusses the two feature extraction methods-RPS and MFCC, Section VI discusses the various classification techniques with emphasis on the Random Forest classifier used in our model, Section VII shows the experimental results obtained from the dataset, Section VIII concludes the rest of the paper by citing important inferences obtained from the two methodologies implemented.

2 Related Works and Background

Speech recognition can be mainly divided into three categories: speech recognition (determining the actual content of the spoken utterance), speaker recognition (identifying the speaker's voice) and language recognition (determining the language of the spoken word)[10]. Speaker recognition, a sub-domain of speech recognition can further be categorized into speaker verification and speaker identification[5]. The former deals with confirmation of the affirmed identity by matching

https://doi.org/10.1051/itmconf/20213701022

his/her voice characteristics with the one already stored in the system (1:1 relationship)[7]. The latter approach aims to determine the correct voice print of the corresponding speaker from a set of known speakers (1:N relationship). SID system can be implemented as per a text-dependent (where the utterances given by the speaker are already known by the system) and a text-independent system (where the utterances are not stored anywhere and the speaker is identified using only his voice features).

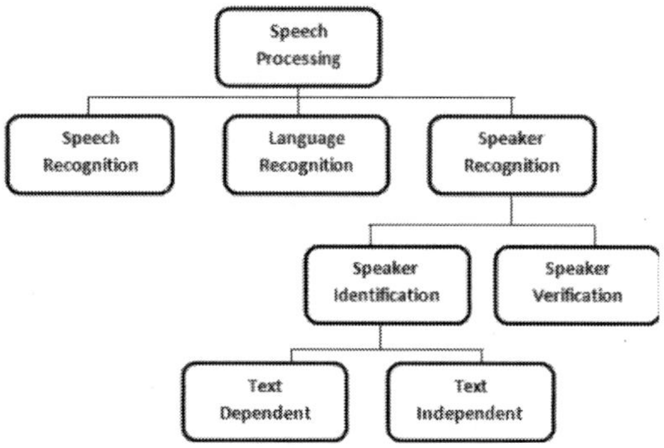

Figure 1. Classification of speaker recognition

3 Dataset

Every machine learning model needs a quality data set in order to operate efficiently. In our implementation of a SID system, we worked with a downloaded dataset, viz. as TIMIT DATASET. TIMIT consists of broadband recordings of 630 speakers of eight major dialects of American English, each further consisting of ten phonetically rich sentences.

We take a set of 38 speakers from the TIMIT dataset. Every speaker's voice notes are contained in a directory named as speaker_dataset that contains speakers labelled as 0 to 37. Each directory contains ten phonetically rich sentences of that particular speaker. So, a total of 380 voice notes will be processed for the extraction of characteristics that are specific to that particular speaker.

4 Pre-processing

Pre-processing is the first phase in the process of recognizing a speaker. It involves the removal of unwanted components such as background noise, unwanted silence in between the uttered words, etc [2]. An efficient speaker identification system must be able to extract the clean speech from the corrupted one. The two most popular methods of pre-processing the speech signal involves spectral subtraction and adaptive noise cancellation[2]. Spectral Subtraction is a method in which non-speech regions in the speech are estimated and are then the average of those regions is subtracted from the speech signal. The efficiency of a pre-processing phase is validated in terms of SNR(Signal-to-Noise Ratio). It refers to the ratio of the power of the correct signal to the noise.

$$SNR = 20 \log_{10}(V_{signal} / V_{noise})$$

where V_{signal} is the voltage of the input signal and V_{noise} is the voltage of the noise.

5 Feature Extraction

Feature extraction is the phase where the pre-processed signal is fed to the model and feature vectors are obtained. Features, in the context of machine learning refers to the numerical characteristics of any given object. In this case the feature vectors can be referred to contain distinct entities of any given speaker such as pitch, amplitude, etc. Feature extraction can be considered as a dimensionality reduction process since the input data fed to the system are too large to be processed and are highly redundant in nature. In our paper, we provide an in depth comparison of two popular feature extraction techniques- MFCC and RPS.

5.1 RPS (Reconstructed Phase Space)

Speaker Identification systems typically uses spectral-based features that reject the state of the phase present in speech signals[18]. Thus, employing some extra features(such as attributing to the turbulence of speech) where the signal phase is not rejected may fill this gap. Moreover, ongoing research on this field have indicated that using nonlinear features may improve the exhibitions of SID systems [16][17]. In one dimensions, the state on any particle can be determined by specifying its position and velocity. In this case, its phase space is a plane. In the case of a voice signal, we have a time series instead of a phase space object. Hence, we need to convert the observations into state vectors. This is phase space reconstruction. RPS method is a multidimensional space where the coordinates are produced by a shift and delay of a sample of a 1-d signal in a time series[6]. In our approach, we embed the speech signal in the reconstructed phase space in order to extract some useful non-linear features. The RPS Distribution plot (scatter graph) in two dimensions is constructed for the vowel /a/ is given in Fig 2.

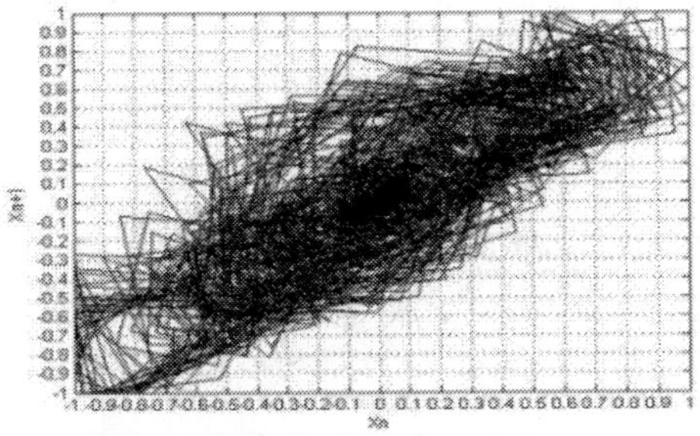

Figure 2. RPS model for the vowel 'a' sound

5.2 MFCC(Mel-Frequency Cepstral Coefficients)

MFCC is one of the most popular and widely used techniques in the field of feature extraction. In our implementation we make use of the librosa library to obtain 40 MFCC features of each speaker's voice sample. This section illustrates the various techniques used to extract features from an audio signal as described in Figure 3.

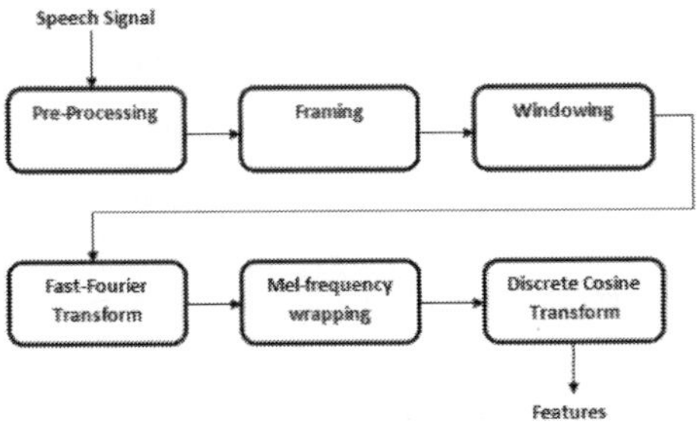

Figure 3. MFCC block diagram

The various phases that are a part of MFCC technique are:

5.2.1 Framing

A speech signals uttered by the user are continuously changing. So, in order to process the signal, we segment it at duration of 25ms such that the signal in a frame becomes quasi-stationary (exhibiting stationary behaviour). This segment often corresponds to a particular phoneme such as "duh" in "duck". Further, to avoid discontinuities among consecutive frames, every two frames are mutually overlapped.

5.2.2 Windowing

Windowing is done in order to minimize the inconsistent fluctuations at the edges of each frame also referred to as spectral distortion [9]. It is implemented by using a window function that lowers down the value to zero at the edges of each frame.

5.2.3 Fast Fourier Transform

This stage converts the time domain signal(Figure 5) into the corresponding frequency domain signal(Figure 4). This phase basically disintegrates the composite signal into the constituent frequencies that make up the composite signal. The output frequencies obtained are referred to as the spectrum of the given frame. The spectrum of all the frames in the given signal form a spectrogram.

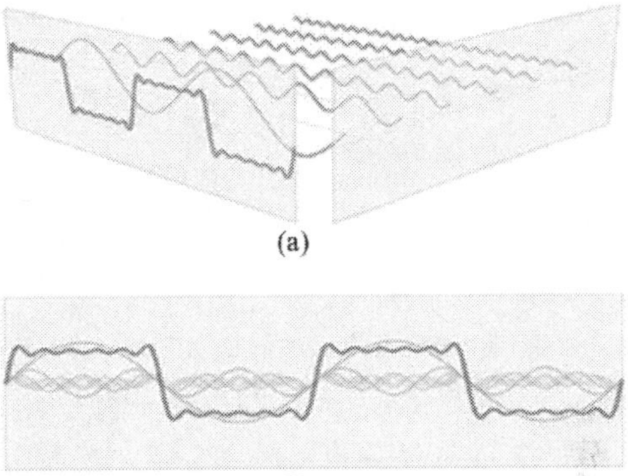

(a)

(b)

Figure 4. Fast-Fourier Transform converts a signal in time domain (a) to a signal in frequency domain (b).

5.2.4 Mel-Frequency wrapping

Several studies have shown that humans perceive frequency not at a linear scale. At low frequencies, the difference in comparing two frequencies as perceived by the human ear is less than the same difference in the frequencies at a higher frequency range [9] as shown in Fig 6. This demands for a different type of scale known as mel scale that sets a threshold frequency below which the mel scale behaves linearly(at low frequencies) and above that logarithmic behaviour(at high frequencies).

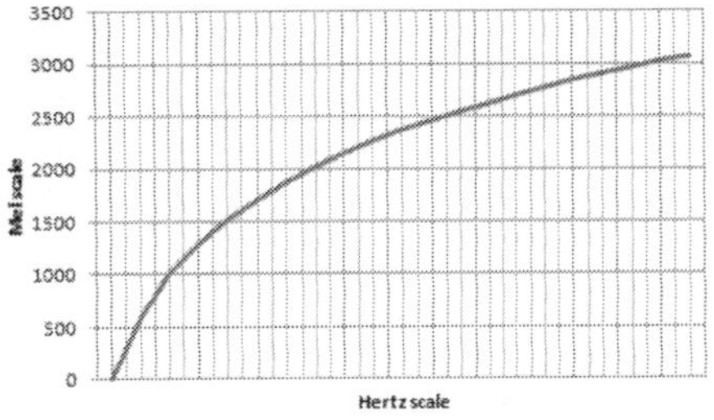

Figure 5. pitch scale vs. mel scale

This is simulated using mel frequency filter bank which has mel frequency scale that converts each derived frequency into its corresponding mel-

https://doi.org/10.1051/itmconf/20213701022

frequency [11]. Mel can be calculated using the below formula:

$$\text{Mel}(f) = 2595 \log_{10}(1+f/700)$$

5.2.5 Discrete Cosine Transform.

The log spectrum thus obtained contains a lot of frequencies. So, in order to remove the whole lot of dimensions, we implement dimensionality reduction in which we combine frequencies close to each other. DCT is basically used to convert the obtained log Mel spectrum into time domain known as cepstrum[3]. The output of DCT gives what is referred to as mel frequency cosine coefficients. The set of these coefficients form the feature vectors are ready for modelling.

6 Classification

The extracted MFCC features are then classified into various classes that corresponds to different speakers. Over the years several classification algorithms have been proposed in the task of speaker identification, all with different rates of success. Most of the popular ones include Support Vector Machine(SVM), K-Nearest Neighbours, Random Forest(RF), K-means and the latest among them Deep Neural Networks (DNN)[8]. In our experiment, we split the features extracted from the speaker dataset as per 75% as training data and 25% as testing data that are then fed onto the Random Forest classifier.

Support Vector Machine is one the most widely known ML algorthim that is used in classification purposes. SVM tries to classify the given set of speakers by trying to come up with a hyperplane that seperates the characteristics the constitutes a particular speaker. One drawback of SVM is it is not particularly suitable for large datasets (size > 10,000). Our feature extraction model of 380 voice wav files generate a massive number of rows around 1,99,000 while using MFCC extraction technique and around 1,89,00,000 in case of features obtained from that of the Reconstructed Phase Space(RPS) technique. So, we opt for a different approach that doesn't really hit the performance issues with larger datasets[15]. In this paper we implemented the Random Forest classification model in order to classify the given speakers' feature. Apart from working efficiently with large datasets, fellow researches in their work [12] proposed RF classifiers as the best performing among 179 classifiers in terms of accuracy and precision. Also, in [14] RF classifiers showed better performances compared to KNN and SVM classifiers in terms of accuracy, precision, recall and f-measure.

Random Forest algorithm is a type of supervised classification algorithm that uses a large combination of decision trees and then trains each of them on a different set of observations. The final predictions are done on the random forest based on the calculations done by averaging each prediction of a single tree. A random forest is considered better than a single decision tree because of the fact that this reduces the over-fitting problem which can be produce really inefficient results among various other classification techniques.

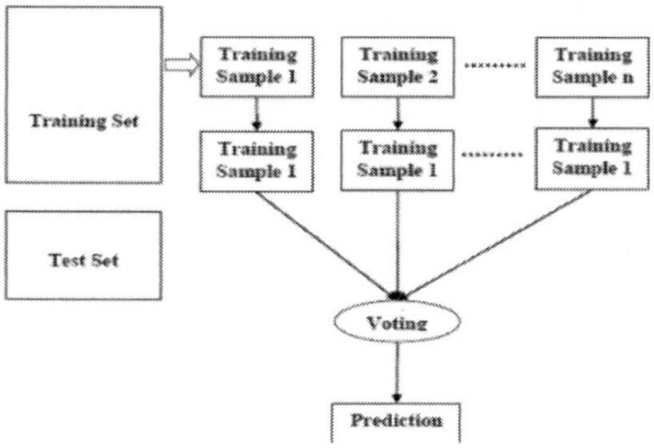

Figure 6. Phases in a Random Forest classifier

7 Experimental results

Table 1. Comparing the two techniques based on accuracy

Speaker Recognition	Training accuracy	Testing accuracy
Random Forest using RPS features	73.6%	71%
Random Forest using MFCC features	100%	97%

Table 1 shows the results obtained from the proposed two methodologies in system identification system using a Random Forest algorithm to classify the features extracted from Mel Frequency Cepstral Coefficient(MFCC) technique and Reconstructed Phase Space(RPS). The MFCC approach shows promising results with an accuracy rate of 100% on training data and 97.4% on testing data as compared to RPS methodology that showed an accuracy of 73.6% on training data and 71% on testing data.

8 Conclusion and Future work

This paper reported the development of a Speaker Identification System using a RF classifier. The paper also briefly described all the stages covered from pre-processing of a speech wav file to eventually identifying the speaker. We infer from the results obtained that spectral features such as MFCC may prove its benefits in showing accurate results and could not be easily replaced by some novel features. But, as a starting point, we propose that future speaker identification software may include a combination of non-linear features(RPS) and the traditional spectral features(MFCC) in order to further improve the performance parameters of speaker identification systems.

References

1. H. S. Jayanna and S. R. Mahadeva Prasanna, May 2009, "Analysis, Feature Extraction, Modeling and Testing Techniques for Speaker Recognition".
2. Yakubu A. Ibrahim, Juliet C. Odiketa, Tunji S. Ibiyemi, 2017, "Preprocessing technique in automatic speech recognition for human computer interaction: An overview", pp186-191
3. Garima Vyas, Barkha Kumari, June 2013, "Speaker Recognition System Based on MFCC and DCT", Vol. 2, Issue 5, pp167-169.
4. Jyoti B. Ramgire, Prof. Sumati M.Jagdale, April 2016, "A Survey on Speaker Recognition With Various Feature Extraction And Classification Techniques", Vol. 03, Issue 04, pp709-712.
5. Satyam P. Todkar, Snehal S. Babar, Rudrendra U. Ambike, Prasad B. Suryakar, April 2018, "Speaker Recognition Techniques: A review", pp1-5.
6. Nisha.V.S , M.Jayasheela, February 2013,"Speaker Identification Using Combined MFCC and Phase Information", Vol. 2, Issue 2, pp1149-1152.
7. Shilpa S. Jagtap and D.G.Bhalke, 2015, "Speaker Verification Using Gaussian Mixture Model".
8. Sreenivas Sremath Tirumala, Seyed Reza Shahamiri, November 2016, "A review on Deep Learning approaches in Speaker Identification".
9. https://shodhganga.inflibnet.ac.in/bitstream/10603/183865/9/09-chapter%203.pdf, "From Speech to Feature Vectors", pp83-128.
10. Joseph P. Campbell, September 1997, "Speaker Recognition: A Tutorial", Vol. 85, no. 9.
11. Fang-Yie Leu, Guan-Liang Lin, 2017, "An MFCC-based Speaker Identification System", pp1055-1062.
12. Manuel Fernández-Delgado, Eva Cernadas, Senén Barro, 2014, "Do we Need Hundreds of Classifiers to Solve Real World Classification Problems?" .
13. Parashar Dhakal, Praveen Damacharla, Ahmad Y. Javaid, Vijay Devabhaktuni, 2019, "A Near Real-Time Automatic Speaker Recognition Architecture for Voice-Based User Interface".
14. Tumisho Billson Mokgonyane, Tshephisho Joseph Sefara, Mercy Mosibudi Mogale, Madimetja Jonas Manamela, 2019, "Automatic Speaker Recognition System based on Machine Learning Algorithms".
15. Mohammed Zakariah ,September 2014, "Classification of large datasets using Random Forest Algorithm in various applications: Survey".
16. Povinelli RJ, Johnson MT, Lindgren AC, Roberts FM, 2006, "Statistical models of reconstructed phase spaces for signal classification", IEEE Trans Signal Process 2006; 54:2178–86.
17. Yasser Shekofteh and Farshad Almasganj, 2013, "Feature Extraction Based on Speech Attractors in the Reconstructed Phase Space for Automatic Speech Recognition Systems".
18. http://shodhganga.inflibnet.ac.in/bitstream/10603/213852/16/16_chapter7.pdf, "Effective Speaker Spotting based on Nonlinear Properties of Vocal Tract" , pp161-180.

Lazy Learning Associative Classification with WkNN and DWkNN Algorithm

Preeti Tamrakar[1], *and S. P. Syed Ibrahim*[1]*

1School of Computing Science and Engineering, VIT Chennai Campus, Tamil Nadu, 600127 India

Abstract. One of the algorithms, which prudently denote better outcomes than the traditional associative classification systems, is the Lazy learning associative classification (LLAC), where the processing of training data is delayed until a test instance is received, whereas in eager learning, before receiving queries, the system begins to process training data. Traditional method assumes that all items within a transaction is same, which is not always true. This paper recommends a new framework called lazy learning associative classification with WkNN (LLAC_WkNN) which uses weighted kNN method with LLAC, that gives a subset of rules when LLAC is applied to the dataset. In order to predict the class label of the unseen test case, the weighted kNN (WkNN) algorithm is then applied to this generated subset. This creates the enhanced accuracy of the classifier. The WkNN also gives an outlier more weight. By applying Dual Distance Weight to LLAC named as LLAC_DWkNN, this limitation of WkNN is resolved. LLAC_DWkNN gives less weightage to outliers, which improve the accuracy of the classifier, further. This algorithm has been applied to different datasets and the experiment results demonstrate that the proposed method is efficient as compared to the traditional and other existing systems.

1 Introduction

Associative classification (AC) [1] has been popularly explored in the last decades and taken full advantage of it in many successful real time applications. It integrates classification and frequent pattern mining into one system. AC can be of two types; Eager associative classification (EAC) and Lazy learning associative classification (LLAC). Two steps are involved in the construction of EAC. Training dataset is used to construct the model in the first step and test dataset is in the second step. The model is validated by using test dataset. EAC generates a large set of rules. During classification, many rules are not useful for classification and some important rules may not be generated. To overcome this issue, LLAC (non- eager) is proposed by Adriano Veloso et al. [2] where focus is given to the attributes of the given test cases. By this way, the chance of generating useful rules is increased. The processing of data is delayed by LLAC until the test instance requires classification. There are two phases in LLAC: 1. Subset Generation and 2. Subset evaluation. Test dataset is used to generate the subsets. In the second phase, subsets generated are assessed. Probability, support and confidence measures are used for the evaluation. For the provided test instance, the highest probability class label is allocated.

Eager associative classification generates a large set of rules. If one instance matches several rules, which classes are different, that is called rule conflict problem. On the other hand, kNN classifies test instance based upon the nearest training samples in the dataset. But it also suffers from two deficiencies: a. Its efficiency is less because it has to compute the similarity/ distance between test and every single training instance and b. Accuracy is totally dependent how to determine the optimum value for k.

To overcome above said issues, Associative classification with kNN (AC-kNN) is proposed by Zaixiang Huang et al. [3]. By integrating these two concepts, it can resolve the rule conflict and low efficiency problem. Rules are developed in AC-kNN method based on the AC system such as CARS discovery, rule ranking, and pruning. kNN applies for the purpose of classification. Haida et al. [4] suggested HAC that would combine the AC algorithm with the NB algorithm to boost classification accuracy and to minimize the memory space and execution time.

Eager Associative Classification with kNN (EAC-kNN) resolves the issue of rule conflict and low efficiency, but Eager associative classification generates many rules. Some of them are not useful and some important rules are not generated. To address this issue, LLAC with Fixed kNN (LLAC_kNN) [5] is proposed in our previous work. Lazy learning associative classification (LLAC) delays the data processing until classification is required for a query. It generates rules specific to each test instance. In LLAC_kNN method, LLAC is applied to the dataset to generate the subset in the first step. In the second step, kNN is applied for the classification. By this way, better accuracy is achieved, but still suffers from the sensitivity problem of choosing k for KNN. To address the sensitivity problem of kNN, Dudani [6] proposed Weighted k Nearest Neighbor (WkNN), in that nearest neighbor gets the heavy weight and farthest neighbor gets the light weight according to their distance from the test instance. This WkNN method is used in the proposed work.

*corresponding author: syedibrahim.sp@vit.ac.in

For the further improvement in performance of LLAC_FkNN, This paper proposes LLAC with Weighed kNN (LLAC_WkNN). In this, subset is generated based on LLAC method. Then the distance is calculated between the generated subset and the test instance. Based on the distance, weight is calculated and assigned to all the k neighbors. In WkNN method, outliers are also getting high weights. To solve this issue Jianping Gou et al. [7] proposed Dual Weighted kNN (DWkNN) method. In that dual weight is assigned in place of the single weight. By applying dual weight, in between outliers get the lesser weight that leads to an accurate classifier. Thus, this paper proposes LLAC_DWkNN method to reduce the weight of outliers. The proposed work has improved the accuracy of the existing system and traditional systems.

At first, Liu B et al. [1] integrated classification and association rule mining (ARM) and referred to as associative classification (AC). AC constructs Class Association Rules (CARS), in that left-hand side is rule and right-hand side is restricted to a class label. AC has been applied fruitfully in several tasks of classification such as detecting fraud, spam filtering, diagnosing cancer, and so on. There are two forms of associative classification, EAC and LLAC.

There are two stages to the Eager Associative Classification (EAC). EAC uses association rule mining in the first step to build the rules on class association. In phase two, the classifier is constructed, based on the CARs created from phase one. The EAC offers enhanced accuracy, but some drawbacks exist.

2 Introduction

Associative classification (AC) [1] has been popularly explored in the last decades and taken full advantage of it in many successful real time applications. It integrates classification and frequent pattern mining into one system. AC can be of two types; Eager associative classification (EAC) and Lazy learning associative classification (LLAC). Two steps are involved in the construction of EAC. Training dataset is used to construct the model in the first step and test dataset is in the second step. The model is validated by using test dataset. EAC generates a large set of rules. During classification, many rules are not useful for classification and some important rules may not be generated. To overcome this issue, LLAC (non- eager) is proposed by Adriano Veloso et al. [2] where focus is given to the attributes of the given test cases. By this way, the chance of generating useful rules is increased. The processing of data is delayed by LLAC until the test instance requires classification. There are two phases in LLAC: 1. Subset Generation and 2. Subset evaluation. Test dataset is used to generate the subsets. In the second phase, subsets generated are assessed. Probability, support and confidence measures are used for the evaluation. For the provided test instance, the highest probability class label is allocated.

Ranking, pruning and producing a huge number of rules are very challenging process.

Lazy learning associative classification is implemented to address these problems. It postpones data processing until the test query requires classification. Merschmann L and Plastino [8] and [9] implemented lazy learning system using the Highest Subset Probability (HiSP) algorithm. Adriano Veloso et al. have proposed numerous lazy classifiers which have enhanced the accuracy of the classification. The attribute, which has scored highest information gain is selected for the construction of rules in LACI as proposed by Syed Ibrahim et al. [10] and Zhang X [11]. Preeti et al. [12] have proposed various ranking learning AC attribute rankings methods and improved classification accuracy. In [13], Variance; a measure of the spread of the data items of number N, from the mean is used. Variance ranking is used to handle the imbalance problem. Rank Order Similarity (ROS) technique is used for similarity measurement. Decision Tree is used in [14] for the early prediction of the Diabetes. The compact subset generation method is used in ChiSC-AC method [15]. Multiple algorithms and methods are presented by multiple authors, but still there is the possibility to improve the accuracy of the classifier.

This paper is divided into five parts, in which the second section addresses a literature survey of the classification and kNN system. Section 3 deals with proposed method and pseudocode. Discussion of the experimental finding is presented in section 4, and the final section addresses the conclusion.

Eager associative classification generates a large set of rules. If one instance matches several rules, which classes are different, that is called rule conflict problem. On the other hand, kNN classifies test instance based upon the nearest training samples in the dataset. But it also suffers from two deficiencies: a. Its efficiency is less because it has to compute the similarity/ distance between test and every single training instance and b. Accuracy is totally dependent how to determine the optimum value for k.

To overcome above said issues, Associative classification with kNN (AC-kNN) is proposed by Zaixiang Huang et al. [3]. By integrating these two concepts, it can resolve the rule conflict and low efficiency problem. Rules are developed in AC-kNN method based on the AC system such as CARS discovery, rule ranking, and pruning. kNN applies for the purpose of classification. Haida et al. [4] suggested HAC that would combine the AC algorithm with the NB algorithm to boost classification accuracy and to minimize the memory space and execution time.

Eager Associative Classification with kNN (EAC-kNN) resolves the issue of rule conflict and low efficiency, but Eager associative classification generates many rules. Some of them are not useful and some important rules are not generated. To address this issue, LLAC with Fixed kNN (LLAC_kNN) [5] is proposed in our previous work. Lazy learning associative classification (LLAC) delays the data processing until classification is required for a query. It generates rules specific to each test instance. In LLAC_kNN method, LLAC is applied to the dataset to generate the subset in the first step. In the second step, kNN is applied for the classification. By this way, better accuracy is achieved, but still suffers from the sensitivity problem of choosing k for KNN. To address the sensitivity problem of kNN, Dudani [6] proposed Weighted k Nearest Neighbor (WkNN), in that nearest neighbor gets the heavy weight and farthest neighbor gets the light weight according to their distance from the test instance. This WkNN method is used in the proposed work.

For the further improvement in performance of LLAC_FkNN, This paper proposes LLAC with Weighed kNN (LLAC_WkNN). In this, subset is generated based on LLAC method. Then the distance is calculated between the generated subset and the test instance. Based on the distance, weight is calculated and assigned to all the k neighbors. In WkNN method, outliers are also getting high weights. To solve this issue Jianping Gou et al. [7] proposed Dual Weighted kNN (DWkNN) method. In that dual weight is assigned in place of the single weight. By applying dual weight, in between outliers get the lesser weight that leads to an accurate classifier. Thus, this paper proposes LLAC_DWkNN method to reduce the weight of outliers. The proposed work has improved the accuracy of the existing system and traditional systems.

At first, Liu B et al. [1] integrated classification and association rule mining (ARM) and referred to as associative classification (AC). AC constructs Class Association Rules (CARS), in that left-hand side is rule and right-hand side is restricted to a class label. AC has been applied fruitfully in several tasks of classification such as detecting fraud, spam filtering, diagnosing cancer, and so on. There are two forms of associative classification, EAC and LLAC.
There are two stages to the Eager Associative Classification (EAC). EAC uses association rule mining in the first step to build the rules on class association. In phase two, the classifier is constructed, based on the CARs created from phase one. The EAC offers enhanced accuracy, but some drawbacks exist. Ranking, pruning and producing a huge number of rules are very challenging process.

Lazy learning associative classification is implemented to address these problems. It postpones data processing until the test query requires classification. Merschmann L and Plastino [8] and [9] implemented lazy learning system using the Highest Subset Probability (HiSP) algorithm. Adriano Veloso et al. have proposed numerous lazy classifiers which have enhanced the accuracy of the classification. The attribute, which has scored highest information gain is selected for the construction of rules in LACI as proposed by Syed Ibrahim et al. [10] and Zhang X [11]. Preeti et al. [12] have proposed various ranking learning AC attribute rankings methods and improved classification accuracy. In [13], Variance; a measure of the spread of the data items of number N, from the mean is used. Variance ranking is used to handle the imbalance problem. Rank Order Similarity (ROS) technique is used for similarity measurement. Decision Tree is used in [14] for the early prediction of the Diabetes. The compact subset generation method is used in ChiSC-AC method [15]. Multiple algorithms and methods are presented by multiple authors, but still there is the possibility to improve the accuracy of the classifier.
This paper is divided into five parts, in which the second section addresses a literature survey of the classification and kNN system. Section 3 deals with proposed method and pseudocode. Discussion of the experimental finding is presented in section 4, and the final section addresses the conclusion.

3 k Nearest Neighbor Algorithm

In data mining, the Nearest Neighbor rule (NN) proposed by Cover [16], is one of the earliest and easiest classification method. In NN, one test instance is given as input along with the training dataset. It has to identify a point from the training dataset, which is closest to the given test instance and assigns its class label to it.

In kNN [17], As input, a training dataset and a test instance are given. The value of k is determined. The algorithm finds the k nearest neighbors of the test instance from the training dataset by calculating the Euclidean distance as given in Eq. (1) between test and all the training instances. The label of the majority class is assigned to the test query given.

Euclidean distance:

$$dist(t,q) = \sqrt{(q_1 - t_1)^2 + (q_2 - t_2)^2 + \cdots + (q_n - t_n)^2} \quad (1)$$

In kNN algorithm, selecting k is difficult task. Classification result directly depends on the selection of k. If k is very small, accuracy may decrease because the prediction will be done based on very few instances. If k is very large, outliers and noise may interrupt the classification result. To overcome this issue, weight is assigned to the neighbors.

Consider the following scenario in Figure 1. Tringle shape is the test instance. There are two classes; circle and star. Task is to assign a class label for the test instance. If k nearest neighbor is applied, then test instance will be assigned as circle class because of the majority voting. But if weight is calculated and assigned, then majority weighted voting is star class. Weight is also an important factor for classification.

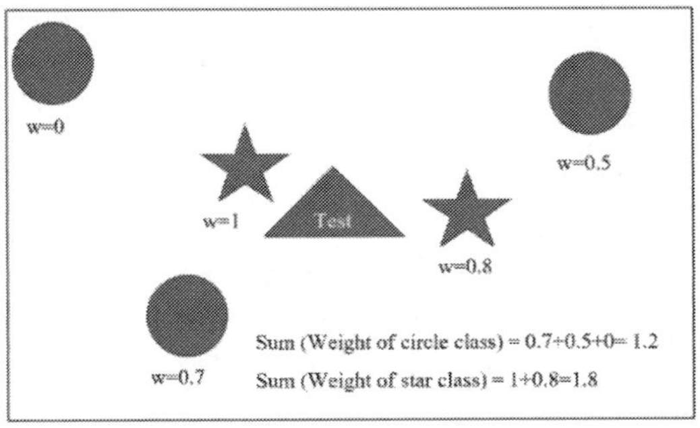

Fig. 1. Illustration of weighted kNN.

Weighted k nearest neighbor (WkNN) is introduced by Dudani, with the concept of assigning more weight to the nearest neighbor and the less weight to the farthest one. After calculating the distance (d1, d2,...,dk) between test instance and training instances, weight (w1, w2,…,wk) is calculated for each neighbor by using the weight formula given in Eq. (2).

$$w_i = \begin{cases} \frac{d_k-d_i}{d_k-d_1}, & if d_k \neq d_i \\ 1, & if d_k = d_i \end{cases} . \quad (2)$$

After calculating the weight for each neighbor, class label is assigned to the given test instance based on the majority weighted voting. If the outliers are present in the neighbors, they are also getting good weights and then the performance of the classifier decreases. To overcome this issue, J. Gou et al. proposed dual distance weighted k nearest neighbor algorithm (DWkNN). In this, dual weight is appointed by multiplying a new weight to the original weight as shown in equation II. By applying dual weight, each neighbor gets lesser weight, so the outliers also get lesser weight and hence the classification performance improves. Dual distance weight formula is given in Eq. (3).

$$\mathbf{w_i} = \begin{cases} \frac{d_k-d_i}{d_k-d_1} \times \frac{d_k+d_i}{d_k+d_1}, & \text{if } d_k \neq d_i \\ \mathbf{1}, & \text{if } d_k = d_i \end{cases} \quad (3)$$

4 Proposed Work

There are two methods proposed in this paper, first is Lazy Learning Associative Classification with Weighted kNN (LLAC_WkNN) and second is Lazy Learning Associative Classification with Dual Weighted kNN (LLAC_DWkNN).

4.1 LLAC_WKNN

The LLAC has been employed by many researchers in various areas and has achieved considerable success when is combined with other algorithms. In machine learning area, this combination or hybridization is to invent new algorithms as a solution necessary in overcoming certain shortcomings observed during the use of classical algorithms. These new algorithms could have new promising features for solving some classification problems. LLAC generates a large number of subsets of rules that degrades the performance of the classifier and kNN is very sensitive to the selection of k. To address these issues, this paper proposes LLAC_WkNN algorithm. In the proposed algorithm, subset is generated based on the LLAC method and the weighted kNN is applied for the classification. The proposed framework is given in Figure 2.

In Algorithm 1, First, subset of test query 't' from the training instances is generated by using information gain attribute. In general, kNN uses Euclidean distance as the distance metric as data is numerical in the dataset, but in this research, data is categorical in the dataset, so similarity function is used to find the similar transaction and arranged in descending order, so similar transaction will be on the top. Similarity is calculated between test query 't' and generated subset 'S' and stored in 'sim'. All the instances of subsets are arranged in descending order according to the similarity value. The more similarity value will be, the more nearest neighbor will be for the test query. Lall and Sharama mentioned that setting a suitable k should satisfy $k = \sqrt{n}$ for training datasets with sample size larger than 100 [18]. Top k nearest neighbors are selected with the class labels and stored in 'x'. Weight 'W= (w1,…,wk)' is calculated for all the k instances. And based on the majority weighted voting, for the given test instance class label is predicted.

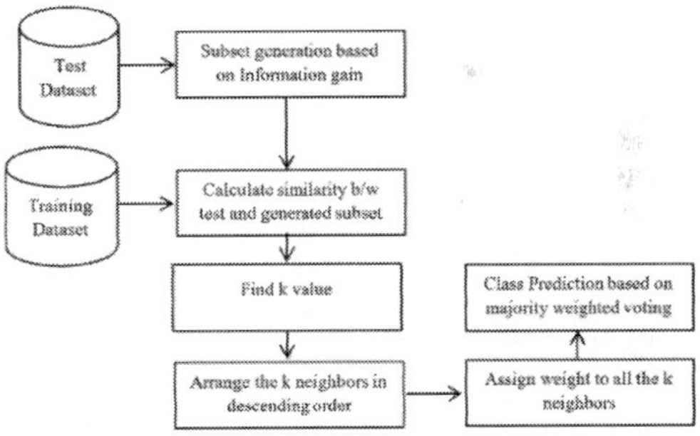

Fig. 2. Proposed Framework for LLAC_WkNN.

Algorithm 1 The proposed LLAC_WkNN algorithm

Input: X = Training Dataset (n X m), Y = Class labels of training dataset, t = test query
Output: c ∈ Y; Predicted class label of the test query.
Step1: Generate subsets for the test query 't' by using information gain attribute.
For i=1 to n do
If (attribute of test query = = attribute of training instance i)

Subset S = training instance i
End if
End for
Step 2: Calculate the similarity between test query 't' and generated subset 'S'

$$Sim(t,s) = \sum_{i=1}^{m} \partial(t,s_i)/_{m}.$$

Where, $\partial(t,s_i) = \begin{cases} = 1, if(t = s_i) \\ = 0, otherwise \end{cases}$.
And, m= number of attributes
Step 3: Sort the similarity 'sim' in descending order
Step 4: Find k and select top k nearset neighbors of the test query 't'.
For i = 1 to k do
xi = si,ci
end for
Step 5: Calculate the weight for all the nearest neighbors. $\boldsymbol{W} = (\boldsymbol{w_1}, \dots \boldsymbol{w_k})$
For i = 1 to k do
If $\boldsymbol{sim_k} = \boldsymbol{sim_1}$
$\boldsymbol{w_i} = \boldsymbol{1}$
else

$$\boldsymbol{w_i} = \frac{\boldsymbol{sim_k} - \boldsymbol{sim_i}}{\boldsymbol{sim_k} - \boldsymbol{sim_1}}$$

end if
end for
Step 6: Assign a majority weighted voting class label 'c' to the test query 't'.

4.2 LLAC_DWKNN

If the outlier exists in the dataset, then they are also getting weights, that leads to poor performance of the system. To overcome this issue of WkNN, Dual distance weighted k nearest neighbor (DWkNN) algorithm is introduced. This paper is also proposing LLAC_DWkNN, in that dual weight is used in place of weight. The procedure is same as shown in algorithm 1 except the weight formula. Another new weight is getting multiplied with the original weight as shown in Eq. (4).

$$w_i = \frac{sim_k - sim_i}{sim_k - sim_1} \times \frac{sim_k + sim_1}{sim_k + sim_i} \tag{4}$$

By assigning dual weight to the nearest neighbors, weight is reduced except the nearest one and the farthest one. It helps improving the accuracy of the classifier by assigning lesser weight to the outliers.

5 Result and Discussion

A total of 9 different datasets are used to test the proposed system. The main datasets are taken from the University of California (UCI Repository) [19]. All experiments are conducted on Intel (R) Core (TM) i3-2120 processor, a 3.3 GHz clock speed and 4 GB RAM. Tenfold cross validation method was used where each 10% of data in turn is used for testing and the remaining 90% for training. Table 1 presents the short illustration of the dataset as well as an abbreviation of dataset names. The dataset names are Balance Scale (BS), Breast Cancer (BC), Breast Wisconsin (BW), Credit Approval (CA), Diabetes, Solar Flare (SF), Glass dataset (GD), Ionosphere (Iono) and Iris Plant (IP).

Table 1. Dataset details.

Dataset	#Transactions	#Attributes	#Classes
BS	625	5	3
BC	286	10	2
BW	699	10	2
CA	690	16	2
Diabetes	768	9	2
SF	1389	13	6
GD	214	10	6
Iono	351	35	2
IP	150	5	3

Accuracy computation: The accuracy is determined using the given equation (5). Comparison of accuracy is provided in Table 2.

$$Accuracy = \frac{Number\ of\ correctly\ predicted\ test\ data}{Total\ no\ of\ test\ data}. \tag{5}$$

The accuracy comparison is shown in Table 2, where dataset name is tabulated in column 1; 2nd column is the traditional method Lazy Associative Classification LAC [2], 3rd, 4th and 5th columns are existing lazy learning methods namely Lazy Associative Classification using Information gain (LACI) [9], Compact Highest Subset Confidence-based Associative Classification (CHiSC-AC) [15] and Attribute ranking based lazy learning AC [12]. The last columns are the proposed Lazy Learning Associative Classification with Weighted kNN (LLAC_WkNN) and Lazy Learning Associative classification with Dual Weighted kNN method (LLAC_DWkNN).

It can be seen in the comparison result that the proposed LLAC-WkNN system is 10.17% better than LAC, 8.23% better than LACI, 3.43% better than CHiSC and 0.40% better than ARBLazyAC respectively. Proposed LLAC-DWkNN is 13.97% better than LAC, 11.97% better than LACI, 7.00%

https://doi.org/10.1051/itmconf/20213701023

better than CHiSC and 3.91% better than ARBLazyAC respectively. Precision, Recall and F1- measure for LLAC_DWkNN is shown in Table 3.

The Win/Draw/Loss table is shown in Table 4. When comparing proposed method with the existing LAC method, the proposed method LLAC_WkNN has improved the classification accuracy for 8 datasets and worse for 1 dataset. When comparing LLAC_DWkNN with LACI, proposed method's accuracy is better for all the 9 datasets.

Table 2. Accuracy comparison with traditional and existing systems.

Datasets	Traditional Classifier	Lazy Learning Associative Classifier (Existing)			Proposed Methods	
	LAC	LACI	CHiSC-AC	ARB Lazy AC	LLAC_ WkNN	LLAC_ DWkNN
BS	69.29	70.32	71.27	76.88	75.58	76.00
BC	66.48	67.86	85.22	73.44	75.86	88.06
BW	93.70	88.57	90.63	94.36	94.28	98.57
CA	76.48	76.81	83.33	89.20	85.50	89.50
Diabetes	69.10	68.83	76.75	75.13	75.32	76.32
SF	81.58	84.71	84.71	85.44	86.99	87.23
GD	57.94	59.09	60.91	65.01	82.50	82.50
Iono	82.29	94.44	96.29	89.56	94.71	95.85
IP	96.67	95.33	87.22	97.40	93.33	96.50

Table 3. Precision, Recall and F-1 Measure

Dataset	Precision	Recall	F1-Measure
BS	0.64	0.71	0.67
BC	0.77	0.76	0.74
BW	0.97	0.94	0.93
CA	0.90	0.90	0.90
Diabetes	0.74	0.75	0.73
SF	0.90	0.90	0.90
GD	0.61	0.73	0.63
Iono	0.90	0.90	0.90
IP	0.95	0.93	0.94

Table 4. Win/Draw/Loss Table

Table Head	Existing Methods			
	LAC	LACI	CHiSC-AC	ARB Lazy AC
LLAC_WkNN	8/0/1	8/0/1	6/0/3	5/0/4
LLAC_DWkNN	8/0/1	9/0/0	7/0/2	7/0/2

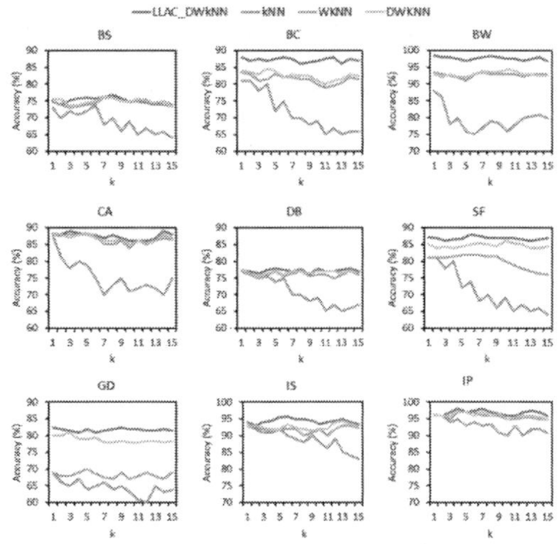

Figure 3 The average accuracy v/s the neighborhood size k for different datasets

Figure 3 shows the average classification result of the proposed LLAC_DWkNN and existing methods DWkNN, WkNN and kNN for the explanation of the sensitivity issue of neighborhood size k. It can be clearly seen that proposed method is performing better than the other related method, mostly when k is large. It is robust to the size of k. Accuracy is not fluctuating when the k changes and showing good performance. The Evaluation results of 9 different datasets from the UC Irvine Machine Learning Repository have proven that the proposed approach achieved higher classification accuracy when compared with competing methods.The average accuracy v/s the neighborhood size k for different datasets is shown in Figure 3. It can be seen that the value of k is not affecting the accuracy for all the 9 datasets. Accuracy is constant irrespective of the value of k. It shows that the system is robust.

Weighted kNN is applied for the classification. Weighted kNN picks only k nearest neighbors among the all generated subsets. By this way, accuracy is improved, and proposed system is robust. The value of k doesn't affect the accuracy. It has been observed that the proposed framework gives the excellent result when compared with the existing one.

6 Conclusion

In this paper, two novel approaches LLAC_WkNN and LLAC_DWkNN are proposed to improve the performance of Lazy learning associative classification. The key idea of the proposed methods is to integrate the two concepts in order to get the advantage of both the methods to improve the classification accuracy and to get rid of the sensitivity issue of neighborhood size k. First lazy learning method is applied to the dataset for subset generation and then weighted kNN is applied for the classification. Weighted kNN picks only k nearest neighbors among the all generated subsets. By this way, accuracy is improved, and proposed system is robust. The value of k doesn't affect the accuracy. The improved outcome directs that the proposed method of uniting LLAC and WkNN is an effective algorithm.

References

1. B. Liu, W. Hsu and M. Yiming, "Integrating Classification and Association Rule Mining," in 4th International Conference on Knowledge Discovery Data Mining (KDD), (1998).
2. A. Veloso, W. Meira Jr. and M. J. Zaki, "Lazy Associative Classification," in Sixth International Conference on Data Mining (ICDM'06), Hong Kong, China, (2006).
3. Z. Huang, Z. Zhou and T. He, "Associative Classification With Knn," Journal of Theoretical and Applied Information Technology, vol. 49, no. 3, pp. 1013-1019, (2013).
4. W. Hadi, Q. A. Al-Radaideh and S. Alhawari, "Integrating associative rule-based classification with Naïve Bayes for text classification," Applied Soft Computing, vol. 69, pp. 344-356, (2018).
5. P. Tamrakar, S. S. Roy, B. Satapathy and I. S. P. Syed, "Integration of lazy learning associative classification with kNN algorithm," in Vision Towards Emerging Trends in Communication and Networking (ViTECoN), Vellore, (2019).
6. S. A. Dudani, "The Distance-Weighted k-Nearest-Neighbor Rule," IEEE Transactions on Systems, Man, and Cybernetics, Vols. SMC-6, no. 4, pp. 325-327, (1976).
7. J. Gou, L. Du, Y. Zhang and T. Xiong, "A New Distance-weighted k-nearest Neighbor Classifier," Journal of Information & Computational Science, vol. 9, no. 6, pp. 1429-1436, (2012).
8. L. Merschmann and A. Plastino, "A Lazy Data Mining Approach for Protein Classification," IEEE Transactions On Nanobioscience, vol. 6, no. 1, pp. 36-42, (2007).
9. L. Merschmann and A. Plastino, "HiSP-GC: A Classification Method Based on Probabilistic Analysis of Patterns," Journal of Information and Data Management, vol. 1, no. 3, pp. 423-438, (2010).
10. S. P. S. Ibrahim, C. K. R. and C. J. K. Kanthasamy, "LACI: Lazy Associative Classification using Information Gain," IACSIT International Journal of Engineering and Technology, vol. 4, no. 1, pp. 1-6, (2012).
11. X. Zhang, G. Chen and Q. Wei , "Building a highly-compact and accurate associative classifier," Applied Intelligence, vol. 34, no. 1, pp. 74-86, (2011).
12. P. Tamrakar and S. P. S. Ibrahim, "Attribute ranking based lazy learning associative classification," ARPN Journal of Engineering and

Applied Sciences, vol. 13, no. 11, pp. 3698-3705, (2018).

13. S. H. Ebenuwa, M. S. Sharif, M. Alazab and A. Al-Nemrat, "Variance Ranking Attributes Selection Techniques for Binary Classification Problem in Imbalance Data," IEEE Access, vol. 7, pp. 24649-24666, (2019).
14. J. Thomas, A. Joseph, I. Johnson and J. Thomas, "Machine Learning Approach For Diabetes Prediction," International Journal of Information Systems and Computer Sciences, vol. 8, no. 2, pp. 55-58, (2019).
15. S. P. S. Ibrahim, K. R. Chandran and C. J. K. Kanthasamy, "Chisc-AC: compact highest subset confidence-based associative classification," Data Science Journal, vol. 13, pp. 127-137, (2014).
16. T. Cover and P. Hart, "Nearest neighbor pattern classification," IEEE Transactions on Information Theory,, vol. 13, no. 1, pp. 21-27, (1967).
17. E. Fix and J. L. Hodges Jr., "Discriminatory Analysis. Nonparametric Discrimination: Consistency Properties," International Statistical Review, vol. 57, no. 3, pp. 238-247, (1989).
18. U. Lall and A. Sharma, "A nearest neighbor bootstrap for resampling hydrologic time series," Water Resources Research, vol. 32, no. 3, pp. 679-693, (1996).
19. D. Dua and C. Graff, "UC Irvine Machine Learning Repository," (2019).

HOME AUTOMATION THROUGH SMART LIGHTING, SMART SECURITY AND OTHER APPLIANCES

Rajarajeswari S [1], Shola Usha Rani [1], Alpanshu Kataria [1], Soumitro Dutta [1]*

[1]School of Computer Science and Engineering, Vellore Institute of Technology, Chennai, Tamilnadu, India

Abstract: The smart home automation system is designed to conveniently manage and monitor household appliances and lighting fixtures remotely, to save time and to use resources effectively. It is a control system for allowing access to home automation devices. The automation devices include lights, fan, camera and doors. It even enhances the main features of the project to control the home appliances from anywhere as remotely. With the camera module, continuous surveillance and protection of the home is accomplished. A significant improvement is achieved in utility costs by an optimal use of energy. Using inexpensive electronic and interactive tools, this initiative can be applied, making it economically, physically and operationally feasible. Basically it involves controlling everything remotely via a visual interface and mobile application. The automation system employs the use of MQTT protocol in order to secure the transmitted device based information to and from in the network. The defense against Man in the Middle attack (MITM) is the main focus of our automation system. The routed messages are checked at the intermediate hub to check for the occurrence of MITM based DoS attacks through manual reset points. .

1. Introduction

There is a huge energy shortage in our world. People are not using the electricity available appropriately. People sometimes fail to turn off their home appliances, such as light fixtures, etc., before leaving home. In these sorts of situations, home automation makes it easy to regulate them conveniently from our mobile from a distant location. People keep running from place to place, working on their never-ending "to-do" list to complete everything. Because of the home automation system, we've never got to think about unlocking the door, shutting off the equipment, and so on. In short, we will save valuable time and maximize everyday efficiency and security.

The convenience, energy management, security are main components initiates the importance for implementing smart home systems. This is new area of research for integrating various components in Smart Home systems, but all the components are not integrated yet. This requires many domains need to communicate through the common interface layer to make the system to work smoother. The further cost will add if other components are connected together to the system. This makes large barrier while doing development to the smart home system as the technical components are getting cheaper the expense cost of smart home system will become cheaper even though it is incorporated with many other components.

*Corresponding author:rajarajeswari.s@vit.ac.in

Smart homes are evolving and becoming more sophisticated, but the goal of giving peace of mind is still there. We believe that the future of home automation will be focused on the continued integration of home devices, and even appliances, to increase personalization, control, comfort, and convenience in the home.

One of the newer fields of science that has not been completely implemented into our everyday lives is Smart Home solutions. This is because research involves the development of a proper smart home in many other research and engineering disciplines. The cost of a smart home is still a major impediment to the advent of smart home solutions on the market. The added cost of the installation is because, while most homes have been installed in the recent past, technology has evolved exponentially. This suggests that before this technology was available, most homes were constructed, and this becomes a hindrance to smart home systems production and sales. However, technology is getting stronger and cheaper, and as new homes are constructed, this will continue to make smart home solutions an investment worth making.

Smart homes are changing and becoming more complicated, but there is still the purpose of ensuring peace of mind. In order to improve personalization, security, and usability in the home, the future of home automation will concentrate on combining home electronics, and even appliances.

Home automation isn't a modern technology at all. Home Automation history goes back further. Since the advent of science fiction, science fiction has centered on smart home technologies, but it is not just that. The best minds in the world have been working for decades on technology that can almost think for itself. There are a lot of highlights in it. 1966: Although never marketed commercially, ECHO IV was the world's first home automation system. The "Electronic Computing Home Operator" (ECHO) was invented by Jim Sutherland to store recipes, relay messages, regulate the temperature of a home and switch appliances on or off. 1969: DARPA launched ARPAnet, the first network in the world, the predecessor to the digital Internet and all our smart technology for the Internet of Things (IoT).

□1981: A predecessor is invented to the cellular (802.11) technology of today. In the 1980s, in the form of garage doors, home surveillance systems, motion-sensing lighting, thermostat controls, and other innovations, home automation became prevalent.

□1991: Ad van Berlo pioneers the gerontechnology-technology sector to better the lives of elderly people. These early innovations formed a solid foundation for the intelligent, life-enhancing characteristics that we enjoy about the smart systems of today.

□1998-2000s: Smart homes have been prevalent. Smart technology evolved in the late 1990s and early 2000s, with gadgets and

□computers becoming more and more available.

Thales is bringing some of the latest home protection IOT solutions, as well as an IOP publication dated 2017, to the industry. The Thales Approach offers an application that is compliant with multiple security systems and sensors, as well as offering a protected network and unique ID for each newly attached device to keep international devices at bay, eliminating many kinds of cyberattacks. The "Home security system using the internet of things" approach in the IOP Publication (2017) talks about using Reed Sensors to track room doorway operation, this is combined with Arduino and Wifi modules, further linked to Blynk Software that the user can connect and receive warning about and unusual behavior tracked by the doorway activity.

2. Existing System

Day by day, energy usage is growing. Remote access to home appliances using IoT is really important to have. That's why the next study field is Smart Home [7][10]. Applications based on IoT have also been very effective for elderly people and persons with any kind of disability [6][11]. This allows the user to monitor equipment such as lamps, fans, etc. without having any physical touches. Some of the Existing Home Security IOT solutions are the ones currently introduced to the market by Thales, and also an IOP Publication dated 2017. The Solution by Thales offers an application that is compatible with various Security devices and Sensors, it also provides a secured network and unique ID for every device that get's newly connected, so as to keep foreign/unlisted devices at bay, preventing many sorts of cyber attacks.

The Solution in the IOP Publication (2017) "Home security system using internet of things", talks about usage of Reed Sensors to track Doorway activity of rooms, this is paired up with Arduino and Wifi modules, further connected to Blynk App that the user can access and be notified about and unusual activity that is monitored by the doorway activity.

The research carried out is recorded in the following sections: [7][9][10][12][14]. The bulk of earlier devices based on these approaches are either Bluetooth-based or DTMF-based [6][7][12][13][15][17]. The fundamental problem is that separate PSTN channels for communication between controlling devices and main supply units are necessary for DTMF-based home automation. In addition, Bluetooth demands that the running appliance be in its range. It is only helpful for short-range engagement.

Home automation in [14] uses MQTT for the sending and receiving of sensor data. Raspberry Pi is used to access sensor information that is used to monitor the room's humidity and temperature as a gateway. Another Raspberry Pi-based home automation system is seen in[9] and the user uses a web-based GUI to monitor the home appliances. The [12] Mobile home automation demonstrates which device is built using ZigBee. We can transform non-smart devices into smart devices with the aid of IOT, which enables us to connect these via the internet. It turns a typical home into a smart home and offers a more efficient way for home appliances to be operated. Protection can also be improved by installing cameras in the home that can be tracked through the internet. Users can also track their home from anywhere and can turn their ON/OFF appliances. This will conserve energy as well as electricity costs as shown in Fig 1

System	Primary Communication	Remote access	Number of Devices	Cost	Speed	Real Time
GSM	SMS messages	Access from anywhere in the world	Unlimited	High cost due to SMS charges	Slow due to delivery issues	No
Bluetooth	Bluetooth and AT commands	Restricted to Bluetooth range- 10 metres	Unlimited	Fast due to proximity	Fast due to proximity	Yes
Phone Based	Phone lines	Anywhere with a phone line	12 due to 12 frequencies of DTMF	Fast	Fast	No
Zigbee	Zigbee and AT commands	Around 10 metres	Unlimited	Fast	Fast	Yes
Wireless	Radio, infrared or other waves	Depending on range and spectrum of waves used	Unlimited	High cost due to licensing and other spectrum issues	Slow due to interferences	Yes

Fig.1. Comparison of energy cost, speed for Phone based and system

3. Proposed System

The proposed system combines the features that an energy saving IOT system can provide with the features of a Secure Home Solution. The solution can even have high cost, high resolution, huge data producing Cameras and Sensors. The system is based on the detection of motion of a human body that is performed by PIR sensors. These PIR Sensors have the ability to work at very low electricity consumption. These PIR's to not only control simple appliances such as lights or fans, but also other security related sensors. Without the presence of objects like Human being the PIR sensor will not sense anything and, the devices will remain in off state and did not produce any data, this will keep the overall database light as well as lower electricity consumption.

The system also consider the situations of MiTM or DoS attacks that can occur at the MQTT broker, hence it'll be secured by TLS. Other than that, QoS is set to 0 as PIR's release data at short intervals of time, and all the messages are not so important to need a guaranteed delivery. QoS 0 makes the part of the MQTT broker easier and faster, also reduces load.

The data created by any security sensors or devices will transmit independent of the system, hence in cases of server malfunction or any malfunction at the microcontroller end can be easily solved by resetting the system and restarting it, this would not cause any loss to precious security data captured by the security sensors or devices.

The microcontroller will also be programmed to get data from MQTT Broker and take actions. Since at this stage too it might be vulnerable to attack, checkpoints that strip the topic of MQTT at which it is subscribing, are present. Abnormal messages getting through to the microcontroller will be filtered out and discarded at the early stages itself (Shown in Fig. 2).

A "Lockdown Mode" is introduced in the system, the same PIR will detects an unusual activity in various rooms, then the sophisticated devices such has high resolution cameras and other sensors will be turned on. This will confuse the burglar because while scoping the house from outside he'll notice the devices in off state, and hence not care about their working at a later stage, and be caught off by the guard. The system connects to mobile devices through LAN. This connection too might be vulnerable to sniffing, so encryption of alerts is supposed to be performed at the microcontroller, and decryption will be done at the end user through APP. This will ensure that alerts is passed to the end user, the one who are using the application. The architecture diagram representing the above proposed system can be seen in the Fig 2.

Fig.2. Architecture diagram of the proposed system

In some applications only simple notification through alarm is sufficient. But some solutions requires more secure and safer applications. To incorporate with simple security system need to be added with more security measures. with the consumption of video surveillance, the user can see who is at their opposite door, even can be alerted is any abnormal even happened at their door steps. And even the system can added an emergency system to add a call to nearby police or sending an SMS to the registered mobile numbers. To add this security feature video cameras are installed in all the rooms inside and outside to improve the security. the sever connected with this cameras will receive the upgraded storage of video and audio streaming. The data related from these cameras is updated into the server based on the infrared sensor data or motion sensor connected to the rooms. This sensor will detect the objects and send the abnormality to the server if find anything. . Light sensors could be included to monitor the light strength during the day and alter the lights in the room accordingly.

The system here it is designed with feature of cameras to note down the entrance and to maintain the log related to the entry. The log includes entrance of object, exit of an object, timestamp and presence of people inside house.

This will automate the feature of alerting the user when some abnormal event occurs. And notified or indicates to the user through the registered mobile number... All the cameras can be set to only activate upon another sensor being tripped in order to keep cameras in a "sleep mode". To control the energy and to increase the efficiency of the system the secure system is incorporated with the feature of central control location. Where this system will invoke the cameras system if it needs otherwise it automatically closes the camera feature. So lockdown mode feature is added, so if the home in lock mode then only the PIR sensor will initiate a signal to camera. Then the camera will start observing the data from the surroundings of the house. If the system make these features to generate an environment that can exchange the common alarm system most households have. With creating an away mode or lockdown mode, it would even be able to be configured to have the household message owner if the away mode is not deactivated within a certain threshold of time when someone enters then house, and alert all household residents via cell phone when someone enters the house.

4. Implementation

The implementation is started with a very simple two-state sensor that could be externally controlled using MQTT. The sensor can be used to simulate real-world objects such as lighting, doors, etc. that have two ON or OFF/Open or close states, etc. The agreed commands are on and off, which, whether acting as a door sensor, switch the lights on and off or open and closed. All other commands are neglected and, using a function node, commands are rendered case-insensitive. As per the instruction, the sensor changes its state and publishes the current state. In node red workspace, a PIR sensor simulating nodes is added. Then the MQTT protocol send the sensor data to the microcontroller and then to the actuators. The model simulates the device such as light bulbs, doors and camera on the dashboard. The system is added different function nodes with the purpose of preventing any type of unauthorized access to the appliances. All the nodes implementation in Node Red can be seen in Fig 3. The system is also configured to send a notification to the user about the bad weather conditions, to alert the user while going out. The notification alert will be sent h by mail or by text message on phone(Fig 4 and Fig 8) Also, manual override switch in the dashboard that will ensure system reset and removal of any private data, in case of malfunction or attack.

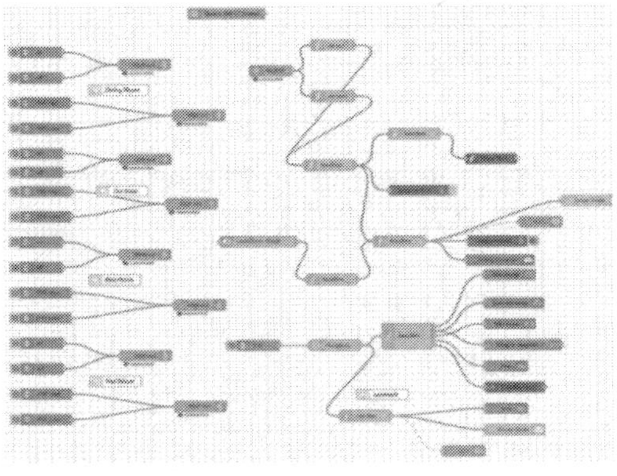

Fig.3. Node red workspace showing all nodes

5. Results and Discussion

Wastage of energy will happen when the usage of is rising. The following pie chart (Fig 5) tells us how much energy is used by various appliances. From the figure 4 it is noticed that the lights and fans absorb full energy. It also observed that means that lamps and fans work excessively; much of the loss of energy arises. The proposed work would aid in the savings of a lot of electricity and thus reducing the electricity bills as well as reducing the harmful impact on environment.

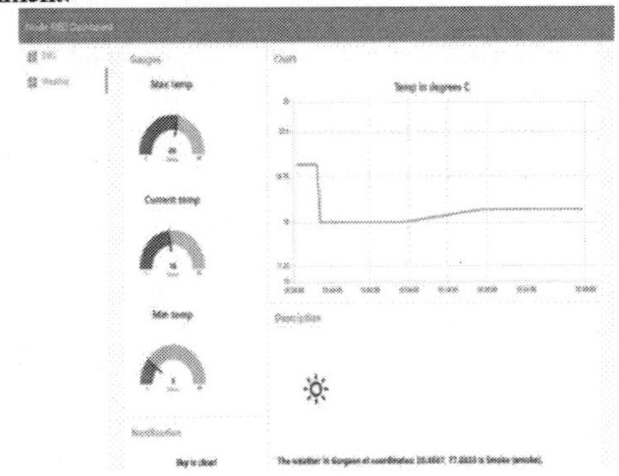

Fig.4. Weather Info on Dashboard

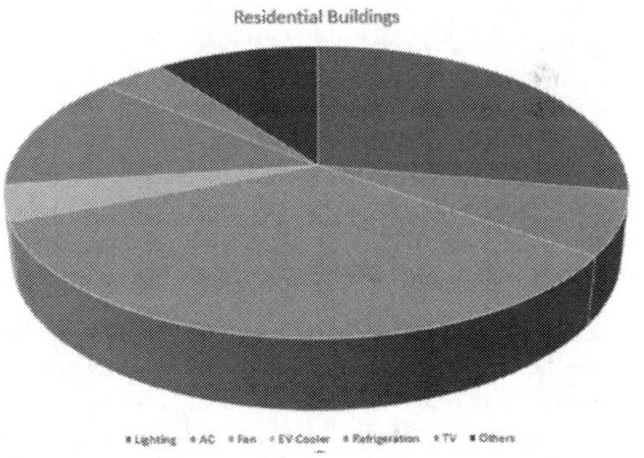

Fig.5. Electricity consumption in India

A basic sound alert system is appropriate for others, while for some, they are not happy until their house is cleaner and safer than anyone else. The device offers protections for those who are interested in a basic security configuration, and may have stronger security measures for those who are looking for more than that. The customer would be able to see who is at their main door by the use of video monitoring, as well as be alerted if anyone stands outside the door. The system can also be programmed to warn the user via text message, in addition to sending notifications to the user from any configuration unit, if the user wishes to. When the device is in away mode and an intrusion gets into the building, the user will set the system to warn 100, and he has been encountered.

In addition, video cameras in all the rooms in the house and outside will be placed on the wall to provide additional protection to the device. However, when they want privacy, users can still quickly switch off the camera. To allow for mass storage and streaming of audio and video, the server will then require an update. The probability of detecting subjects will be obtained by infrared motion sensors on walls. To track the light intensity during the day and change the lighting in the room accordingly, light sensors may be mounted.

A surveillance camera to track the main entrance is included with the current function which can be updated to retain a record on whether someone has entered and left the home, what time, and whether or not someone is present in the house at the moment. A major piece of security is the installation of surveillance cameras around the residence in numerous locations around the home. They can be automated with these cameras to warn the user when anyone is detected in an area, and to inform the user by text message or some other way. Only when another sensor is activated will all cameras be programmed to activate in order to hold cameras in "sleep mode" in order to save electricity and prevent the cameras from wasting extraneous quantities of power in order to observe nothing of significance. For these cameras, another security feature we should introduce is to be able to wake them up and monitor them from a central control spot. In order to create an ecosystem that can replace the common alarm system that many households have the group would like to create these features. Additional Lockdown mode is configured, which notifies the user about any unknown activity in the house.(Fig 6 and Fig 7). A switch kind of function is built to activate the lockdown mode. Finally the burglar alerts and weather notifications are intimated to the user through mail of the registered user as shown in Fig 8.

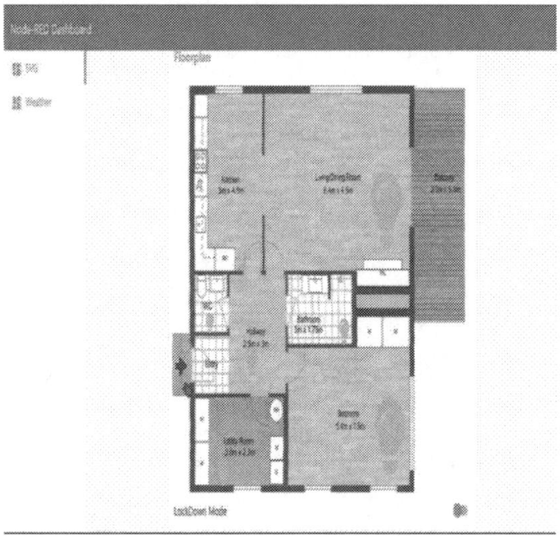

Fig.6. House Plan Showing all devices when Lockdown Mode OFF

Fig.7. House Plan Showing all devices when Lockdown Mode ON

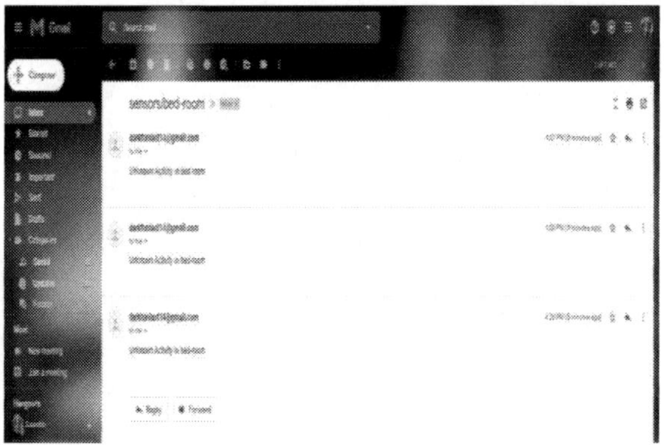

Fig.8. E-Mail Notification

6. Conclusion and Future work

For this project, there are several potential future iterations, incorporating new functionality to boost the user's performance and easier methods of interfacing the device, its individual components and the user. Increased protection capabilities, phone apps, audio/video features, touch screen control, wireless connectivity, speech recognition, faster on-screen navigation, and easy user download are some of these potential implementations. Users will customize the device, whether they want it to be armed in a remote mode, in a standard mode or in a low security mode. In addition to all this, users can watch their home, a record of comings and goings, and a live camera feed at the front door. This can be achieved by broadcasting the video via Wi-Fi over the internet. A sensor will watch when the sun shines through a window and when it is, close the drapes. This would minimize the incoming heat and hence the air conditioner's power demand would be minimized. When the exterior lights that are used hit the stage where the electrical lights produce the same amount of energy, the electrical lights may be switched off. The machine should take advantage of the natural energy provided by the atmosphere. Using the cold air from the outside to cool the house when the weather is cooler rather than depending on the AC when it is not required, the device can operate according to outside temperatures. In addition, it will help to use natural capital in the most productive manner around the building. A graphics-based UI panel in the house is another addition which could be made.

Although the device addresses much of the needs of the customer as being able to use the system without a computer, the system need to support for the mobile phone devices from remote access.

References

1. 1.N. Sriskanthan and Tan Karand. Bluetooth Based Home Automation System. Journal of Microprocessors and Microsystems
2. 2.Muhammad Izhar Ramli, Mohd Helmy Abd Wahab, Nabihah, Towards Smart Home: Control Electrical Devices Online, Nornabihah Ahmad International Conference on Science and Technology: Application in Industry and Education.
3. 3.E. Yavuz, B. Hasan, I. Serkan and K. Duygu. Safe and Secure PIC Based Remote Control Application for Intelligent Home. International Journal of Computer Science and Network Security.
4. Amul Jadhav, S. Anand, Nilesh Dhangare, K.S. Wagh Universal Mobile Application Development (UMAD) On Home Automation Marathwada Mitra Mandal's Institute of Technology, University of Pune, India Network and Complex Systems (2012)
5. 5.Pratik Gadtaula, Home Automation, Telemark University College, Faculty of Technology, Master's Thesis
6. 6.A. J. Jara, Wearable Internet: Powering Personal Devices with the Internet of Things Capabilities 2014 International Conference on Identification, Information and Knowledge in the Internet of Things, Beijing,, pp. 7-7(2014)
7. Y. Kung, S. Liou, G. Qiu, B. Zu, Z. Wang and G. Jong, Home monitoring system based internet of things 2018 IEEE International Conference on Applied System Invention (ICASI), Chiba, pp. 325-327(2018)
8. Y. Sun, Y. Xia, H. Song and R. Bie, Internet of Things Services for Small Town 2014 International Conference on Identification, Information and Knowledge in the Internet of Things, Beijing, , pp. 92-95(2014).
9. D. Pavithra and R. Balakrishnan, IoT based monitoring and control system for home automation, Global Conference on Communication Technologies (GCCT), Thuckalay, , pp. 169-173.(2015)
10. H. V. Bhatnagar, P. Kumar, S. Rawat and T. Choudhury, Implementation model of Wi-Fi based Smart Home System, International Conference on Advances in Computing and Communication Engineering (ICACCE), Paris, pp. 23-28(2018).
11. M. C. Domingo, An overview of the Internet of Things for people with disabilities, Journal of Network and Computer Applications, vol. 35, Issue 2, pp. 584-596, (2012).
12. A. Olteanu, G. Oprina, N. Tapus and S. Zeisberg, Enabling Mobile Devices for Home Automation Using ZigBee,19th International Conference on Control Systems and Computer Science, Bucharest, pp. 189-195(2018)
13. R. Piyare and M. Tazil, Bluetooth based home automation system using cell phone, 2011 IEEE 15th International Symposium on Consumer Electronics (ISCE), Singapore, pp. 192-195(2011)
14. Y. Upadhyay, A. Borole and D. Dileepan, MQTT based secured home automation system Symposium on Colossal Data Analysis and Networking (CDAN), Indore, pp. 1-4 (2016)
15. T. Wang, Y. Li and H. Gao, The smart home system based on TCP/IP and DTMF technology 2008 7th World Congress on Intelligent Control and Automation, Chongqing, pp. 7686-7691(2008).
16. T. M. Ladwa, S. M. Ladwa, R. S. Kaarthik, A. R. Dhara and N. Dalei, Control of remote domestic system using DTMF, International Conference on Instrumentation, Communication, Information Technology, and Biomedical Engineering 2009, Bandung, pp. 1-6 (2009)
17. N. M. Morshed, G. M. Muid-Ur-Rahman, M. R. Karim and H. U. Zaman, Microcontroller based home automation system using Bluetooth, GSM, Wi-Fi and DTMF, 2015 International Conference on Advances in Electrical Engineering (ICAEE), Dhaka,pp. 101-104(2015)
18. H. Brooke Stauffer Smart Enabling System for Home automation, IEEE Transactions on Consumer Electronics, Vol. 37(2) , pp. 29-35.(1991)

19. Faisal Baig, Saira Beg and Muhammad Fahad Khan, ZigBee Based Home Appliances Controlling Through Spoken Commands Using Handheld Devices International Journal of Smart Home, Vol. 7(1), pp 19 -26. (2013)
20. Pooja S Chinchansure, Charudatta V Kulkarni, Home Automation System based on FPGA and GSM, International Conference on Computer Communication and Informatics (ICCCI), Coimbatore, India, pp. 1 - 5 (2014)
21. A. Alheraish, Design and Implementation of Home Automation System, IEEE Transactions on Consumer Electronics ,Vol. 50(4) , pp. 1087-1092(2004).
22. Manoj Kumar Singh, Sadhan Mahapatra, Atreya Kumar Sudhir Green Building Design: A step towards sustainable habitat Conference Paper • (2012),
23. Home Automation Systems - A Study. Article in International Journal of Computer Applications • DOI: 10.5120/20379-2601(2015)

ITM Web of Conferences **37**, 01025 (2021) https://doi.org/10.1051/itmconf/20213701025
ICITSD-2021

Early Detection of LDoS Attack using SNMP MIBs

Gayathri Rajakumaran[1*], *Neelanarayanan Venkataraman*[1], *Abdul Quadir MD*[1]

[1] School of Computer Science and Engineering, Vellore Institute of Technology, Chennai, India
[1*] Corresponding author: gayathri.r@vit.ac.in

Abstract. Early detection of Denial of Service (DoS) attacks are given more emphasizing due to its adverse effects on disrupting the services of legitimate users. LDoS attack is one among the DoS category which floods the target at ideal rate to keep the connections open for longer duration. Traditional defense measures are inadequate to filter due to its less traffic volume. The current works focus on either empirical studies or signal processing models to capture the behavioural characteristics of LDoS based on TCP's congestion control and timeout mechanism but none carries out detection at a faster timestamp. Early detection solutions are the main focus as it could scale up the revenue losses in today's online application issues. Hence our model is based on Simple Network Management Protocol (SNMP), through which the early detection of LDoS attacks is carried out. The relevant detection metrics are identified through theoretical validation of SNMP MIBs and existing dataset analysis. Experimental simulations illustrate the LDoS detection efficiency and the same has been validated for theoretically.

1 Introduction

Internet plays the vital role due to advanced computing technologies and digitized environment. It acts as the backbone without which the day-to-day activities turn out to be zero. The growth of technologies goes hand-by-hand with the security disruptions. Distributed Denial of Service (DDoS) attack is a critical threat as it makes complete disruptions to the internet community by keeping its traces in the availability feature. During DDoS, the target is flooded with continuous bogus requests which occupy victim's memory. The exhausted victim is unable to accept or respond to any kind of requests from legitimate users. Resources exhausted during such attacks include bandwidth of network, CPU cycles, and Server's memory; interrupt processing capability and protocol structures. The current applications also face low spike attacks which produce the effect similar to DDoS high spike. LDoS attack is one among the category of DDoS family as it makes the TCP sessions engaged for longer duration by sending low spikes of traffic. A large traffic spike originating from attacker reaches the target at a slower rate meanwhile very few connections are established to escape from the time out mechanism and traditional defense strategies.

LDoS attack exploits the vulnerability in the congestion control mechanism of TCP by either periodically or continuously sending attack requests in short term or at constant rate. The time out mechanism of TCP congestion control is based on the time taken for receiving a complete request from the source. In general the Round Trip Time (RTT) is constant and ranges from 10 to 100s of milliseconds. In case of delayed request from the source, the Retransmission Time Out (RTO) operates on a longer time-scale.

Since the LDoS attacks poses very less volume of incoming traffic which mostly goes undetected by the traditional security defenses. Due to large incubation periods, the attack traffic mixes with the normal ones which make the traffic segregation and analysis part most difficult. If an online service gets disrupted either due to high spike or low spike DDoS, the trust among legitimate users and regular customers will be vanished. Uninterrupted service is the success behind big online giants like Amazon, Flipkart, Snapdeal, etc. According to the latest Amazon report [1], even 100 millisecond disruptions of its online services causes 1% drop in the overall sales. Hence early detection is the expected and needy solution to better address the issues faced in the current scenario. We aim to offer such solution by employing the features of Simple Network Management Protocol.

Numerous researches are ongoing to address the DDoS issues and to provide better solutions. The existing solutions are categorized under detection measures in general and detection measures based on hypothesis testing.

2 Technical Backgrounds

Low rate attack detection through information metrics [8] helps in measuring the difference in distance between the normal and attack traffic cases. Information metrics can quantify the differences in network traffic concerning various probability distributions. The generalized entropy metric and information distance metric are utilized in the process of detecting low rate DoS attacks. The entropy metric can detect the attack seven hops earlier than the traditional Shannon metric. This approach outperforms the famous Kullback-Leibler divergence approach as it enlarges the adjudication distance and yields the optimal detection sensitivity. The information metric can effectively reduce the low rate DoS attacks with a clear reduction in the false-positive rate. The IP traceback algorithm can find all attacks as well as attackers within the own Local Area Network (LAN) and discards the incoming traffic. It also considers the attacks based on the category of insiders.

Low rate shrew based DoS attacks [9] are detected through the TCP congestion control window behavior. The shrew based DoS attacks are threatening for real-time applications as it can easily throttle TCP flows through a very low attack cost. By capturing the adjustment behaviors of the TCP's congestion control window, the combined effect of the attack pattern concerning the network environment is realized.

Empirical evaluation of the information metrics [10] attempts to detect the low and high rate DoS attacks in the networked environment. The empirical evaluation is carried out for the metrics namely Hartley entropy, Shannon entropy, Renyi's entropy, generalized entropy, kullback-Leibler divergence and generalized information measure in the process of detecting both low and high rate DoS attacks. These metrics help greatly in the differentiation of network traffic data and facilitates the process of building an optimal model. For illustrating the efficiency and effectiveness of the metrics related to DoS through the data sets MIT Lincoln Laboratory, CAIDA, and TUIDS.

Robust RED algorithm [11] is applied for detecting low rate DoS attacks. The RED algorithm maximizes the TCP throughput and attempts to filter, detect attack packets before the adoption of a normal RED algorithm for attack flows. The RRED algorithm claims the incoming flow as an attack only if the majority packets in the flow are sent within the short duration after a packet drop. RRED is efficient in suspecting both TCP and UDP based flow in the case of detecting low rate DoS attacks.

CPR based approach is used for the detection and filtering of the LDoS attack as they intend to cause network congestion. An incoming flow with Congestion Participation Rate (CPR) higher than the expected threshold is declared as suspicious and all the subsequent packets are dropped. The effectiveness of CPR is quantified through the average CPR distance for the normal and attack flows. This approach is more effective in terms of comparison with the existing Discrete Fourier Transform (DCT) technique in the process of detecting the LDoS attacks. A major difference in differentiating the TCP flows is that the normal TCP flows usually avoid the network congestion as it poses the TCP congestion control mechanism whereas the LDoS attack traffic introduces network congestion to degrade the network performance. In extreme cases, the LDoS attack throttles all normal TCP incoming flows and the aggregate value of the attack is very closer to the bandwidth of the network. CPR approach works well as it doesn't drop any packet and no network congestion is observed.

Low rate DoS detection based on network multifractal [12] considers the characteristics of network traffic in the process of detecting DoS attack. LDoS sends periodic pulse sequences with a low-frequency relative to form aggregation flows at the victim side. LDoS attacks, in general, are harder to detect as it poses the low rate property. For characterizing and analysing the network traffic, mathematical models are used for exploring the complex multifractal structure. Even though the LDoS attacks are slow, it contributes to the multifractal characteristics of network traffic.

The Multifractal Detrended Fluctuation Analysis Algorithm (MF-DFA) identifies the changes in the multifractal characteristics in small quantity for detecting the LDoS. Through the wavelet analysis process, the singularity, bursty nature of network traffic is captured and estimated using the Holder exponent. The difference values of Holder exponent between the normal and LDoS attack traffic are distinguished. The difference value is used as the basis for differentiating normal flow with the attack ones.

The dynamic time warping approach is used for robust and accurate identification of DDoS attacks. When the affected TCP flow enters into timeout and starts to retransmit the packets, the LDoS attack will send a small burst to force the TCP flows to enter into the RTO again. This results in very low transmission bandwidth for the TCP flows. When an attack is identified through the dynamic warping approach, the count of affected TCP flow is minimized, sufficient resource protection is done for the affected flows and behavioural analysis based prediction is carried out. This method has very low false positives and false negatives and efficient in the process of isolating legitimate users with the attacker.

3 Characterization of LDoS Attack

In LDoS, the attacker sends low traffic spikes at a very low frequency in order to hide its presence in the network. Complete analysis of attack pattern and characteristics plays a vital role for yielding a prominent and proactive solution. Representation of LDoS is depicted as in below Fig 1. LLDoS indicates the duration of attack pulse, SLDoS is the beginning of attack, RLDoS is the rate of requests received during attack and FLDoS represents the frequency of attack. The attack strength ASLDoS is represented as below.

$$ASLDoS = RLDoS \text{ X } LLDoS/FLDoS \quad (1)$$

The range of ASLDoS is very small in the case of normal traffic whereas it varies significanlty during attack traffic. The general behvarioural inhibitence of DDoS least helps in the LDoS detection procedure.

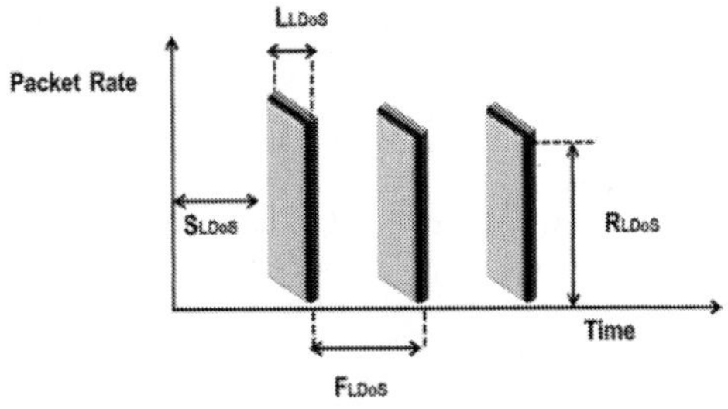

Fig. 1. Representation of LDoS

On receiving the low traffic spikes, the TCP's timeout delay gets increased from its initial value. In general, the delay is increased to facilitate proper connectivity and to prevent connection breakdown for the requesting client.

As per RFC 6298 standard [2], the RTO calculation for TCP is achieved through its two states SRTT (Smoothed Round-Trip Time) and RTTVAR (Round Trip Time Variation). The clock granuarity is assumed as G seconds. Initially the RTO is updated as 1 second till the RTO measurement is estimated.

$$RTO = SRTT + max(G, K*RTTVAR) \quad (2)$$

For the first measurement of RTO value, RTTVAR=R and SRTT=R/2. For the subsequent RTO measurements the values :(1-β)*RTTVAR+β*|SRTT-R| . In case of abnormal behaviour, where the ACK is not received from the sender within the depicted RTO value, then the RT gradual increase in RTO is depicted in Fig 2.

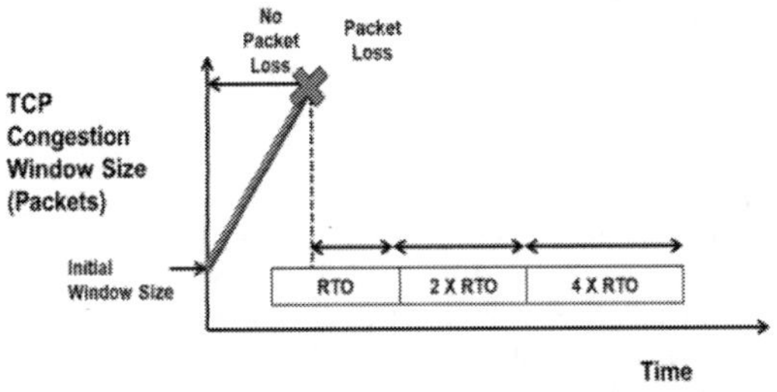

Fig. 2. TCP Retransmission Strategy

For better understanding, TCP's deterministic re-transmission strategy is exhibited. The RTO will be set for each and every incoming request.

$$Total(t) = \begin{cases} RTOmin \text{ where } LDoS = 0 \\ 2* RTOmin \text{ where } LDoS = 1 \end{cases} \quad (3)$$

In general, the requests arises in a network is modeled as a Poisson distribution where the events are random [3]. 'n' number of arrivals in a time interval 't' is considered. The arrival rate is represented as 'λ'.

$$P(n\_t) = (\lambda t)^\wedge n\ e^\wedge(-\lambda t)/n! \quad (4)$$

A modulation of the above assumption could be done from the little's formula as few days network will be congested where it will not be the case for all times. In such cases, a network system could be modelled as (R_a,R_t). 'Ra' is the average number of reqeusts in the system and 'Rt' is the amount of time spent by each request in the system and 'λ' is the arrival rate.

$$R\_a = \lambda\ R\_t \quad (5)$$

The little's theorem [4] could be applied to either a whole system or part of a system as both serves the purpose. In the case of LDoS attack, the parameters affected network metrics could be packet transmission time, propagation delay, average queueing delay and average number of packets received. Estimating all these metrics values helps in faster attack detection as well helps to analyse the performance issues raised during LDoS. All the packets are considered as requests as we are employing the SNMP for anlaysing the incoming traffic. SNMP captures all incoming requests which are further considered for a detailed analysis. The various system parts considered for a bried analysis are listed as below

- Transmitter / Sender
- Transmission Line
- Buffer/ Memory
- Transmitter / Sender + Buffer/ Memory

'D_tr' is the request transmision time. The average number of requests at transmitter is depicted as 'D_tr' or þ for link utilization. 'D_r' is the propagation delay. The average number of requests on the fly is λ D_r. 'Dq' is the average queuing delay. The average number of requests in memory is λ Dq. Finally the average number of requests is illustrated by the equation þ + λ Dq.

4 Attack Modelling of TCP Variants

Detection solution for LDoS is achieved by considering the current variant of TCP adopted in today's Internet scenario and majority of websites. Among the other prevalent TCP variants, BBR [4],[5], [6], [7] is the dominant as it is adopted in most of the websites inclusive of Google cloud and Amazon Web Services (AWS). Our solution is modelled by assuming the constraints specific to behav-ioural characteristics of BBR. It attempts to provide solu-tions on the basis of traffic delivery and latency of roundtrips. BBR adopted in Amazon CloudFront effectively increased the performance gain upto 22% on aggregate throughput across various networks and regions. The performance gains rely on quality, capacity and distance of the connectivity. The congestion indicators of TCP BBR are Current Bandwidth Estimate (BWE) and RTTmin which is depicted in below cases. Rrate indicates the response rate of BBR TCP connections,

B_{qu} is the bottleneck queue utilization, tcp is the transit capacity, B_{asr} is the base sending rate, B_a is available bandwidth.

Case 1: Primary Congestion Response

$$\downarrow BWE \rightarrow \downarrow R_{rate} \quad (6)$$

Case 2: Recovery Mode

$$Count\ (R_{data}) = Count\ (REQ_{ack}) \quad (7)$$

Case 3: Core State (PROBE_BW) B_{qu} = tcp

$$BWE = B_{asr} \quad (8)$$

$$C_{wnd} = 2X\ BWE\ X\ RTT_{min} \quad (9)$$

Case 4: Exhausted Bandwidth

$$\downarrow Ba \rightarrow \downarrow Rrate\ |\uparrow Bqu\ |\ \downarrow BWE \quad (10)$$

$$\uparrow Ba \rightarrow \downarrow Bqu \quad (11)$$

TCP BBR addresses the bottleneck adopting the solutions based on the above computations. If the Ba falls, BWE will not represent it in the first 10 RTT's. The increased sending rate of BBR raises the B_{qu} and fall in BWE to match the returning ACK's rate. Similarly, the B_{qu} falls with the rise in Ba. Adopting SNMP in the above context, attempts to eliminate the congestion at primary level. The Rrate, REQ_{ack} could be obtained based on the TCP specific SNMP MIB's tcpPassiveOpens, tcpCurrEstab and tcpActiveOpens respectively. Through SNMP, by comparing the incoming and response counter values, LDoS attack traffic could be detected. To enhance the detection accuracy and minimize time constraint, the additional required parameters could be better validated based on outcomes of real time data set analysis.

5 Detection metrics based on Dataset Analysis

To understand the behavioural characteristics and traffic pattern of Denial of Service attacks, a complete analysis of the existing data set is important. To carry out the same, the KDD-99, NSL-KDD data set are chosen as it is one of the standard bench-marked data set. Out of the overall 41 features, the incoming frequency count is a derived one which plays a vital role in DoS detection. In order to achieve a complete traffic distinction between attack and normal traffic, additional parameters need to be explored. Hence the EDGAR (Electronic Data Gathering, Analysis, and Retrieval) data set is considered. Differentiation of normal traffic with Denial of Service is the expected outcome of the analysis, based on which the mitigation measures could be tested by generating the synthetic data set.

The Division of Economic and Risk Analysis (DERA) has assembled information on internet search traffic for EDGAR filings through SEC.gov generally covering the period February 14, 2003 through June 30, 2017.The data is intended to provide insight into the usage of publicly accessible EDGAR company filings in a simple but extensive manner.

The attack features are exactly understandable through the traffic patterns of the KDD data set but the normal requests which could be received per second from vari-ous IP addresses could not be retrieved from it as the data set does not contain the IP address column. The data set utilized for analysis is the most the recent year 2017 log record data set where it contains numerous factors particularly IP address, data and incoming time of request entry in seconds. These properties are utilized for further detailed analysis. Hence the U.S government data set EDGAR is used for observing the patterns of normal traffic. The detailed steps in the process of extracting requests based on the IP address and mapping it to the time scale are illustrated as below steps:

- The dataset is isolated with the factors IP ad-dress, date and time in seconds
- Normalize the obtained data set for removing duplicates
- Plot the incoming frequency of incoming IP ad-dress Vs Time to identify the number of requests which arrives from the same IP address per second
- Plot the distribution pattern based on the IP ad-dress which in turn helps to distinguish the normal traffic with the attack one

EDGAR dataset collection is achieved through 11 variables which provides complete picture about the IP ad-dress and status code of incoming requests and few other important metrics which are chosen through the feature selection process. The frequency of incoming requests from the same IP address is calculated by observing the IP address of incoming requests concerning time. The various variables of EDGAR dataset are illustrated as below:

- IP address of the incoming request
- Date
- Time
- Zone
- cik - index key associated with the requested document
- Accession number of the document
- Document file size
- Apache status code for the incoming request
- Referrer header
- Crawler information
- Browser information

The identified detection metrics are mapped with the TCP BBR modes for estimating the current state of incoming connection requests.

6 Experimental Testbed for LDoS

In order to validate the finding from the real time EDGAR data set analysis, experimental test bed set up is done to generate the synthetic data set based on the input metrics Arrival Rate and Request Size. To carry out the validation procedure, TCP specific SNMP MIBs are monitored by performing the experimental set up of SNMP in a controlled environment as per the study. The set up involved 1 PC as Attacker, 1 PC as Normal user, 1 L2 switch and 1 PC as SNMP manager. SNMP agent is installed in both the attacker PC and Normal user PC for collecting the statistics. A test bed is set up for simulating the TCP-SYN attack and normal traffic requests. Any Web servers such as either Apache or XAMPP server will be handling requests in the victim system.

The test bed setup is connected to the D-Link DES-3528 switch. The switch could be managed through the serial port, telnet or web based management agent. The Com-mand Line Interface (CLI) is utilized for configuring and managing the switches through the serial port or telnet interfaces. This type is designed to provide the features fault tolerance, flexibility, port density, robust security and maximum throughput by providing the user-friendly management interface for the users. The test bed consists of one web server, an attacker, legitimate user, SNMP agent and SNMP manager system is depicted in Fig 3.

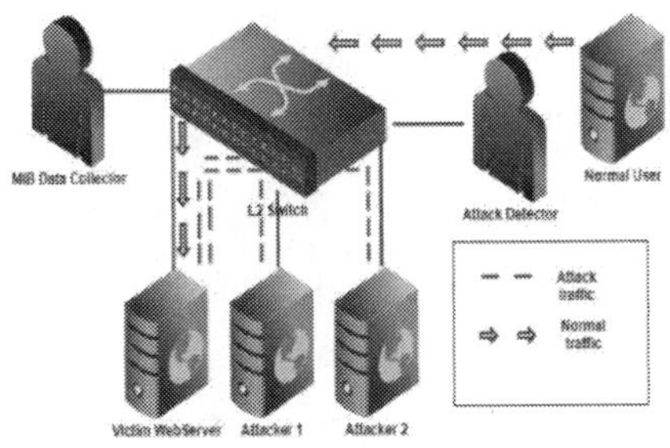

Fig 3: LDoS Testbed

6.1 Traffic Differentiation Based EDGAR Metrics on IP Frequency

The frequency of incoming requests from the same host is an important metric for differentiating the DoS traffic with the normal one which is derived from the EDGAR data set. During the first observation of per second arrival rate, the incoming requests from the same IP address are not exceeding the maximum threshold of 10. The observations are repeated for 30 days of the data set and the arrival rate of incoming IP addresses is monitored randomly to estimate the rough arrival rate of requests from the same IP address per second. All the 30 days results illustrate that the maximum requests which arise from the same IP address per second range between 10 to 40.

Graphs are constructed for a peak day of traffic from 9 AM to 5 PM in order to analyse the maximum requests which arise from the same IP addresses in the interval of 1 hour from the time period of 9 AM to 5 PM in a randomly chosen day of data. The obtained results are plotted with respect to the IP address and number of incoming requests depicted in Fig 4.

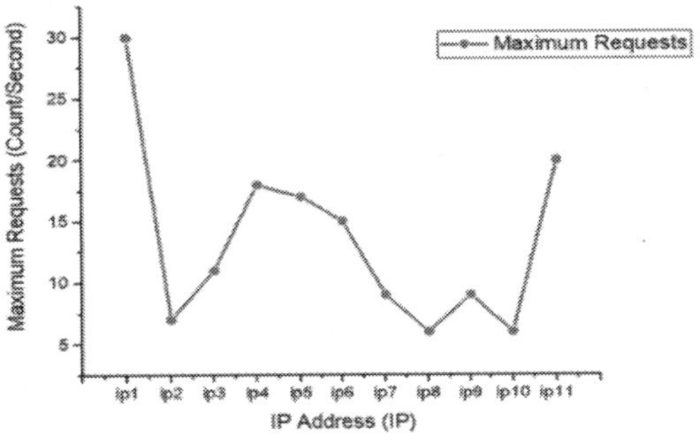

Fig 4: Request frequency 9 AM to 10 AM

Based on the incoming frequency analysis of the graphs based on the data collected from 9 AM to 5 PM, it is inferred that the count of requests which arrive from the same IP address reaches the maximum point of 34 in all the graphs. The analysis is done for the random 1-hour traffic data chosen for the 30 days statistics of the EDGAR data set and all the graphs are showing the same variation. Another interesting feature that has to be observed in the EDGAR analysis is that, even if the repeated requests arise from the same IP address, the request size is different for every individual request which is evident from the below graph. The overall incoming requests from the same IP address from 9 AM to 5 PM are represented in Fig 5.

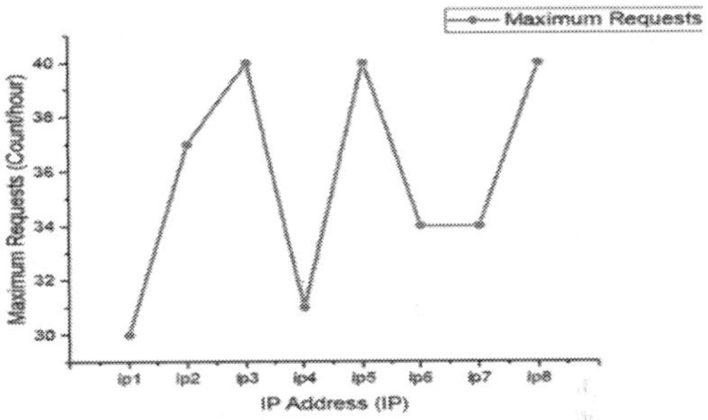

Fig 5: Request frequency from 9 AM to 5 PM

From the above graphs, it can be concluded that if the incoming request is normal, as per human behavioural and EDGAR analysis, a request of same size cannot be executed more than 5 by a normal user even if he tries to access the same file, for any scenario the request size will vary and it doesn't remains the same. This contradicts the attack scenario, as the incoming attack request sizes follows the same size and the arrival rates of attack varies from the range of 100 to 1 million as it merely depends on the capacity of the attacker system.

On observing the arrival rates of normal pattern of EDGAR, it ranges between 1 to maximum of 40 requests and not exceeding the mentioned range. From this analysis, conclusion is drawn for differentiating the normal traffic with the attack one. The analysis of each 1 hour traffic on every day is captured and analysis for a day is illustrated in the below Fig 6. It illustrates the

 https://doi.org/10.1051/itmconf/20213701025

number of IP's with the same number of request count for the randomly chosen 1 hour time period in a day of EDGAR data set.

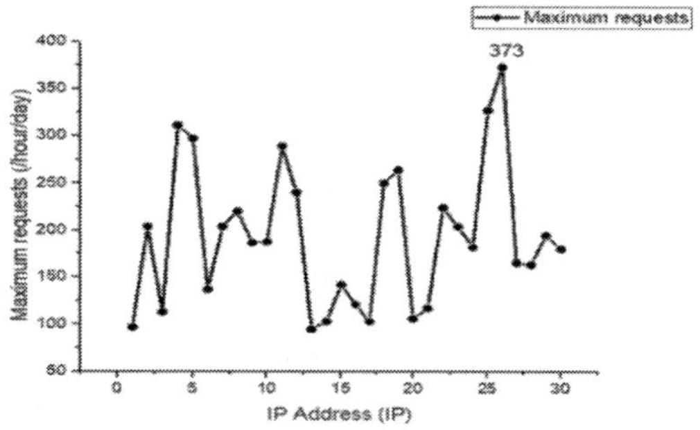

Fig 6: Maximum Request Count from various IP's

The maximum request value arriving from multiple IP addresses irrespective of the requested content are based on the 78 observation in the stipulated time period. Repeating this observation for all 30 days of the randomly chosen time period for each day, the maximum incoming request pattern and behaviour of normal IP's can be clearly understood. According to the observation, 96 are the maximum requests which arise from 107 IP addresses depicted in below figures. The graph is plotted by considering the IP address with respect to maximum request count for each day observed from the chosen time period.

It is concluded based on the observation of complete 30 days of EDGAR traffic, that if a request comes from a normal IP address, the maximum threshold from various IP addresses are not exceeding the range 373 and the number of requests per second from the same IP address is not exceeding the range of 40. In order to strongly conclude the incoming traffic as either attack or normal, additional parameter request size also needs to be considered as according to the EDGAR analysis, even maximum of 40 requests from the same IP address is handled as it belongs to different request sizes.

From the overall observation, only 2 requests originate from the same IP for the same request size which is termed as normal as it is within the threshold 5. Hence for traffic to be normal, the incoming request from the same IP should contain the various size of request else it will be distinguished as DoS attack traffic. The conclusions from the EDGAR dataset for attack distinction are carried out for the detection of LDoS attack.

6.2 Traffic Differentiation based on SNMP Metrics

We aim to address the above research gaps in the existing measures through the Simple Network Management Protocol (SNMP) [17][18]. The characterization of SNMP MIBs should be done initially which helps greatly to identify the purpose and importance of each. Based on the reference, the below SNMP MIBs are chosen.

The SNMP components, basic structure and the way of retrieving Management Information Base (MIB) variables are observed from the references. The MIBs relevant for LDoS detection are identified through the techniques theoretical validation and Linear Regression.

The metrics Request Size and Arrival Rate are fetched based on the real time data behaviour analysis of the KDD-99 dataset. To derive relation between the various attack distinction metrics for the LDoS attack traffic, the below ones are formulated. To have a detailed in depth analysis, the overall traffic, normal traffic, statistic traffic and attack traffic need to be represented which are denoted as ov(t), n(t), s(t) and a(t). The ov (t) could be expressed as

$$ov\,(t) = n(t)+a(t) \quad (12)$$

If the server is under normal traffic, then the representation is

$$a(t) = 0 \quad (13)$$

Therefore $ov(t) = n(t)$

If the server is under attack, then there is a rapid increase in the value of a(t) to larger levels. For easy attack identification, a(t) value needs to be captured. Our proposed method helps to capture the value of a(t) through the Management Information Base (MIB) variables of Simple Network Management Protocol (SNMP). The related MIB's are tcpActiveOpens, tcpPassiveOpens and tcpCurrEstab [14], [15], [16]. The attack detection algorithm is formulated as below.

Algorithm 1:Attack detection

```
Input: Analysis based on SNMP MIB's
        User requests: (R1, R2,..Rn)
        Sent user requests: IP_ tcpActiveOpensi
         Received user requests: IP_ tcpPassiveOpensi
        Failed user requests: IP_ tcpAttemptFailsi
        Established user requests: IP_ tcpCurrEstabi
    Threshold: T
        Sent user requests: IP_ ActiveOpensi
Output: Categorization of traffic 'LDoS' or 'Normal'
  begin
     if tcpActiveOpensi >= T
       "Differerentiate incoming Traffic"
     if tcpActiveOpensi != tcpPassiveOpensi &&
     tcpActiveOpensi != tcpCurrEstabi
Perform IP similarity comparison
     Compare consecutive IP's (IPi, IPj) at T
     if (IPi== IPj)
       "LDoS traffic"
     else
       "Normal traffic"
     endif
     endif
end
```

The captured results for SNMP MIB's from the experimental testbed are illustrated in the Fig 7, 8 and 9.

https://doi.org/10.1051/itmconf/20213701025

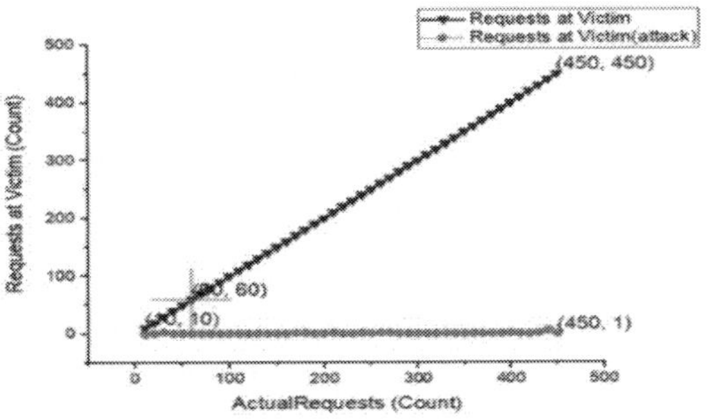

Fig 7: Variation of Actual Requests Vs tcpPassiveOpens

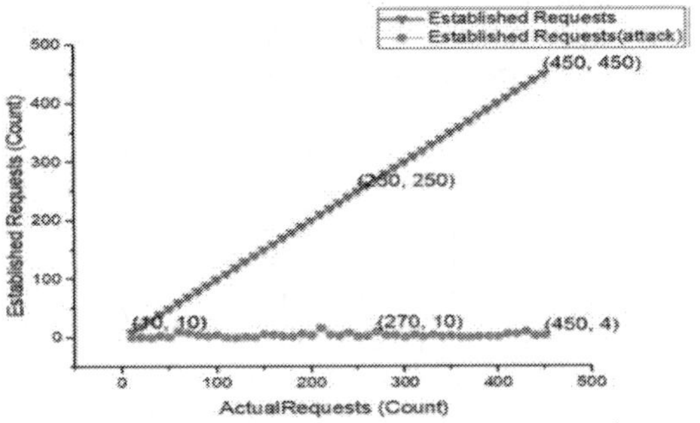

Fig 8: Variation of Actual Requests Vs tcpCurrEstab

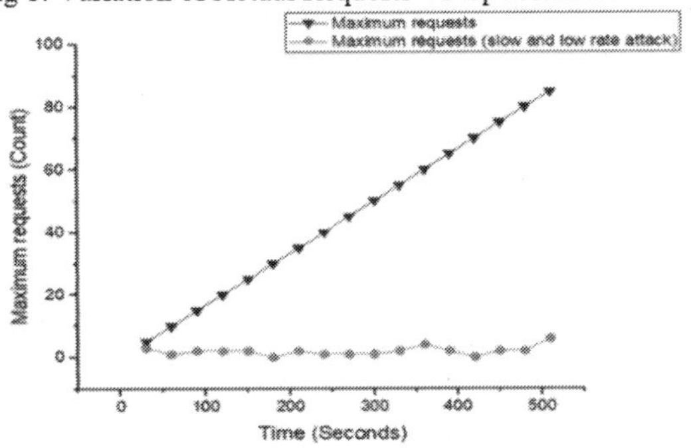

Fig 9: Variation of Time Vs Maximum requests

From the above, it is inferred that the during the nor-mal traffic, minimum variation exists between the SNMP MIB's tcpActiveOpens and tcpCurrEstab as 30 connections are established in 30 seconds. During attack traffic, there exists huge deviation between the chosen MIB's as only single connection is established in 30 seconds. Hence by combining the detection measures tcpActiveOpens, tcpCurrEstab, Time and maximum requests validation, 98.7 % detection accuracy is achieved. The general features and characteristics of DoS and LDoS [20], [21], [22], [23], [24] are analyzed based on the literature to arrive at a conclusion of the defensive measures.

7. Conclusion

The SNMP based detection measure is accompanied with the metrics based on the real time analysis of EDGAR dataset which helps in accurate detection of LDoS attacks and yields 98.7%. Early detection is another important criterion which is achieved in 6.9 seconds. The reseach gaps mentioned in the existing literature are examined carefully which paved way for the identification of an important metric tcpCurrEstab which helped to boost the accuracy. The chosen TCP specific SNMP MIB's are effective in distinguishing the LDoS attack from the normal one which is demonstrated through the simulation tools deployed in the experimental testbed. Theoretical validation is done is incorporated for the the aforementioned SNMP MIB's. The future work aims to analyse the real time log patterns to enhance the detection accuracy.

References

1. G. Linden, "Make Data Useful", Presentation, Amazon, November, 2006.
2. V. Paxson, M. Allman, J. Chu and M. Sargent, "Computing TCP's Retransmission Timer", https://tools.ietf.org/html/rfc6298, 2011.
3. Al-Haidari, F., Salah, K., Sqalli, M. et al. , "Performance Modeling and Analysis of the EDoS-Shield Mitigation", Arab J Sci Eng , Springer, 42, 793–804 (2017). https://doi.org/10.1007/s13369-016-2331-z.
4. Eytan Modiano , "Introduction to Queueing Theory ",http://web.mit.edu/modiano/www/6.263/lec5-6.pdf
5. Mishra, Ayush & Sun, Xiangpeng & Jain, Atishya & Pande, Sameer & Joshi, Raj & Leong, Ben. (2019). The Great Internet TCP Congestion Control Census. Proceedings of the ACM on Measurement and Analysis of Computing Systems. 3. 1-24. 10.1145/3366693.
6. AWS Admin, "TCP BBR Congestion Control With Amazon CloudFront", https://aws.amazon.com/blogs/networking-and-content-delivery/tcp-bbr-congestion-control-with-amazon-cloudfront/.
7. Colt McAnlis, "TCP BBR: Magic dust for network performance", https://medium.com/google-cloud/tcp-bbr-magic-dust-for-network-performance-57a5f1ccf437.
8. Xiang, Y., Li, K. and Zhou, W. (2011), 'Low-rate ddos attacks detection and traceback by using new information metrics', IEEE transactions on information forensics and security 6(2), 426–437.
9. Luo, J., Yang, X., Wang, J., Xu, J., Sun, J. and Long, K. (2014), 'On a mathematical model for low-rate shrew ddos', IEEE Transactions on Information Forensics and Security 9(7), 1069–1083.
10. Bhuyan, M. H., Bhattacharyya, D. and Kalita, J. K. (2015), 'An empirical evaluation of information

metrics for low-rate and high-rate ddos attack detection', Pattern Recognition Letters 51, 1–7.

11. Zhang, C., Yin, J., Cai, Z. and Chen, W. (2010), 'Rred: robust red algorithm to counter low-rate denial-of-service attacks', IEEE Communications Letters 14(5), 489–491.
12. Wu, Z., Zhang, L. and Yue, M. (2015), 'Low-rate dos attacks detection based on network multifractal', IEEE Transactions on Dependable and Secure Computing 13(5), 559–567.
13. Yu, H. (2007), Dos-resilient secure aggregation queries in sensor networks, in 'Proceedings of the twenty-sixth annual ACM symposium on Principles of distributed computing', ACM, pp. 394–395.
14. Gayathri R and Neelanarayanan V, "Denial of Service Attack prediction using Gradient Descent Algorithm", Advances in Internet Research and Engineering (Springer Nature), Vol:1(45), pp.1-8,2020
15. Gayathri R and Neelanarayanan V, "Identification of Regression function and distribution model for Denial of Service attack in Second Life online community using Simple Network Management Protocol", International Journal of Web Based Online Communities , Inderscience, Vol:15(3), pp.225-237, August 2019
16. Gayathri R and Neelanarayanan V, "DoS detection solution for cloud platform using SNMP", International Journal of Pure and Applied Mathematics , vol.119 (11), pp.175-183, 2018.
17. Xiao, L., Wei, W., Yang, W., Shen, Y. and Wu, X. (2017), 'A protocol-free detection against cloud oriented reflection dos attacks', Soft Computing 21(13), 3713–3721.
18. Thapngam, T., Yu, S., Zhou, W. and Makki, S. K. (2014), 'Distributed denial of service (ddos) detection by traffic pattern analysis', Peer-to-peer networking and applications 7(4), 346–358.
19. Stavrou, A., Fleck, D. and Kolias, C. (2016), 'On the move: Evading distributed denialof- service attacks', Computer 49(3), 104–107.
20. Shevtekar, A., Anantharam, K. and Ansari, N. (2005), 'Low rate tcp denial-of-service attack detection at edge routers', IEEE Communications Letters 9(4), 363–365.
21. Luo, X., Chan, E.W. and Chang, R. K. (2006), Vanguard: a new detection scheme for a class of tcp-targeted denial-of-service attacks, in '2006 IEEE/IFIP Network Operations and Management Symposium NOMS 2006', IEEE, pp. 507–518.
22. Liu, Z. and Guan, L. (2010), Attack simulation and signature extraction of low-rate dos, in '2010 Third International Symposium on Intelligent Information Technology and Security Informatics', IEEE, pp. 544–548.
23. Kline, E., Afanasyev, A. and Reiher, P. (2011), Shield: Dos filtering using traffic deflecting, in '2011 19th IEEE International Conference on Network Protocols', IEEE, pp. 37–42.
24. Kiuchi, T., Hori, Y. and Sakurai, K. (2010), A design of history based traffic filtering with probabilistic packet marking against dos attacks, in '2010 10th IEEE/IPSJ International Symposium on Applications and the Internet', IEEE, pp. 261–264.
25. Jingle, I. D. J., Rajsingh, E. B. and Paul, P. M. (n.d.), 'Distributed detection of dos using clock values in wireless broadband networks', International Journal of Engineering and Advanced Technology (IJEAT) ISSN pp. 2249–8958.
26. Hoque, M. A. and Chakraborty, B. (2015), 'Anomaly based intrusion detection systems using snmp data', International Journal of Science, Engineering and Computer Technology 5(3), 44.
27. Habib, A., Hefeeda, M. and Bhargava, B. K. (2003), Detecting service violations and dos attacks., in 'NDSS'.
28. Gao, Z. and Ansari, N. (2005), 'Tracing cyber attacks from the practical perspective', IEEE Communications Magazine 43(5), 123–131.
29. Elleithy, K. M., Blagovic, D., Cheng, W. K. and Sideleau, P. (2005), 'Denial of service attack techniques: Analysis, implementation and comparison'.

AUTHOR INDEX